NINTH EDITION

Rachna Sachdev (handwritten)

39,40 (handwritten)

Benson's
Microbiological
Applications

Laboratory Manual in General Microbiology

Alfred E. Brown
Auburn University

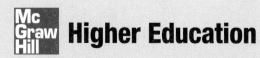

Mc Graw Hill **Higher Education**

Boston   Burr Ridge, IL   Dubuque, IA   Madison, WI   New York   San Francisco   St. Louis
Bangkok   Bogotá   Caracas   Kuala Lumpur   Lisbon   London   Madrid   Mexico City
Milan   Montreal   New Delhi   Santiago   Seoul   Singapore   Sydney   Taipei   Toronto

The McGraw·Hill Companies

Higher Education

BENSON'S MICROBIOLOGICAL APPLICATIONS SHORT VERSION:
LABORATORY MANUAL IN GENERAL MICROBIOLOGY, NINTH EDITION

This book is printed on acid-free paper.

1 2 3 4 5 6 7 8 9 0 QPD/QPD 0 9 8 7 6 5 4

ISBN 0-07-282397-6

Publisher: *Martin J. Lange*
Senior sponsoring editor: *Patrick E. Reidy*
Senior developmental editor: *Jean Sims Fornango*
Marketing manager: *Tami Petsche*
Project manager: *Joyce Watters*
Senior production supervisor: *Sherry L. Kane*
Designer: *Rick D. Noel*
Cover/interior design: *Maureen McCutcheon*
Cover image: *©Firefly Productions/Corbis*, Agar plates in a Microbiology Laboratory, SC-020-0190
Senior photo research coordinator: *Lori Hancock*
Compositor: *Electronic Publishing Services Inc., NYC*
Typeface: *11/12 Times Roman*
Printer: *Quebecor World Dubuque, IA*

Some of the laboratory experiments included in this text may be hazardous if materials are handled improperly or if procedures are conducted incorrectly. Safety precautions are necessary when you are working with chemicals, glass test tubes, hot water baths, sharp instruments, and the like, or for any procedures that generally require caution. Your school may have set regulations regarding safety procedures that your instructor will explain to you. Should you have any problems with materials or procedures, please ask your instructor for help.

www.mhhe.com

ABOUT THE AUTHOR

McGraw-Hill is pleased to introduce the new author of Benson's Microbiological Applications: Laboratory Manual in General Microbiology, ninth edition

Alfred E. Brown
Professor, Auburn University
Ph.D., UCLA

Research focus: Physiology of photosynthetic bacteria; mode of action of herbicides on photosynthetic bacteria and soil microbial populations.

"The primary focus of my research has been the mode of action of certain herbicides on photosynthetic bacteria and on soil bacteria. For the photosynthetic bacteria, I have characterized the mechanism for how atrazine and other **s**-triazines inhibit photosynthetic electron transport in the purple nonsulfur bacteria and how resistance to these inhibitors develops in this group of organisms. The **s**-triazines act as quinone antagonists whereby they replace ubiquinone on a binding site associated with the L-subunit of the photosynthetic reaction center. Resistance to these herbicides occurs when a mutation in the L-subunit gene gives rise to an amino acid change in the binding protein thereby diminishing the affinity of the herbicide for the quinone binding site. Studies in my lab have also shown that atrazine induces the production of stress proteins in herbicide-sensitive strains of these bacteria. More recently, the research focus has shifted to another class of herbicides, the sulfonylureas, which inhibit the enzyme acetolactacte synthase(ALS), the first enzyme in the biosynthetic pathway for valine/isoleucine. This work has concentrated on soil bacteria that are resistant to the sulfonylureas and a characterization of ALS in these bacteria. Thus far, investigations have shown that the ALS in the soil isolates and in photosynthetic bacteria is different in its response to these herbicides when compared to the model system described for *Escherichia coli*."

Dr. Brown's teaching experience includes General Microbiology with laboratory sections, Medical Microbiology, Methods in Microbiology, Special Problems, Applied and Environmental Microbiology, Photosynthesis, and Biomembranes.

CONTENTS

CONTENTS ∎

New Beginnings!

Benson's Microbiological Applications has been the "gold standard" of microbiology lab manuals for over 30 years. The manual has a number of attractive features that resulted in its adoption in universities, colleges, and community colleges for a wide variety of microbiology courses. These features include "user friendly" diagrams that students can easily follow, clear instructions, and an excellent array of reliable exercises suitable for beginning or advanced courses.

Needless to say, taking over as the author of the manual was a daunting task. In revising the manual for the ninth edition, I have tried to maintain these important strengths and enhance the manual by up-dating the art and exercises with more modern figures, photographs, and information that more appropriately reflect our current times. However, in most cases the procedures in the exercises have remained relatively unchanged because they have stood the test of time.

ORGANIZATIONAL CHANGES

The organization of the manual has undergone some significant changes. A number of older lab exercises have been eliminated. In some cases several exercises repeated essentially the same procedures with different medium. For example, one strong exercise on microbial spoilage of food-stuff has been included, rather than three exercises on the same procedure.

- **The second section in the manual concerns the manipulation of microorganisms** because students usually must manipulate and transfer bacterial cultures early in a microbiology course. This section was created by combining aseptic technique, and the pour and streak plate exercises.
- **The standard plate count** is used in several exercises in the manual but the general procedure is introduced in **Culture Methods** rather than repeating it for each individual exercise.
- **A simple procedure for the endospore stain has been added** to the staining exercises in the manual. This does not require the prolonged heating required in conventional procedures and I find that it works extremely well in labs taught at Auburn University.
- **The section on Virology has been expanded** to include a phage titer determination and the phage typing exercise has been moved from medical microbiology to this section. Although the virology exercises do not cover all kinds of viruses, for ex-

ample animal and plant viruses are not included, the exercises serve to introduce basic virology through the use of bacteriophage.

NEW ART PROGRAM

The majority of illustrations have been re-rendered in a **modern four color style** to reflect equipment and practices (ie, wearing gloves) common to today's microbiology labs.

A number of **new full color photos** have been included illustrating new equipment, experiment reactions, and proper procedures. This visual enhancement will help today's student succeed in the laboratory experience and reflect the environment they encounter in their own microbiology labs.

SAFETY FIRST

New information on safety has been added, including information on the **Biosafety Level** ratings of organisms used in the manual and the recommendations for safe handling of those materials. Although no organisms above a BSL rating of 2 are included in this manual, the safety precautions through level 4 are outlined to provide students with the full scope of knowledge about lab safety. This material appears in the front of the manual before students begin working on exercises.

ACKNOWLEDGEMENTS

The updates and improvements in this edition were guided by the helpful reviews of the following instructors. Their input was critical to the decisions that shaped this edition of *Benson's Microbiological Applications*.

Emmanuel Brako, *Winona State University*
Carolyn Dabirsiaghi, *Baltimore City Community College*
Kathleen Dannelly, *Indiana State University*
Johnny El-Rady, *University of South Florida*
David Essar, *Winona State University*
Kathy Foreman, *Moraine Valley Community College*
Joseph Gauthier, *University of Alabama*
Kathy Germain, *Southwest Tennessee Community College*
Janice Haggart, *North Dakota State University*
Laraine Powers, *East Tennessee State University*
Susan Skelly, *Rutgers University*

Goeffrey Smith, *New Mexico State University*
Carolyn Thompson, *SUNY-Morrisville*
James Urban, *Kansas State University*

Patricia Vary, *Northern Illinois University*
Lori Zeringue, *Louisiana State University*

BASIC MICROBIOLOGY
LABORATORY SAFETY

Every student and instructor must focus on the need for safety in the microbiology laboratory. While the lab is a fascinating and exciting learning environment, there are hazards that must be acknowledged and guarded against. The following outline of basic lab safety will give every member of the lab section the information needed to guarantee a safe learning environment.

Microbiological laboratories are special, often unique work environments that may pose identifiable infectious disease risks to persons in or near them. Infections have been con-

The "Biohazard" symbol must be affixed to any container or equipment used to store or transport potentially infectious materials.

tracted in the laboratory throughout the history of microbiology. Published reports around the turn of the century described laboratory-associated cases of typhoid, cholera, glanders, brucellosis, and tetanus.

The term "containment" is used in describing safe methods for managing infectious materials in the laboratory environment where they are being handled or maintained. The purpose of containment is to reduce or eliminate exposure of laboratory workers, other persons, and the outside environment to potentially hazardous agents. Primary containment, the protection of personnel and the immediate laboratory environment from exposure to infectious agents is provided by both good microbiological technique and the use of appropriate safety equipment. The use of vaccines may provide an increased level of personal protection. Secondary containment, the protection of the environment external to the laboratory from exposure to infectious materials, is provided by a combination of facility design and operational practices. Therefore, the three elements of containment include laboratory practice and technique, safety equipment, and facility design. The risk assessment of the work to be done with a specific agent will determine the appropriate combination of these elements.

BIOSAFETY LEVEL (BSL)

The recommended biosafety level(s) for the organisms represent those conditions under which the agent ordinarily can be safely handled. The laboratory director is specifically and primarily responsible for assessing the risks and appropriately applying the recommended biosafety levels. When specific information is available to suggest that virulence,

pathogenicity, antibiotic resistance patterns, vaccine and treatment availability, or other factors are significantly altered, more (or less) stringent practices may be specified.

1. BSL1- work with agents not known to cause disease in healthy adults; "standard microbiological practices SMP)" apply; no safety equipment required; sinks required.
2. BSL2- work with agents associated with human disease; SMP apply plus limited access, biohazard signs, sharps precautions, and biosafety manual required; BSC used for aerosol/splash generating operations; lab coats, gloves, face protection required; contaminated waste is autoclaved.
3. BSL3- work with indigenous/exotic agents which may have serious or lethal consequences and with potential for aerosol transmission; BSL2 practices plus controlled access; decontamination of all waste and lab clothing before laundering; determination of baseline serums; BSC used for all specimen manipulations; respiratory protection used as needed; physical separation from access corridors; double door access; negative airflow into lab; exhaust air not recirculated.
4. BSL4- work with dangerous/exotic agents of life threatening nature or unknown risk of transmission; BSL3 practices plus clothing change before entering lab; shower required for exit; all materials are decontaminated on exit; positive pressure personnel suit required for entry; separated/isolated building; dedicated air supply/exhaust and decon systems.

Each of the four biosafety levels (BSLs) consist of combinations of laboratory practices and techniques, safety equipment, and laboratory facilities. Each combination is specifically appropriate for the operations performed, the documented or suspected routes of transmission of the infectious agents, and the laboratory function or activity. However, common to all four biosafety levels are the "Standard Practices" which remain the same from BSL1 to BSL4.

STANDARD PRACTICES

1. Access to lab is limited or restricted by the lab director when work with infectious agents is in progress.
2. Persons wash their hands after handling viable material and animals, after removing gloves, and before leaving lab.

3. Eating, drinking, smoking, handling contact lenses, and applying cosmetics are not permitted in work areas.
4. Mouth pipetting is prohibited.
5. All procedures are performed to minimize aerosol or splash production.
6. Work surfaces are decontaminated daily and after any spill of viable material.
7. All cultures, stocks and other regulated wastes are decontaminated before disposal by an approved decontamination method such as autoclaving.
8. An insect and rodent control program is in effect.
9. BSCs are used whenever there is a potential for aerosol/splash creation or when high concentrations/large volumes of infectious agents are used.
10. Face protection is used for anticipated splashes/sprays to the face.
11. Lab coats/gowns/smocks/uniforms are worn while in the lab.
12. Gloves are worn when handling infected animals and when hands may come in contact with infectious materials, contaminated equipment, or surfaces.
13. All infectious/regulated waste is decontaminated via autoclave, chemical disinfection, incinerator, or other approved method.

Safe Operations

- Biohazard warning signs listing responsible laboratory personnel and infectious agents are posted on all laboratory access doors.
- OSHA requires the wearing of personal protective safety glasses whenever working with or around hazardous materials.
- Sandals and open-toe shoes are not appropriate footwear in the laboratory.
- All laboratory materials (lab coats, gloves, eyewear, etc.) remain in the laboratory unless properly decontaminated.

CENTRIFUGES

Procedure

1. Check centrifuge tubes for cracks/chips before use.
2. Do not fill centrifuge tubes to the very top of the tube.
3. Tightly seal all centrifuge tubes or use safety cups/buckets to prevent aerosol escape.
4. Take care that matched sets of buckets, adapters, and plastic inserts are kept together.
5. Ensure that rotors are "locked" to the spindle and buckets are "sealed" before operation.
6. Use a biological safety cabinet (BSC) to load and open tubes, safety cups, and buckets when work-

ing with biohazardous materials. Decontaminate tubes, safety cups, and buckets before removal from the BSC and transport to the centrifuge.
7. Close the centrifuge top during operation.
8. Allow the centrifuge to come to a complete stop before opening.
9. Disinfect weekly and immediately following any spill or breakage of the surfaces of the centrifuge head, bowl, trunnions, and buckets. Use 70% alcohol, 2% glutaraldehyde, or any registered mycobactericidal. For radioactive contamination, use equal parts of 70% ethanol, 10% SDS, and water, followed by water rinses and drying with a soft cloth. Dupont COUNT-OFF and other radioactive decontaminates must not be used on aluminum rotors as they will remove the anodized coatings.

BIOLOGICAL WASTE DISPOSAL*

Procedure:

1. Line discard pan with appropriately sized *clear* autoclave bag.
2. Fold upper part of bag over the side of the pan.
3. Add waste- keep pan covered when not in use and do not overfill.
4. When the pan is ≈3/4 full, carefully add 250-500 ml of water or germicidal solution. Avoid splashing!
5. Twist the bag closed. Do not tie!
6. Replace pan lid and tape closed with autoclave tape.
7. Affix identifying label to end of pan and transport to autoclave room. Leave pan in appropriate area for autoclaving.

Shipping Containers:

1. Decontaminate if necessary (autoclave or wipe with disinfectant)
2. Deface biohazard sticker, and
3. Mark outer cardboard container as "TRASH".

Laboratory Equipment:

1. Decontaminate with an appropriate disinfectant, and
2. Affix "signed" CDC form 0.593 (Decontamination Sticker) to outside.

Safe Operations:

- **NEVER** place lab waste in office waste containers.
- Place **all** sharps into "sharps" containers.
- Place all lab waste (pipettes, pipette tips, pipette wrappers, tissue cultures flasks, Kimwipes, etc.) into appropriate discard pans or discard/autoclave bag.
- **DECONTAMINATE** discard pans before leaving lab:
  a. Disinfect outside

    b. Label
    c. Tape ends with autoclave tape
    d. Secure for transport to autoclave

BIOLOGICAL EMERGENCIES

Surface Contamination

1. Define/isolate contaminated area.
2. Alert co-workers.
3. Put on appropriate PPE.
4. Remove glass/lumps with forceps or scoop.
5. Apply absorbent towel(s) to spill; remove bulk and reapply if needed.
6. Apply disinfectant* to towel surface.
7. Allow adequate contact time (20 minutes).
8. Remove towel, mop up; clean with alcohol or soap/water.
9. Properly dispose of materials.
10. Notify supervisor and OHS.

Personnel Contamination

1. Clean exposed surface with soap/water, eyewash (eyes), or saline (mouth).
2. Apply first aid and treat as an emergency.
3. Notify supervisor or security desk (after hours).
4. Follow "On-the-Job Injury/Illness Reporting Procedures."

5. Report to the Occupational Health Clinic for treatment/counseling.

More Information?

Phillips, C. B. 1986. Human Factors in Microbiological Laboratory Accidents, p. 43-48, *in* Laboratory Safety: Principles and Practices. ASM, Washington, D. C.

Harding, L. and D. F. Liberman. 1995. Epidemiology of Laboratory Associated Infections, p. 7-15, in Laboratory Safety: Principles and Practices. ASM, Washington, D. C.

Biosafety in Microbiological and Biomedical Laboratories (1999), CDC/NIM publication.

Primary Containment for Biohazards: Selection, Installation, and Use of Biological Safety Cabinets (1995), CDC/NIH publication.

Rayburn, Stephen R. 1990. The Foundations of Laboratory Safety: A Guide for the Biomedical Laboratory. pp. 14-19, pp. 69-71. Springer-Verlag, New York.

Veseley, D., Lauer, J. 1986. Decontamination, Sterilization, Disinfection, and Antisepsis in the Microbiology Laboratory, pp. 187-190. *In* Briton M. Miller (ed.), Laboratory Safety: Principles and Practices. American Society for Microbiology, Washington, D. C.

*1:100 solution of household bleach (hypochlorite) for most spills, 1/10 solution for spills containing large amounts of organic material, or any EPA registered mycobactericidal. Alcohols are not recommended as surface decontaminates because of their evaporative nature which decreases "contact time."

Biosafety Levels for Selected Infectious Agents

Biosafety Level (BSL)	Typical Risk	Organism
BSL 1	Not likely to pose a disease risk to healthy adults.	*Alicaligenes denitrificans* *Alicaligenes faecalis* *Bacillus cereus* *Bacillus subtillis* *Corneybacterium pseudodiphtheriticum* *Enterobacter aerogenes* *Enterococcus faicalis* *Micrococcus luteus* *Neisseria sicca* *Proteus vulgaris* *Psuedomonas aeruginose* *Staphylococcus epidermidis* *Staphylococcus saprophyticus*
BSL 2	Poses a moderate risk to healthy adults; unlikely to spread throughout community; effective treatment readily available.	*Escherichia coli* *Klebsiella pneumoniae* *Mycobacterium phlei* *Salmonella typhimurium* *Shigella flexneri* *Staphylococcus aureus* *Staphylococcus pneumoniae* *Staphylococcus pyogenes*
BSL 3	Can cause disease in healthy adults; may spread to community; effective treatment readily available.	*Chlamydia trachomatis* *Francisella tularensis* *Mycobacterium bovis* *Mycobacterium tuberculosis* *Pseudomas mallei* *Yersinia pestis* *Blastomyces dermatitidis* *Coccidiodies immitisj* *Histoplasma capsulatum* *Coxiella burnetii* *Richettsia Canada* *Rickettsia prowazekii*
BSL 4	Can cause disease in healthy adults; poses a lethal risk and does not respond to vaccines or antimicrobial therapy.	Filovirus *Herpesvirus simiae* Lassa virus Marburg virus

Microorganisms Used in These Lab Exercises and Their BSL Ratings

Organisms	Lab Exercise	BSL Rating
Alcaligenes faecalis	31	1
Azotobacter insignis	52	1
Azotobacter nigricans	52	1
Azotobacter vinlandii	52	1
Bacillus cereus (mycoides)	59	1
Bacillus coagulans	64	1
Bacillus megaterium	13, 15, 16, 19, 30, 33	1
Bacillus stearothermophilus	64	1
Bacillus subtilis	16, 21, 42	1
Candida glabrata	31	1
Clostridium rubrum	21	2
Clostridium sporogenes	21, 57, 64	2
Clostridium thermosacharolyticum	64	2
Corneybacterium diphtheriae	12	2
Desulfovibrio desulfuricans	56	1
Enterobacter aerogenes	41, 43, 61	1
Escherichia coli	9, 19, 21, 22, 26, 27, 29, 30, 31, 32, 34, 36, 41, 42, 58, 59, 61, 64, 66, 67, 68	2
Halobacterium salinarium	32	1
Kelbsiella pneumoniae	14	2
Micrococcus luteus	18, 78	1
Moraxella (Branhamella) catarrhalis	15	1
Mycobacterium smegmatis	15, 17	2
Paracoccus denitrificans	53	1
Penicillum notatum	59	1
Physarum polycephalum	23	1
Proteus vulgaris	18, 36, 42, 43, 58	1
Pseudomonas aeruginosa	15, 36, 37, 41	1
Pseudomonas fluorescens	59	
Saccharomyces cerevisia	31	1
Salmonella typhimurium	73	2
Serratia marcescens	10, 29, 78	1
Staphylococcus aureus	10, 13, 15, 17, 21, 28, 30, 31, 32, 33, 34, 36, 37, 41, 42, 43, 57, 58, 59, 70, 71, 73	1

Microorganisms Used in These Lab Exercises and Their BSL Ratings (continued)

Organisms	Lab Exercise	BSL Rating
Staphylococcus epidermidis	49, 70	1
Staphylococcus saprophyticus	70	1
Streptococcus faecalis	21	2
Streptococcus mutans	14	2
Streptococcus pneumoniae	14	2

MICROSCOPY

Although there are many kinds of microscopes available to the micro-biologist today, only four types will be described here for our use: the brightfield, darkfield, phase-contrast, and fluorescence microscopes. If you have had extensive exposure to microscopy in previous courses, this unit may not be of great value to you; however, if the study of microorganisms is a new field of study for you, there is a great deal of information that you need to acquire about the proper use of these instruments.

Microscopes in a college laboratory represent a considerable investment and require special care to prevent damage to the lenses and mechanical parts. A microscope may be used by several people during the day and moved from the work area to storage; which results in a much greater chance for damage to the instrument than if the microscope were used by only a single person.

The complexity of some of the more expensive microscopes also requires that certain adjustments be made periodically. Knowing how to make these adjustments to get the equipment to perform properly is very important. An attempt is made in the five exercises of this unit to provide the necessary assistance for getting the most out of the equipment.

Microscopy should be as fascinating to the beginner as it is to the professional of long standing; however, only with intelligent understanding can the beginner approach the achievement that occurs with years of experience.

Brightfield Microscopy

A microscope that allows light rays to pass directly through to the eye without being deflected by an intervening opaque plate in the condenser is called a *brightfield microscope.* This is the conventional type of instrument encountered by students in beginning courses in biology; it is also the first type to be used in this laboratory.

All brightfield microscopes have certain things in common, yet they differ somewhat in mechanical operation. Similarities and differences of various makes are discussed in this exercise so that you will know how to use the instrument that is available to you. Before attending the first laboratory session in which the microscope is used, read over this exercise and answer all the questions on the Laboratory Report. Your instructor may require that the Laboratory Report be handed in prior to doing any laboratory work.

CARE OF THE INSTRUMENT

Microscopes represent considerable investment and can be damaged easily if certain precautions are not observed. The following suggestions cover most hazards.

Transport   When carrying your microscope from one part of the room to another, use both hands to hold the instrument, as illustrated in figure 1.1. If it is carried with only one hand and allowed to dangle at your side, there is always the danger of collision with furniture or some other object. And, *under no circumstances should one attempt to carry two microscopes at one time.*

Clutter   Keep your workstation uncluttered while doing microscopy. Keep unnecessary books and other materials away from your work area. A clear work area promotes efficiency and results in fewer accidents.

Electric Cord   Microscopes have been known to tumble off of tabletops when students have entangled a foot in a dangling electric cord. Don't let the light cord on your microscope dangle in such a way as to risk foot entanglement.

Lens Care   At the beginning of each laboratory period, check the lenses to make sure they are clean. At the

FIGURE 1.1   The microscope should be held firmly with both hands while being carried.

end of each lab session, be sure to wipe any immersion oil off the immersion lens if it has been used. More specifics about lens care are provided on page 5.

Dust Protection   In most laboratories dustcovers are used to protect the instruments during storage. If one is available, place it over the microscope at the end of the period.

COMPONENTS

Before we discuss the procedures for using a microscope, let's identify the principal parts of the instrument as illustrated in figure 1.2.

Framework   All microscopes have a basic frame structure, which includes the **arm** and **base.** To this

FIGURE 1.2 The compound microscope
(Courtesy of the Olympus Corporation, Lake Success, NY)

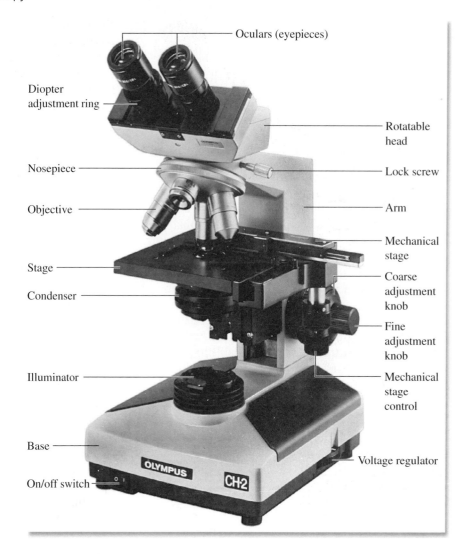

framework all other parts are attached. On many of the older microscopes the base is not rigidly attached to the arm as is the case in figure 1.2; instead, a pivot point is present that enables one to tilt the arm backward to adjust the eyepoint height.

Stage The horizontal platform that supports the microscope slide is called the **stage**. Note that it has a clamping device, the **mechanical stage,** which is used for holding and moving the slide around on the stage. Note, also, the location of the **mechanical stage control** in figure 1.2.

Light Source In the base of most microscopes is positioned some kind of light source. Ideally, the lamp should have a **voltage control** to vary the intensity of light. The microscope in figure 1.2 has a knurled wheel on the right side of its base to regulate the voltage supplied to the light bulb.

Most microscopes have some provision for reducing light intensity with a **neutral density filter**. Such a filter is often needed to reduce the intensity of light below the lower limit allowed by the voltage control. On microscopes such as the Olympus CH-2, one can simply place a neutral density filter over the light source in the base. On some microscopes a filter is built into the base.

Lens Systems All microscopes have three lens systems: the oculars, the objectives, and the condenser. Figure 1.3 illustrates the light path through these three systems.

The **ocular,** or eyepiece, is a complex piece, located at the top of the instrument, that consists of two or more internal lenses and usually has a magnification of 10×. Most modern microscopes (figure 1.2) have two ocular (binocular) lenses.

Three or more **objectives** are usually present. Note that they are attached to a rotatable **nosepiece,**

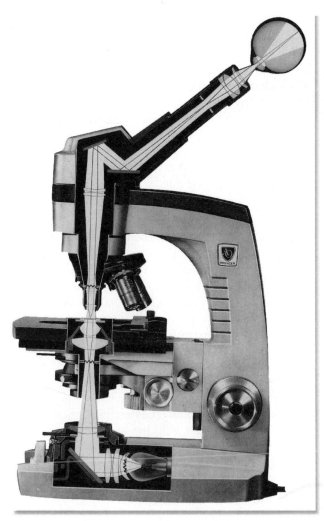

FIGURE 1.3   The light pathway of a microscope

which makes it possible to move them into position over a slide. Objectives on most laboratory microscopes have magnifications of 10×, 45×, and 100×, designated as **low power, high-dry,** and **oil immersion,** respectively. Some microscopes will have a fourth objective for rapid scanning of microscopic fields that is only 4×.

The third lens system is the **condenser,** which is located under the stage. It collects and directs the light from the lamp to the slide being studied. The condenser can be moved up and down by a knob under the stage. A **diaphragm** within the condenser regulates the amount of light that reaches the slide. Microscopes that lack a voltage control on the light source rely entirely on the diaphragm for controlling light intensity. On the Olympus microscope in figure 1.2, the diaphragm is controlled by turning a knurled ring. On some microscopes, a diaphragm lever is present. Figure 1.3 illustrates the location of the condenser and diaphragm.

Focusing Knobs   The concentrically arranged **coarse adjustment** and **fine adjustment knobs** on the side of the microscope are used for bringing objects into focus when studying an object on a slide. On some microscopes, these knobs are not positioned concentrically as shown here.

Ocular Adjustments   On binocular microscopes, one must be able to change the distance between the oculars and to make diopter changes for eye differences. On most microscopes, the interocular distance is changed by simply pulling apart or pushing together the oculars.

To make diopter adjustments, one focuses first with the right eye only. Without touching the focusing knobs, diopter adjustments are then made on the left eye by turning the knurled **diopter adjustment ring** (figure 1.2) on the left ocular until a sharp image is seen. One should now be able to see sharp images with both eyes.

RESOLUTION

The resolution limit, or **resolving power,** of a microscope lens system is a function of its numerical aperture, the wavelength of light, and the design of the condenser. The optimum resolution of the best microscopes with oil immersion lenses is around 0.2 μm. This means that two small objects that are 0.2 μm apart will be seen as separate entities; objects closer than that will be seen as a single object.

To get the maximum amount of resolution from a lens system, the following factors must be taken into consideration:

- A **blue filter** should be in place over the light source because the short wavelength of blue light provides maximum resolution.
- The **condenser** should be kept at its highest position where it allows a maximum amount of light to enter the objective.
- The **diaphragm** should not be stopped down too much. Although stopping down improves contrast, it reduces the numerical aperture.
- **Immersion oil** should be used between the slide and the 100× objective.

Of significance is the fact that, as magnification is increased, the resolution must also increase. Simply increasing magnification by using a 20× ocular won't increase the resolution.

LENS CARE

Keeping the lenses of your microscope clean is a constant concern. Unless all lenses are kept free of dust, oil, and other contaminants, they are unable to achieve the degree of resolution that is intended. Consider the following suggestions for cleaning the various lens components:

Cleaning Tissues Only lint-free, optically safe tissues should be used to clean lenses. Tissues free of abrasive grit fall in this category. Booklets of lens tissue are most widely used for this purpose. Although several types of boxed tissues are also safe, *use only the type of tissue that is recommended by your instructor* (figure 1.4).

Solvents Various liquids can be used for cleaning microscope lenses. Green soap with warm water works very well. Xylene is universally acceptable. Alcohol and acetone are also recommended, but often with some reservations. Acetone is a powerful solvent that could possibly dissolve the lens mounting cement in some objective lenses if it were used too liberally. When it is used it should be used sparingly. Your instructor will inform you as to what solvents can be used on the lenses of your microscope.

Oculars The best way to determine if your eyepiece is clean is to rotate it between the thumb and forefinger as you look through the microscope. A rotating pattern will be evidence of dirt.

If cleaning the top lens of the ocular with lens tissue fails to remove the debris, one should try cleaning the lower lens with lens tissue and blowing off any excess lint with an air syringe or gas cannister. *Whenever the ocular is removed from the microscope, it is imperative that a piece of lens tissue be placed over the open end of the microscope as illustrated in figure 1.4.*

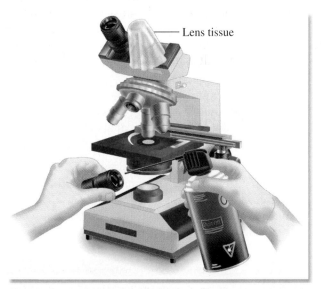

Lens tissue

FIGURE 1.4 When oculars are removed for cleaning, cover the ocular opening with lens tissue. A blast from an air syringe or gas cannister removes dust and lint.

Objectives Objective lenses often become soiled by materials from slides or fingers. A piece of lens tissue moistened with green soap and water, or one of the acceptable solvents mentioned above, will usually remove whatever is on the lens. Sometimes a cotton swab with a solvent will work better than lens tissue. At any time that the image on the slide is unclear or cloudy, assume at once that the objective you are using is soiled.

Condenser Dust often accumulates on the top surface of the condenser; thus, wiping it off occasionally with lens tissue is desirable.

PROCEDURES

If your microscope has three objectives you have three magnification options: (1) low-power, or 100× total magnification, (2) high-dry magnification, which is 450× total with a 45× objective, and (3) 1000× total magnification with a 100× oil immersion objective. Note that the total magnification seen through an objective is calculated by simply multiplying the power of the ocular by the power of the objective.

Whether you use the low-power objective or the oil immersion objective will depend on how much magnification is necessary. Generally speaking, however, it is best to start with the low-power objective and progress to the higher magnifications as your study progresses. Consider the following suggestions for setting up your microscope and making microscopic observations.

Low-Power Examination The main reason for starting with the low-power objective is to enable you to explore the slide to look for the object you are planning to study. Once you have found what you are looking for, you can proceed to higher magnifications. Use the following steps when exploring a slide with the low-power objective:

1. Position the slide on the stage with the material to be studied on the *upper* surface of the slide. Figure 1.5 illustrates how the slide must be held in place by the mechanical stage retainer lever.

FIGURE 1.5 The slide must be properly positioned as the retainer lever is moved to the right.

2. Turn on the light source, using a *minimum* amount of voltage. If necessary, reposition the slide so that the stained material on the slide is in the *exact center* of the light source.
3. Check the condenser to see that it has been raised to its highest point.
4. If the low-power objective is not directly over the center of the stage, rotate it into position. Be sure that as you rotate the objective into position it clicks into its locked position.
5. Turn the coarse adjustment knob to lower the objective until it stops. A built-in stop will prevent the objective from touching the slide.
6. While looking down through the ocular (or oculars), bring the object into focus by turning the fine adjustment focusing knob. Don't readjust the coarse adjustment knob. If you are using a binocular microscope, it will also be necessary to adjust the interocular distance and diopter adjustment to match your eyes.
7. Manipulate the diaphragm lever to reduce or increase the light intensity to produce the clearest, sharpest image. Note that as you close down the diaphragm to reduce the light intensity, the contrast improves and the depth of field increases. Stopping down the diaphragm when using the low-power objective does not decrease resolution.
8. Once an image is visible, move the slide about to search out what you are looking for. The slide is moved by turning the knobs that move the mechanical stage.
9. Check the cleanliness of the ocular, using the procedure outlined earlier.
10. Once you have identified the structures to be studied and wish to increase the magnification, you may proceed to either high-dry or oil immersion magnification. However, before changing objectives, *be sure to center the object you wish to observe.*

High-Dry Examination   To proceed from low-power to high-dry magnification, all that is necessary is to rotate the high-dry objective into position and open up the diaphragm somewhat. It may be necessary to make a minor adjustment with the fine adjustment knob to sharpen up the image, but *the coarse adjustment knob should not be touched.*

Good quality modern microscopes are **parfocal.** This means that the image will remain in focus when changing from a lower power objective lens to a higher power lens. Only minimal focusing should be necessary with the fine focus adjustment.

When increasing the lighting, be sure to open up the diaphragm first instead of increasing the voltage

on your lamp; the reason is that *lamp life is greatly extended when used at low voltage.* If the field is not bright enough after opening the diaphragm, feel free to increase the voltage. A final point: Keep the condenser at its highest point.

Oil Immersion Techniques   The oil immersion lens derives its name from the fact that a special mineral oil is interposed between the lens and the microscope slide. The oil is used because it has the same refractive index as glass, which prevents the loss of light due to the bending of light rays (diffraction) as they pass through air. The use of oil in this way enhances the resolving power of the microscope. Figure 1.6 reveals this phenomenon.

With parfocal objectives one can go directly to oil immersion from either low power or high-dry. On some microscopes, however, going from low power to high power and then to oil immersion is better. Once the microscope has been brought into focus at one magnification, the oil immersion lens can be rotated into position without fear of striking the slide.

Before rotating the oil immersion lens into position, however, a drop of immersion oil must be placed on the slide. An oil immersion lens should never be used without oil. Incidentally, if the oil appears cloudy, it should be discarded.

When using the oil immersion lens, it is best to open the diaphragm as much as possible. Stopping down the diaphragm tends to limit the resolving power of the optics. In addition, the condenser must be kept at its highest point. If different colored filters are available for the lamp housing, it is best to use blue or greenish filters to enhance the resolving power.

Since the oil immersion lens will be used extensively in all bacteriological studies, it is of paramount importance that you learn how to use this lens properly. Using this lens takes a little practice due to the difficulties usually encountered in manip-

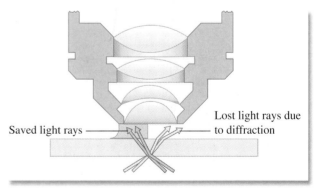

Saved light rays ────

Lost light rays due to diffraction

FIGURE 1.6   Immersion oil, having the same refractive index as glass, prevents light loss due to diffraction.

TABLE 1.1   Relationship of Working Distance to Magnification

Lens	Magnification	Focal length (mm)	Working distance (mm)
Low power	10X	16.0	7.7
High dry	40X	4.0	0.3
Oil immersion	100X	1.8	0.12

ulating the lighting. It is important for all beginning students to appreciate that the working distance of a lens, the distance between the lens and microscope slide, decreases significantly as the magnification of the lens increases (table 1.1). Hence, the potential for damage to the oil immersion because of a collision with the microscope slide is very great. A final comment of importance: At the end of the laboratory period remove all immersion oil from the lens tip with lens tissue.

PUTTING IT AWAY

When you take a microscope from the cabinet at the beginning of the period, you expect it to be clean and in proper working condition. The next person to use the instrument after you have used it will expect the same consideration. A few moments of care at the end of the period will ensure these conditions. Check over

the following list of items at the end of each period before you return the microscope to the cabinet.

1. Remove the slide from the stage.
2. If immersion oil has been used, wipe it off the lens and stage with lens tissue. (Do not wipe oil off slides you wish to keep. Simply put them into a slide box and let the oil drain off.)
3. Rotate the low-power objective into position.
4. If the microscope has been inclined, return it to an erect position.
5. If the microscope has a built-in movable lamp, raise the lamp to its highest position.
6. If the microscope has a long attached electric cord, wrap it around the base.
7. Adjust the mechanical stage so that it does not project too far on either side.
8. Replace the dustcover.
9. If the microscope has a separate transformer, return it to its designated place.
10. Return the microscope to its correct place in the cabinet.

LABORATORY REPORT

Before the microscope is to be used in the laboratory, answer all the questions on Laboratory Report 1 that pertain to brightfield microscopy. Preparation on your part prior to going to the laboratory will greatly facilitate your understanding. Your instructor may wish to collect this report at the *beginning of the period* on the first day that the microscope is to be used in class.

Darkfield Microscopy

Delicate transparent living organisms can be more easily observed with darkfield microscopy than with conventional brightfield microscopy. This method is particularly useful when one is attempting to identify spirochaetes in an exudate from a syphilitic lesion. Figure 2.1 illustrates the appearance of these organisms under such illumination. This effect may be produced by placing a darkfield stop below the regular condenser or by replacing the condenser with a specially constructed one.

Another application of darkfield microscopy is in the fluorescence microscope. Although fluorescence may be seen without a dark field, it is greatly enhanced with this application.

To achieve the darkfield effect it is necessary to alter the light rays that approach the objective in such a way that only oblique rays strike the objects being viewed. The obliquity of the rays must be so extreme that if no objects are in the field, the background is completely light-free. Objects in the field become brightly illuminated, however, by the rays that are reflected up through the lens system of the microscope.

Although there are several different methods for producing a dark field, only two devices will be described here: the star diaphragm and the cardioid condenser. The availability of equipment will determine the method to be used in this laboratory.

THE STAR DIAPHRAGM

One of the simplest ways to produce the darkfield effect is to insert a star diaphragm into the filter slot of the condenser housing as shown in figure 2.2. This device has an opaque disk in the center that blocks the central rays of light. Figure 2.3 reveals the effect of this stop on the light rays passing through the condenser. If such a device is not available, one can be made by cutting round disks of opaque paper of different sizes that are cemented to transparent celluloid disks that will fit into the slot. If the microscope normally has a diffusion disk in this slot, it is best to replace it with rigid clear celluloid or glass.

An interesting modification of this technique is to use colored celluloid stops instead of opaque paper. Backgrounds of blue, red, or any color can be produced in this way.

In setting up this type of darkfield illumination, it is necessary to keep these points in mind:

1. Limit this technique to the study of large organisms that can be seen easily with low-power magnification. *Good resolution with higher powered objectives is difficult with this method.*

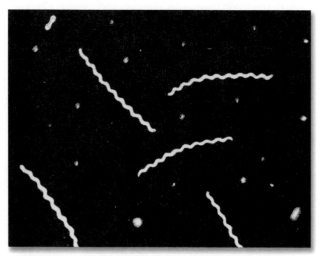

FIGURE 2.1 Transparent living microorganisms, such as the syphilis spirochaete, can be seen much more easily when observed in a dark field.

FIGURE 2.2 The insertion of a star diaphragm into the filter slot of the condenser will produce a dark field suitable for low magnifications. © The McGraw-Hill Companies/Auburn University Photographic Service

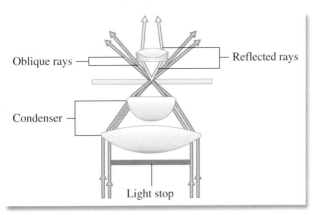

FIGURE 2.3 The star diaphragm allows only peripheral light rays to pass up through the condenser. This method requires maximum illumination.

2. Keep the diaphragm wide open and use as much light as possible. If the microscope has a voltage regulator, you will find that the higher voltages will produce better results.
3. Be sure to center the stop as precisely as possible.
4. Move the condenser up and down to produce the best effects.

THE CARDIOID CONDENSER

The difficulty that results from using the star diaphragm or opaque paper disks with high-dry and oil immersion objectives is that the oblique rays are not as carefully metered as is necessary for the higher magnifications. Special condensers such as the cardioid or paraboloid types must be used. Since the cardioid type is the most frequently used type, its use will be described here.

Figure 2.4 illustrates the light path through such a condenser. Note that the light rays entering the lower element of the condenser are reflected first off a convex mirrored surface and then off a second concave surface to produce the desired oblique rays of light.

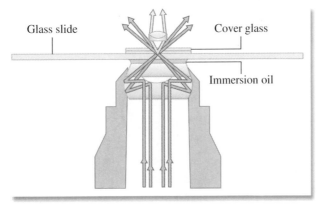

FIGURE 2.4 A cardioid condenser provides greater light concentration for oblique illumination than the star diaphragm.

Once the condenser has been installed in the microscope, the following steps should be followed to produce ideal illumination.

MATERIALS

- slides and cover glasses of excellent quality (slides of 1.15–1.25 mm thickness and No. 1 cover glasses)

1. Adjust the upper surface of the condenser to a height just below stage level.
2. Place a clear glass slide in position over the condenser.
3. Focus the 10× objective on the top of the condenser until a bright ring comes into focus.
4. Center the bright ring so that it is concentric with the field edge by adjusting the centering screws on the darkfield condenser. If the condenser has a light source built into it, it will also be necessary to center it as well to achieve even illumination.
5. Remove the clear glass slide.
6. If a funnel stop is available for the oil immersion objective, remove this object and insert this unit. (This stop serves to reduce the numerical aperture of the oil immersion objective to a value that is less than the condenser.)
7. Place a drop of immersion oil on the upper surface of the condenser and place the slide on top of the oil. The following preconditions in slide usage must be adhered to:
   - Slides and cover glasses should be optically perfect. Scratches and imperfections will cause annoying diffractions of light rays.
   - Slides and cover glasses must be free of dirt or grease of any kind.
   - A cover glass should always be used.
8. If the oil immersion lens is to be used, place a drop of oil on the cover glass.
9. If the field does not appear dark and lacks contrast, return to the 10× objective and check the ring concentricity and light source centration. If contrast is still lacking after these adjustments, the specimen is probably too thick.
10. If sharp focus is difficult to achieve under oil immersion, try using a thinner cover glass and adding more oil to the top of the cover glass and bottom of the slide.

LABORATORY REPORT

This exercise may be used in conjunction with Part 2 when studying the various types of organisms. After reading over this exercise and doing any special assignments made by your instructor, answer the questions in Laboratory Report 1 about darkfield microscopy.

LABORATORY REPORT

Student: _____

Date: _____ Section: _____

EXERCISE 2   Darkfield Microscopy

A. Questions

1. What characteristic of living bacteria makes them easier to see with a darkfield condenser than with a regular brightfield condenser?

    _____

    _____

2. If a darkfield condenser causes all light rays to bypass the objective, where does the light come from that makes an object visible in a dark field?

    _____

    _____

3. What advantage does a cardioid condenser have over a star diaphragm?

    _____

    _____

Phase-Contrast Microscopy

In order to visualize cells in the brightfield microscope, it is necessary to stain cells. If one tries to observe cells without the benefit of staining, very little contrast or detail is present and the cells appear transparent in the surrounding medium in which they are suspended. Staining increases the contrast between the cell and its surrounding medium and allows the observer to see greater detail, including cellular organelles. However, most staining procedures usually lead to the death of the cells to be observed and, as a result, the observation of stained cells can give rise to artifacts.

A microscope that is able to differentiate the transparent protoplasmic structures and enhance the contrast between the cell and its surroundings, without the necessity of staining, is the ***phase-contrast microscope.*** The first phase-contrast microscope was developed in 1933 by Frederick Zernike and was originally called the *Zernike microscope.* Today, it is the microscope of choice for observing living cells and activities of cells such as motility. Figure 3.1 illustrates the contrast differences between brightfield and phase-contrast images. In this exercise, you will learn to use the phase-contrast microscope and observe the activities of living cells.

IMAGE CONTRAST

Objects in a microscopic field may be categorized as being either amplitude or phase objects. **Amplitude objects** (illustration 1, figure 3.2) show up as dark objects under the microscope because the amplitude (intensity) of light rays is reduced as the rays pass through the objects. **Phase objects** (illustration 2, figure 3.2), on the other hand, are completely transparent since light rays pass through them unchanged with respect to amplitude. As some of the light rays pass through phase objects, however, they are retarded by $\frac{1}{4}$ wavelength.

This retardation, known as *phase shift,* occurs with no amplitude diminution; thus, the objects appear transparent rather than opaque. Since most biological specimens are phase objects, lacking in contrast, it becomes necessary to apply dyes of various kinds to cells that are to be studied with a brightfield microscope. To understand how Zernike took advantage of the $\frac{1}{4}$ wavelength phase shift in developing his microscope, we must understand the difference between direct and diffracted light rays.

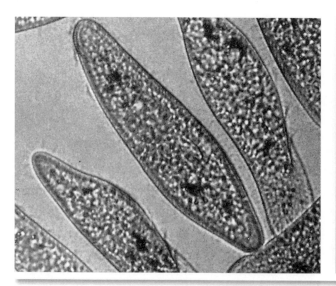

(a) Brightfield

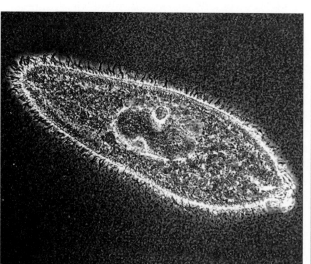

(b) Phase contrast

FIGURE 3.1 Comparison of brightfield and phase-contrast images

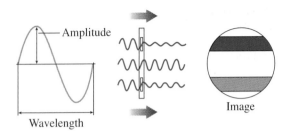

AMPLITUDE OBJECTS

(1) The extent to which the amplitude of light rays is diminished determines the darkness of an object in a microscopic field.

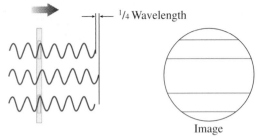

PHASE OBJECTS

(2) Note that the retardation of light rays without amplitude diminution results in transparent phase objects.

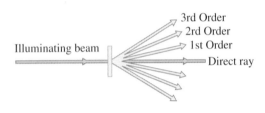

DIRECT AND DIFFRACTED RAYS

(3) A light ray passing through a slit or transparent object emerges as a direct ray with several orders of diffracted rays. The diffracted rays are $\frac{1}{4}$ wavelength out of phase with the direct ray.

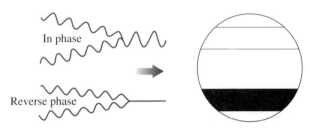

COINCIDENCE AND INTERFERENCE

(4) Note that when two light rays are in phase they will unite to produce amplitude summation. Light rays in reverse phase, however, cancel each other (interference) to produce dark objects.

FIGURE 3.2   The utilization of light rays in phase-contrast microscopy

TWO TYPES OF LIGHT RAYS

Light rays passing through a transparent object emerge as either direct or diffracted rays. Those rays that pass straight through unaffected by the medium are called **direct rays.** They are unaltered in amplitude and phase. The balance of the rays that are bent by their slowing through the medium (due to density differences) emerge from the object as **diffracted rays.** It is these rays that are retarded $\frac{1}{4}$ wavelength. Illustration 3, figure 3.2, illustrates these two types of light rays.

An important characteristic of these light rays is that if the direct and diffracted rays of an object can be brought into exact phase, or *coincidence,* with each other, the resultant amplitude of the converged rays is the sum of the two waves. This increase in amplitude will produce increased brightness of the object in the field. On the other hand, if two rays of equal amplitude are in reverse phase ($\frac{1}{2}$ wavelength off), their amplitudes cancel each other to produce a dark object. This phenomenon is called *interference.* Illustration 4, figure 3.2, shows these two conditions.

THE ZERNIKE MICROSCOPE

In constructing his first phase-contrast microscope, Zernike experimented with various configurations of diaphragms and various materials that could be used to retard or advance the direct light rays. Figure 3.3 illustrates the optical system of a typical modern phase-contrast microscope. It differs from a conventional brightfield microscope by having (1) a different type of diaphragm and (2) a phase plate.

The diaphragm consists of an **annular stop** that allows only a hollow cone of light rays to pass up through the condenser to the object on the slide. The **phase plate** is a special optical disk located at the rear focal plane of the objective. It has a **phase ring** on it that advances or retards the direct light rays $\frac{1}{4}$ wavelength.

Note in figure 3.3 that the direct rays converge on the phase ring to be advanced or retarded $\frac{1}{4}$ wavelength. These rays emerge as solid lines from the object on the slide. This ring on the phase plate is coated with a material that will produce the desired phase shift. The diffracted rays, on the other hand, which have already been retarded $\frac{1}{4}$ wavelength by the phase object on the slide, completely miss the phase ring and are not affected by the phase plate. It should be clear, then, that depending on the type of phase-contrast microscope, the convergence of diffracted and direct rays on the image plane will result in either a brighter image (*amplitude summation*) or a darker image (*amplitude interference* or *reverse phase*).

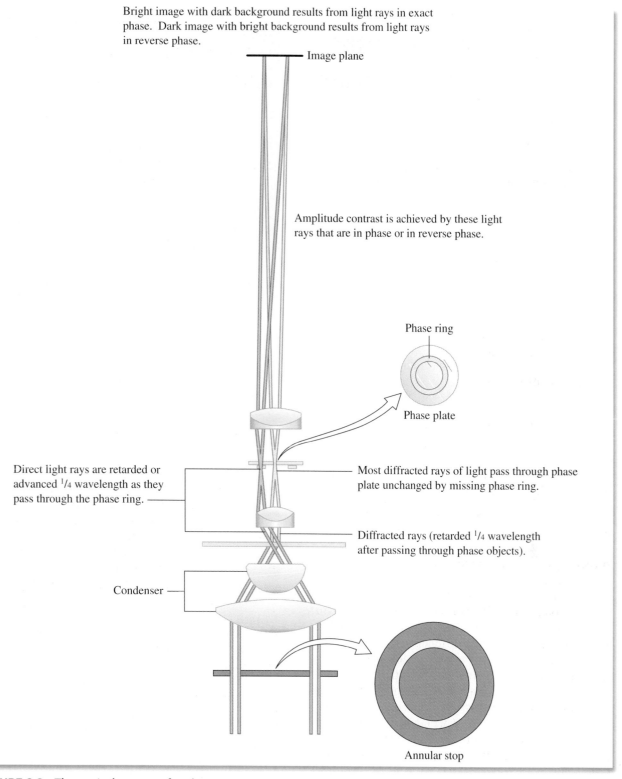

Bright image with dark background results from light rays in exact phase. Dark image with bright background results from light rays in reverse phase.

Image plane

Amplitude contrast is achieved by these light rays that are in phase or in reverse phase.

Phase ring

Phase plate

Direct light rays are retarded or advanced ¼ wavelength as they pass through the phase ring.

Most diffracted rays of light pass through phase plate unchanged by missing phase ring.

Diffracted rays (retarded ¼ wavelength after passing through phase objects).

Condenser

Annular stop

FIGURE 3.3 **The optical system of a phase-contrast microscope**

The former is referred to as ***bright phase*** microscopy; the latter as ***dark phase*** microscopy. The apparent brightness or darkness, incidentally, is proportional to the square of the amplitude; thus, the image will be four times as bright or dark as seen through a bright-field microscope.

It should be added here that the phase plates of some microscopes have coatings to change the phase

17

of the diffracted rays. In any event, the end result will be the same: to achieve coincidence or interference of direct and diffracted rays.

MICROSCOPE ADJUSTMENTS

If the annular stop under the condenser of a phase-contrast microscope can be moved out of position, this instrument can also be used for brightfield studies. Although a phase-contrast objective has a phase ring attached to the top surface of one of its lenses, the presence of that ring does not impair the resolution of the objective when it is used in the brightfield mode. It is for this reason that manufacturers have designed phase-contrast microscopes in such a way that they can be quickly converted to brightfield operation.

To make a microscope function efficiently in both phase-contrast and brightfield situations, one must master the following procedures:

- lining up the annular ring and phase rings so that they are perfectly concentric,
- adjusting the light source so that maximum illumination is achieved for both phase-contrast and brightfield usage, and
- being able to shift back and forth easily from phase-contrast to brightfield modes. The following suggestions should be helpful in coping with these problems.

Alignment of Annulus and Phase Ring

Unless the annular ring below the condenser is aligned perfectly with the phase ring in the objective, good phase-contrast imagery cannot be achieved. Figure 3.4 illustrates the difference between nonalignment and alignment. If a microscope has only one phase-contrast objective, there will be only one annular stop that has to be aligned. If a microscope has two or more phase objectives, there must be a substage unit with separate annular stops for each phase objective, and alignment procedure must be performed separately for each objective and its annular stop.

Since the objective cannot be moved once it is locked in position, all adjustments are made to the annular stop. On some microscopes the adjustment may be made with tools, as illustrated in figure 3.5. On other microscopes, figure 3.6, the annular rings are moved into position with special knobs on the substage unit. Since the method of adjustment varies from one brand of microscope to another, one has to follow the instructions provided by the manufacturer. Once the adjustments have been made, they are rigidly set and needn't be changed unless someone inadvertently disturbs them.

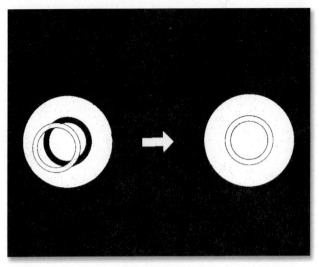

FIGURE 3.4 The image on the right illustrates the appearance of the rings when perfect alignment of phase ring and annulus diaphragm has been achieved.

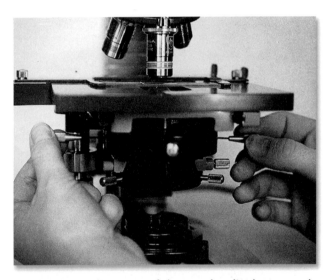

FIGURE 3.5 Alignment of the annulus diaphragm and phase ring is accomplished with a pair of Allen-type screwdrivers on this American Optical microscope.

To observe ring alignment, one can replace the eyepiece with a **centering telescope** as shown in figure 3.7. With this unit in place, the two rings can be brought into sharp focus by rotating the focusing ring on the telescope. Refocusing is necessary for each objective and its matching annular stop. Some manufacturers, such as American Optical, provide an aperture viewing unit (figure 3.8), which enables one to observe the rings without using a centering telescope. Zeiss microscopes have a unit called the ***Optovar,*** which is located in a position similar to the American Optical unit that serves the same purpose.

FIGURE 3.6 Alignment of the annulus and phase ring is achieved by adjusting the two knobs as shown. ©The McGraw-Hill Companies/Auburn University Photographic Service

FIGURE 3.8 Some microscopes have an aperture viewing unit that can be used instead of a centering telescope for observing the orientation of the phase ring and annular ring.

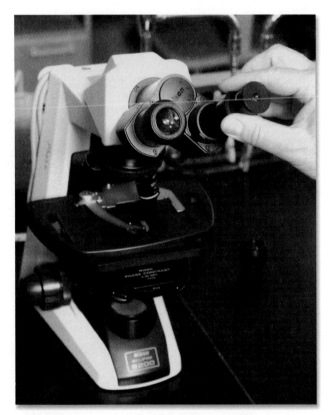

FIGURE 3.7 If the ocular of a phase-contrast microscope is replaced with a centering telescope, the orientation of the phase ring and annular ring can be viewed. ©The McGraw-Hill Companies/Auburn University Photographic Service

Light Source Adjustment

For both brightfield and phase-contrast modes, it is essential that optimum lighting be achieved. This is no great problem for a simple setup such as the American Optical instrument shown in figure 3.9. For multiple phase objective microscopes, however, there are many more adjustments that need to be made. A few suggestions that highlight some of the problems and solutions follow:

1. Since blue light provides better images for both phase-contrast and brightfield modes, make certain that a blue filter is placed in the filter holder that is positioned in the light path. If the microscope has no filter holder, placing the filter over the light source on the base will help.
2. Brightness of field under phase-contrast is controlled by adjusting the voltage or the iris diaphragm on the base. Considerably more light is required for phase-contrast than for brightfield since so much light is blocked out by the annular stop.
3. The evenness of illumination on some microscopes, seen on these pages, can be adjusted by removing the lamp housing from the microscope and focusing the light spot on a piece of translucent white paper. For the detailed steps in this procedure, one should consult the instruction manual that comes with the microscope. Light source adjustments of this nature are not necessary for the simpler types of microscopes.

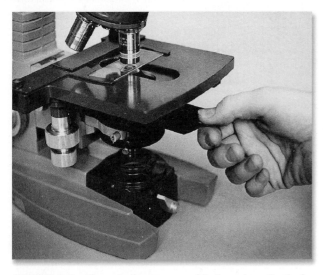

FIGURE 3.9 The annular stop on this American Optical microscope has the annular stop located on a slideway. When pushed in, the annular stop is in position.

WORKING PROCEDURES

Once the light source is correct and the phase elements are centered you are finally ready to examine slide preparations. Keep in mind that from now on most of the adjustments described earlier should not be altered; however, if misalignment has occurred due to mishandling, it will be necessary to refer back to alignment procedures. The following guidelines should be adhered to in all phase-contrast studies:

• Use only optically perfect slides and cover glasses (no bubbles or striae in the glass).
• Be sure that slides and cover glasses are completely free of grease or chemicals.
• Use wet mount slides instead of hanging drop preparations. The latter leave much to be desired.

Culture broths containing bacteria or protozoan suspensions are ideal for wet mounts.
• In general, limit observations to living cells. In most instances, stained slides are not satisfactory.

The first time you use phase-contrast optics to examine a wet mount, follow these suggestions:

1. Place the wet mount slide on the stage and bring the material into focus, *using brightfield optics* at low-power magnification.
2. Once the image is in focus, switch to phase optics at the same magnification. Remember, it is necessary to place in position the matching annular stop.
3. Adjust the light intensity, first with the base diaphragm and then with the voltage regulator. In most instances, you will need to increase the amount of light for phase-contrast.
4. Switch to higher magnifications, much in the same way you do for brightfield optics, except that you have to rotate a matching annular stop into position.
5. If an oil immersion phase objective is used, add immersion oil to the top of the condenser as well as to the top of the cover glass.
6. Don't be disturbed by the "halo effect" that you observe with phase optics. Halos are normal.

LABORATORY REPORT

This exercise may be used in conjunction with Part 2 in studying various types of organisms. Organelles in protozoans and algae will show up more distinctly than with brightfield optics. After reading this exercise and doing any special assignments made by your instructor, answer the questions in Laboratory Report 3.

LABORATORY REPORT

Student: _____

Date: _____ Section: _____

EXERCISE 3   Phase-Contrast Microscopy

A. Questions

1. Which rays (*direct* or *diffracted*) are altered by the phase ring on the phase plate?

2. How much phase shift occurs in the light rays that emerge from a transparent object?

3. Differentiate:

    Bright phase microscope: _____

    Dark phase microscope: _____

4. List two items that can be used for observing the concentricity of the annulus and phase ring:

    a. _____   b. _____

B. Multiple Choice

Select the answer that best completes the following statements.

1. If direct rays passing through an object are advanced $\frac{1}{4}$ wavelength by the phase ring, the diffracted rays are
    1. in phase with the direct rays.
    2. $\frac{1}{2}$ wavelength out of phase with the direct rays.
    3. $\frac{3}{4}$ wavelength out of phase with direct rays.
    4. in reverse phase with the direct rays.
    5. Both 2 and 4 are correct.

2. Amplitude summation occurs in phase-contrast optics when both direct and diffracted rays are
    1. in phase.
    2. in reverse phase.
    3. off $\frac{1}{4}$ wavelength.
    4. None of these are correct.

3. The phase-contrast microscope is best suited for observing
    1. living organisms in an uncovered drop on a slide.
    2. stained slides with cover glasses.
    3. living organisms in hanging drop slide preparations.
    4. living organisms on a slide with a cover glass.

ANSWERS

Multiple Choice

1. _____

2. _____

3. _____

4. _____

5. _____

4. A phase centering telescope is used to
    1. improve the resolution of the ocular.
    2. increase magnification with the oil immersion objective.
    3. observe the relationship of the annular diaphragm to the phase ring.
    4. None of the above are correct.

5. The visible spectrum of light is between
    1. 200 and 800 nanometers.
    2. 400 and 780 nanometers.
    3. 300 and 800 nanometers.
    4. None of these are correct.

Microscopic Measurements

FIGURE 4.1 The ocular micrometer with retaining ring is inserted into the base of the eyepiece.

FIGURE 4.2 Stage micrometer is positioned by centering the small glass disk over the light source.

With an ocular micrometer properly installed in the eyepiece of your microscope, it is a simple matter to measure the size of microorganisms that are seen in the microscopic field. An **ocular micrometer** consists of a circular disk of glass that has graduations engraved on its upper surface. These graduations appear as shown in illustration B, figure 4.4. On some microscopes one has to disassemble the ocular so that the disk can be placed on a shelf in the ocular tube between the two lenses. On most microscopes, however, the ocular micrometer is simply inserted into the bottom of the ocular, as shown in figure 4.1. Before one can use the micrometer, it is necessary to calibrate it for each of the objectives by using a stage micrometer.

The principal purpose of this exercise is to show you how to calibrate an ocular micrometer for the various objectives on your microscope.

CALIBRATION PROCEDURE

The distance between the lines of an ocular micrometer is an arbitrary value that has meaning only if the ocular micrometer is calibrated for the objective that is being used. A **stage micrometer** (figure 4.2), also known as an *objective micrometer,* has lines scribed on it that are exactly 0.01 mm (10 μm) apart. Illustration C, figure 4.4, shows these graduations.

To calibrate the ocular micrometer for a given objective, it is necessary to superimpose the two scales

and determine how many of the ocular graduations coincide with one graduation on the scale of the stage micrometer. Illustration A in figure 4.4 shows how the two scales appear when they are properly aligned in the microscopic field. In this case, seven ocular divisions match up with one stage micrometer division of 0.01 mm to give an ocular value of 0.01/7, or 0.00143 mm. Since there are 1000 micrometers in 1 millimeter, these divisions are 1.43 μm apart.

With this information known, the stage micrometer is replaced with a slide of organisms (figure 4.3) to

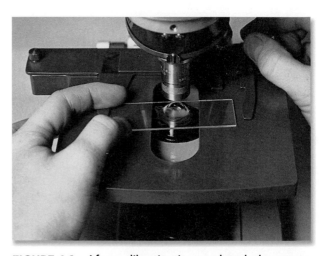

FIGURE 4.3 After calibration is completed, the stage micrometer is replaced with a slide for measurements.

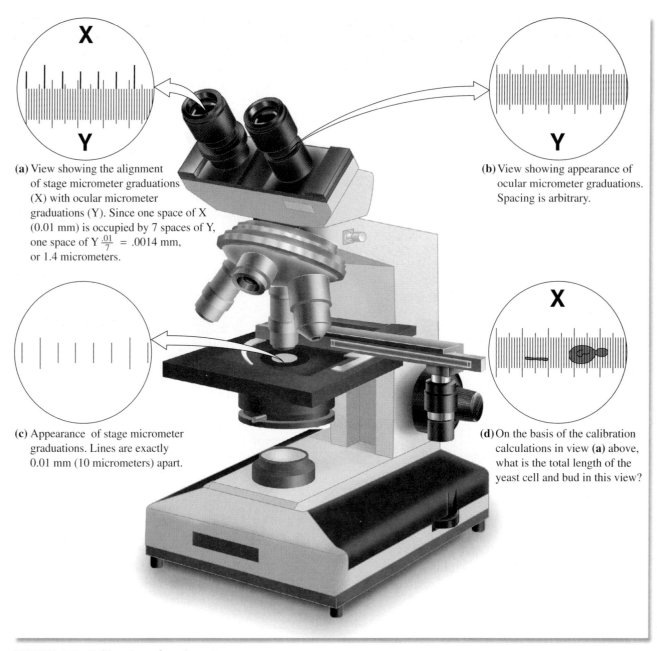

(a) View showing the alignment of stage micrometer graduations (X) with ocular micrometer graduations (Y). Since one space of X (0.01 mm) is occupied by 7 spaces of Y, one space of $Y \frac{.01}{7} = .0014$ mm, or 1.4 micrometers.

(b) View showing appearance of ocular micrometer graduations. Spacing is arbitrary.

(c) Appearance of stage micrometer graduations. Lines are exactly 0.01 mm (10 micrometers) apart.

(d) On the basis of the calibration calculations in view **(a)** above, what is the total length of the yeast cell and bud in this view?

FIGURE 4.4 Calibration of ocular micrometer.

be measured. Illustration D, figure 4.4, shows how a field of microorganisms might appear with the ocular micrometer in the eyepiece. To determine the size of an organism, then, it is a simple matter to count the graduations and multiply this number by the known distance between the graduations. When calibrating the objectives of a microscope, proceed as follows.

MATERIALS

- ocular micrometer or eyepiece that contains a micrometer disk
- stage micrometer

1. If eyepieces are available that contain ocular micrometers, replace the eyepiece in your microscope

with one of them. If it is necessary to insert an ocular micrometer in your eyepiece, find out from your instructor whether it is to be inserted below the bottom lens or placed between the two lenses within the eyepiece. In either case, great care must be taken to avoid dropping the eyepiece or reassembling the lenses incorrectly. *Only with your instructor's prior approval shall eyepieces be disassembled.* Be sure that the graduations are on the upper surface of the glass disk.

2. Place the stage micrometer on the stage and center it exactly over the light source.

3. With the low-power (10×) objective in position, bring the graduations of the stage micrometer into focus, *using the coarse adjustment knob. Reduce the lighting.* Note: If the microscope has an automatic stop, do not use it as you normally would for regular microscope slides. The stage micrometer slide is too thick to allow it to function properly.

4. Rotate the eyepiece until the graduations of the ocular micrometer lie parallel to the lines of the stage micrometer.

5. If the **low-power objective** is the objective to be calibrated, proceed to step 8.

6. If the **high-dry objective** is to be calibrated, swing it into position and proceed to step 8.

7. If the **oil immersion lens** is to be calibrated, place a drop of immersion oil on the stage micrometer, swing the oil immersion lens into position, and bring the lines into focus; then, proceed to the next step.

8. Move the stage micrometer laterally until the lines at one end coincide. Then look for another line on the ocular micrometer that coincides *exactly* with one on the stage micrometer. Occa-

sionally, one stage micrometer division will include an even number of ocular divisions, as shown in illustration A of figure 4.4. In most instances, however, several stage graduations will be involved. In this case, divide the number of stage micrometer divisions by the number of ocular divisions that coincide. The figure you get will be that part of a stage micrometer division that is seen in an ocular division. This value must then be multiplied by 0.01 mm to get the amount of each ocular division.

Example: 3 divisions of the stage micrometer line up with 20 divisions of the ocular micrometer.

$$\text{Each ocular division} = \tfrac{3}{20} \times 0.01$$
$$= 0.0015 \text{ mm}$$
$$= 1.5 \text{ μm}$$

Replace the stage micrometer with slides of organisms to be measured.

MEASURING ASSIGNMENTS

Organisms such as protozoans, algae, fungi, and bacteria in the next few exercises may need to be measured. If your instructor requires that measurements be made, you will be referred to this exercise.

Later on you will be working with unknowns. In some cases measurements of the unknown organisms will be pertinent to identification.

If trial measurements are to be made at this time, your instructor will make appropriate assignments. **Important:** Remove the ocular micrometer from your microscope at the end of the laboratory period.

LABORATORY REPORT

Answer the questions in Laboratory Report 4.

LABORATORY REPORT

Student: _____

Date: _____ Section: _____

EXERCISE 4   Microscopic Measurements

Questions

1. What is the distance between each of the graduations on the stage micrometer?_____ mm

2. Why must the entire calibration procedure be performed for each objective?_____

_____

SURVEY OF MICROORGANISMS

Microorganisms abound in the environment. Eukaryotic microbes such as protozoa, algae, diatoms, and amoebas are plentiful in ponds and lakes. Bacteria are found associated with animals, occur abundantly in the soil and in water systems, and have even been isolated from core samples taken from deep within the earth's crust. Bacteria are also present in the air where they are distributed by convection currents that transport them from other environments. The Archaea, modern day relatives of early microorganisms, occupy some of the most extreme environments such as acidic-volcanic hot springs, anaerobic environments devoid of any oxygen, and lakes and salt marshes excessively high in sodium chloride. Cyanobacteria are photosynthetic prokaryotes that can be found growing in ponds and lakes, on limestone rocks, and even on the shingles that protect the roofs of our homes. Fungi are a very diverse group of microorganisms that are found in most common environments. For example, they degrade complex molecules in the soil, thus contributing to its fertility. Sometimes, however, they can be nuisance organisms; they form mildew in our bathroom showers and their spores cause allergies. The best way of describing the distribution of microorganisms is to say that they are ubiquitous, or found everywhere.

Intriguing questions to biologists are how are the various organisms related to one another and where do the individual organisms fit in an evolutionary scheme? Molecular biology techniques have provided a means to analyze the genetic relatedness of the organisms that comprise the biological world and determine where the various organisms fit into an evolutionary scheme. By comparing the sequence of ribosomal RNA molecules, coupled with biochemical data, investigators have developed a phylogenetic tree that illustrates the current thinking on the placement of the various organisms into such a

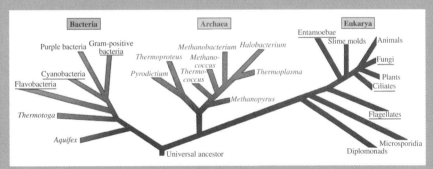

From: Extremophiles. Michael T. Madigan and Barry L. Marrs in *Scientific American* Vol. 276, Number 4, pages 82–87, April 1997.

scheme. This evolutionary scheme divides the biological world into three domains.

Domain Bacteria   These organisms have a prokaryotic cell structure. They lack organelles such as mitochondria and chloroplasts, are devoid of an organized nucleus with a nuclear membrane, and possess 70S ribosomes that are inhibited by many broad spectrum antibiotics. The vast majority of organisms are enclosed in a cell wall composed of peptidoglycan. The bacteria and cyanobacteria are members of this domain.

Domain Eukarya   Organisms in this domain have a eukaryotic cell structure. They contain organelles such as mitochondria and chloroplasts, an organized nucleus enclosed in a nuclear membrane, and 80S ribosomes that are not inhibited by broad spectrum antibiotics. Plants, animals and microorganisms such as protozoa, algae, and the fungi belong to this domain.

Domain Archaea   The Archaea exhibit the characteristics of both the bacteria and Eukarya. These organisms are considered to be the relatives of ancient microbes that existed during Archean times. Like their bacterial counterparts, they possess a simple cell structure that lacks organelles and an organized nucleus. They have 70S ribosomes but the latter are more similar to 80S ribosomes and they are not sensitive to antibiotics. They have a cell wall but its structure is not composed of peptidoglycan. The principle habitats of these organisms are extreme environments such as volcanic hot springs, environments with excessively high salt, and environments devoid of oxygen. Thus, they are referred to as "extremophiles." The acido-thermophiles, the halobacteria, and the methanogens (methane bacteria) are examples of the Archaea.

In the exercises of Part 2, you will have the opportunity to study some of these organisms. In pond water, you may see amoebas, protozoans, various algae, diatoms, and cyanobacteria. You will sample for the presence of bacteria by exposing growth media to various environments. The fungi will be studied by looking at cultures and preparing slides of these organisms. Because the Archaea occur in extreme conditions and also require specialized culture techniques, it is unlikely that you will encounter any of these organisms.

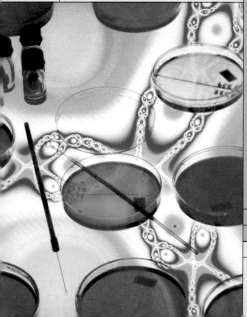

Protozoans, Algae, and Cyanobacteria

In this exercise, a study will be made of protozoans, algae, and cyanobacteria that are found in pond water. Bottles that contain water and bottom debris from various ponds will be available for study. Illustrations and text provided in this exercise will be used to assist you in an attempt to identify the various types that are encountered. Unpigmented, moving microorganisms will probably be protozoans. Greenish or golden-brown organisms are usually algae. Organisms that appear blue-green will be cyanobacteria. Supplementary books on the laboratory bookshelf will also be available for assistance in identifying organisms that are not described in the short text of this exercise.

The purpose of this exercise is, simply, to provide you with an opportunity to become familiar with the differences between the three groups by comparing their characteristics. The extent to which you will be held accountable for the names of various organisms will be determined by your instructor. The amount of time available for this laboratory exercise will determine the depth of scope to be pursued.

To study the microorganisms of pond water, it will be necessary to make wet mount slides. The procedure for making such slides is relatively simple. All that is necessary is to place a drop of suspended organisms on a microscope slide and cover it with a cover glass. If several different cultures are available, the number of the bottle should be recorded on the slide with a marking pen. As you prepare and study your slides, observe the guidelines below.

MATERIALS

- bottles of pond-water samples
- microscope slides and cover glasses
- rubber-bulbed pipettes and forceps
- Sharpie marking pen
- reference books

1. Clean the slide and cover glass with soap and water, rinse thoroughly, and dry. Do not attempt to study a slide that lacks a cover glass.

2. When using a pipette, insert it into the bottom of the bottle to get a maximum number of organisms. Very few organisms will be found swimming around in middepth of the bottle.
3. To remove filamentous algae from a specimen bottle, use forceps. Avoid putting too much material on the slides.
4. Explore the slide first with the low-power objective. Reduce the lighting with the iris diaphragm. Keep the condenser at its highest point.
5. When you find an organism of interest, swing the high-dry objective into position and adjust the lighting to get optimum contrast. If your microscope has phase-contrast elements, use them.
6. Refer to figures 5.1 through 5.6 and the text on these pages to identify the various organisms that you encounter.
7. Record your observations on the Laboratory Reports.

THE PROTISTS

Single-celled eukaryons that lack tissue specialization are called *protists*. Protozoologists group all protists in **Kingdom Protista**. Those protists that are animal-like are put in **Subkingdom Protozoa** and the protists that are plantlike fall into **Subkingdom Algae.** This system of classification includes all colonial species as well as the single-celled types.

SUBKINGDOM PROTOZOA

Externally, protozoan cells are covered with a cell membrane, or pellicle; cell walls are absent; and distinct nuclei with nuclear membranes are present. Specialized organelles, such as contractile vacuoles, cytostomes, mitochondria, ribosomes, flagella, and cilia, may also be present.

All protozoans produce **cysts,** which are resistant dormant stages that enable them to survive drought, heat, and freezing. They reproduce asexually by cell division and exhibit various degrees of sexual reproduction.

The Subkingdom Protozoa is divided into three phyla: Sarcomastigophora, Ciliophora, and

Apicomplexa. Type of locomotion plays an important role in classification here. A brief description of each phylum follows:

Phylum Sarcomastigophora

Members of this phylum have been subdivided into two subphyla: Sarcodina and Mastigophora.

Sarcodina (Amoebas) Members of this subphylum move about by the formation of flowing protoplasmic projections called *pseudopodia*. The formation of pseudopodia is commonly referred to as *amoeboid movement*. Illustrations 5 through 8 in figure 5.1 are representative amoebas.

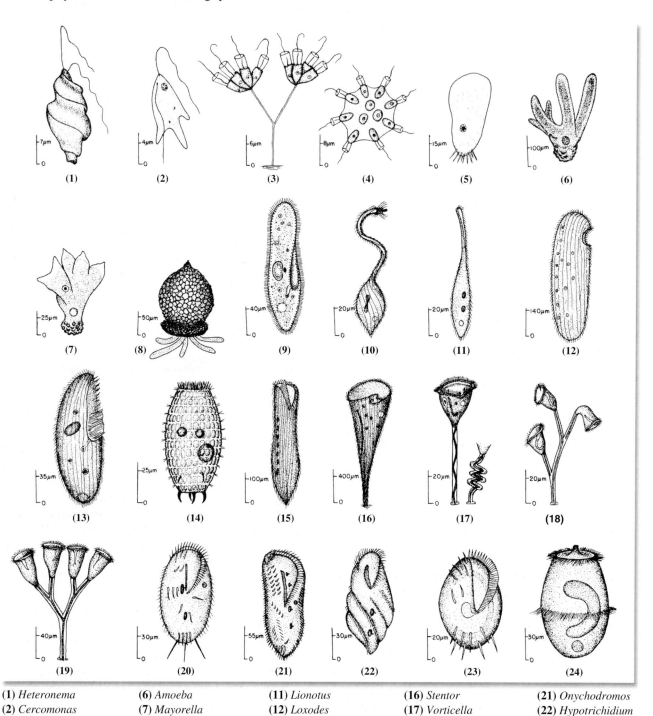

(1) Heteronema	(6) Amoeba	(11) Lionotus	(16) Stentor	(21) Onychodromos
(2) Cercomonas	(7) Mayorella	(12) Loxodes	(17) Vorticella	(22) Hypotrichidium
(3) Codosiga	(8) Diffugia	(13) Blepharisma	(18) Carchesium	(23) Euplotes
(4) Protospongia	(9) Paramecium	(14) Coleps	(19) Zoothamnium	(24) Didinium
(5) Trichamoeba	(10) Lacymaria	(15) Condylostoma	(20) Stylonychia	

FIGURE 5.1 Protozoans

Mastigophora (*Zooflagellates*) These protozoans possess whiplike structures called *flagella*. There is considerable diversity among the members of this group. Only a few representatives (illustrations 1 through 4) are seen in figure 5.1.

Phylum Ciliophora

These microorganisms are undoubtedly the most advanced and structurally complex of all protozoans. Evidence seems to indicate that they have evolved from the zooflagellates. Movement and food getting are accomplished with short hairlike structures called *cilia*. Illustrations 9 through 24 are typical ciliates.

Phylum Apicomplexa

This phylum has only one class, the *Sporozoa*. Members of this phylum lack locomotor organelles and all are internal parasites. As indicated by their class name, their life cycles include spore-forming stages. *Plasmodium,* the malarial parasite, is a significant pathogenic sporozoan of humans.

SUBKINGDOM ALGAE

The Subkingdom Algae includes all the photosynthetic eukaryotic organisms in Kingdom Protista. Being true protists, they differ from the plants (*Plantae*) in that tissue differentiation is lacking.

The algae may be unicellular, as those shown in the top row of figure 5.2; colonial, like the four in the lower right-hand corner of figure 5.2; or filamentous, as those in figure 5.3. The undifferentiated algal structure is often referred to as a *thallus*. It lacks the stem, root, and leaf structures that result from tissue specialization.

These microorganisms are universally present where ample moisture, favorable temperature, and sufficient sunlight exist. Although a great majority of them live submerged in water, some grow on soil. Others grow on the bark of trees or on the surfaces of rocks.

Algae have distinct, visible nuclei and chloroplasts. **Chloroplasts** are organelles that contain **chlorophyll a** and other pigments. Photosynthesis takes place within these bodies. The size, shape, distribution, and number of chloroplasts vary considerably from species to species. In some instances, a single chloroplast may occupy most of the cell space.

Although there are seven divisions of algae, only five will be listed here. Since two groups, the cryptomonads and red algae, are not usually encountered in freshwater ponds, they have not been included here.

Division 1 Euglenophycophyta
(Euglenoids)

Illustrations 1 through 6 in figure 5.2 are typical euglenoids, representing four different genera within this relatively small group. All of them are flagellated and appear to be intermediate between the algae and protozoans. Protozoanlike characteristics seen in the euglenoids are (1) the absence of a cell wall, (2) the presence of a gullet, (3) the ability to ingest food but not through the gullet, (4) the ability to assimilate organic substances, and (5) the absence of chloroplasts in some species. In view of these facts, it becomes readily apparent why many zoologists often group the euglenoids with the zooflagellates.

The absence of a cell wall makes these protists very flexible in movement. Instead of a cell wall, they possess a semirigid outer **pellicle,** which gives the organism a definite form. Photosynthetic types contain **chlorophylls a** and **b,** and they always have a red **stigma** (eyespot), which is light sensitive. Their characteristic food-storage compound is a lipopolysaccharide, **paramylum.** The photosynthetic euglenoids can be bleached experimentally by various means in the laboratory. The colorless forms that develop, however, cannot be induced to revert back to phototrophy.

Division 2 Chlorophycophyta
(Green Algae)

The majority of algae observed in ponds belong to this group. They are grass-green in color, resembling the euglenoids in having **chlorophylls a** and **b.** They differ from euglenoids in that they synthesize **starch** instead of paramylum for food storage.

The diversity of this group is too great to explore its subdivisions in this preliminary study; however, the small flagellated *Chlamydomonas* (illustration 8, figure 5.2) appears to be the archetype of the entire group and has been extensively studied. Many colonial forms, such as *Pandorina, Eudorina, Gonium,* and *Volvox* (illustrations 14, 15, 19, and 20, figure 5.2), consist of organisms similar to *Chlamydomonas.* It is the general consensus that from this flagellated form all the filamentous algae have evolved.

Except for *Vaucheria* and *Tribonema,* all of the filamentous forms in figure 5.3 are Chlorophycophyta. All of the nonfilamentous, nonflagellated algae in figure 5.4 also are green algae.

A unique group of green algae is the **desmids** (illustrations 16 through 20, figure 5.4). With the exceptions of a few species, the cells of desmids consist of two similar halves, or semicells. The two halves usually are separated by a constriction, the *isthmus*.

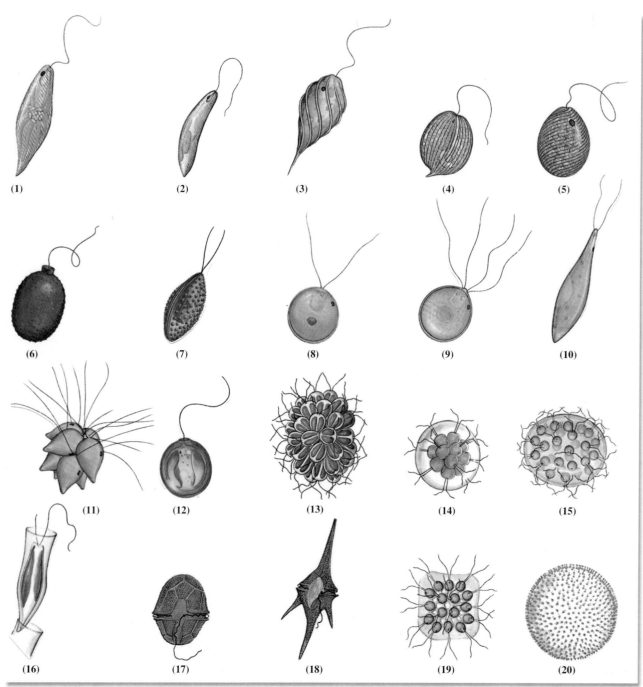

(1) *Euglena* (700X)
(2) *Euglena* (700X)
(3) *Phacus* (1000X)
(4) *Phacus* (350X)
(5) *Lepocinclis* (350X)

(6) *Trachelomonas* (1000X)
(7) *Phacotus* (1500X)
(8) *Chlamydomonas* (1000X)
(9) *Carteria* (1500X)
(10) *Chlorogonium* (1000X)

(11) *Pyrobotrys* (1000X)
(12) *Chrysococcus* (3000X)
(13) *Synura* (350X)
(14) *Pandorina* (350X)
(15) *Eudorina* (175X)

(16) *Dinobyron* (1000X)
(17) *Peridinium* (350X)
(18) *Ceratium* (175X)
(19) *Gonium* (350X)
(20) *Volvox* (100X)

FIGURE 5.2 Flagellated algae (Courtesy of the U.S. Environmental Protection Agency, Office of Research & Development, Cincinnati, OH 45268)

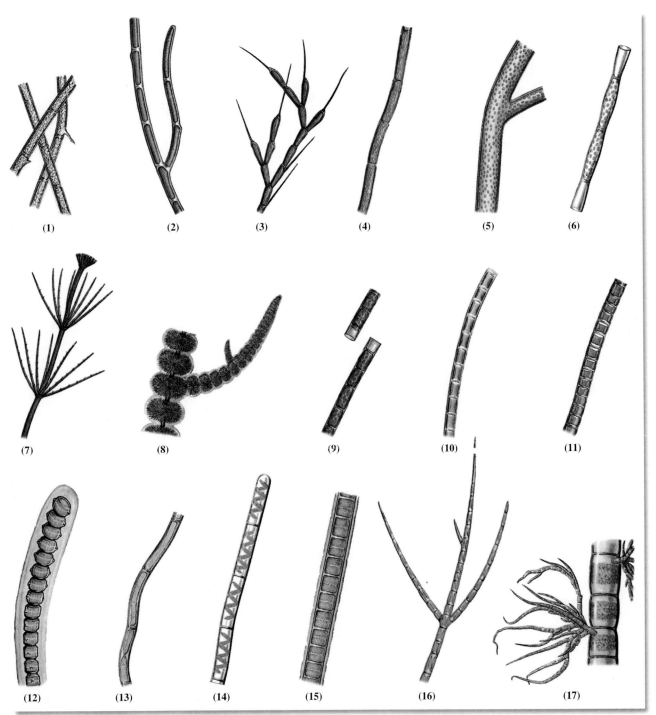

FIGURE 5.3   Filamentous algae (Courtesy of the U.S. Environmental Protection Agency, Office of Research & Development, Cincinnati, OH 45268)

(1) *Rhizoclonium* (175X)
(2) *Cladophora* (100X)
(3) *Bulbochaete* (100X)
(4) *Oedogonium* (350X)

(5) *Vaucheria* (100X)
(6) *Tribonema* (300X)
(7) *Chara* (3X)
(8) *Batrachospermum* (2X)

(9) *Microspora* (175X)
(10) *Ulothrix* (175X)
(11) *Ulothrix* (175X)
(12) *Desmidium* (175X)

(13) *Mougeotia* (175X)
(14) *Spirogyra* (175X)
(15) *Zygnema* (175X)
(16) *Stigeoclonium* (300X)
(17) *Draparnaldia* (100X)

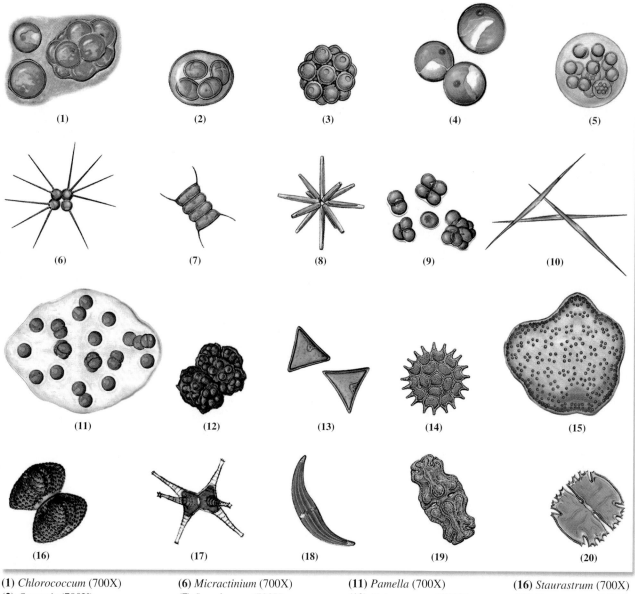

(1) *Chlorococcum* (700X)   (6) *Micractinium* (700X)   (11) *Pamella* (700X)   (16) *Staurastrum* (700X)
(2) *Oocystis* (700X)   (7) *Scendesmus* (700X)   (12) *Botryococcus* (700X)   (17) *Staurastrum* (350X)
(3) *Coelastrum* (350X)   (8) *Actinastrum* (700X)   (13) *Tetraedron* (1000X)   (18) *Closterium* (175X)
(4) *Chlorella* (350X)   (9) *Phytoconis* (700X)   (14) *Pediastrum* (100X)   (19) *Euastrum* (350X)
(5) *Sphaerocystis* (350X)   (10) *Ankistrodesmus* (700X)   (15) *Tetraspora* (100X)   (20) *Micrasterias* (175X)

FIGURE 5.4   Nonfilamentous and nonflagellated algae (Courtesy of the U.S. Environmental Protection Agency, Office of Research & Development, Cincinnati, OH 45268)

Division 3 Chrysophycophyta

(Golden Brown Algae)

This large diversified division consists of over six thousand species. They differ from the euglenoids and green algae in that (1) food storage is in the form of **oils** and **leucosin,** a polysaccharide; (2) **chlorophylls a** and **c** are present; and (3) **fucoxanthin,** a brownish pigment, is present. It is the combination of fucoxanthin, other yellow pigments, and the chlorophylls that causes most of these algae to appear golden brown.

Representatives of this division are seen in figures 5.2, 5.3, and 5.5. In figure 5.2, *Chrysococcus, Synura,* and *Dinobyron* are typical flagellated chrysophycophytes. *Vaucheria* and *Tribonema* are the only filamentous chrysophycophytes shown in figure 5.3.

All of the organisms in figure 5.5 are chrysophycophytes and fall into a special category of algae

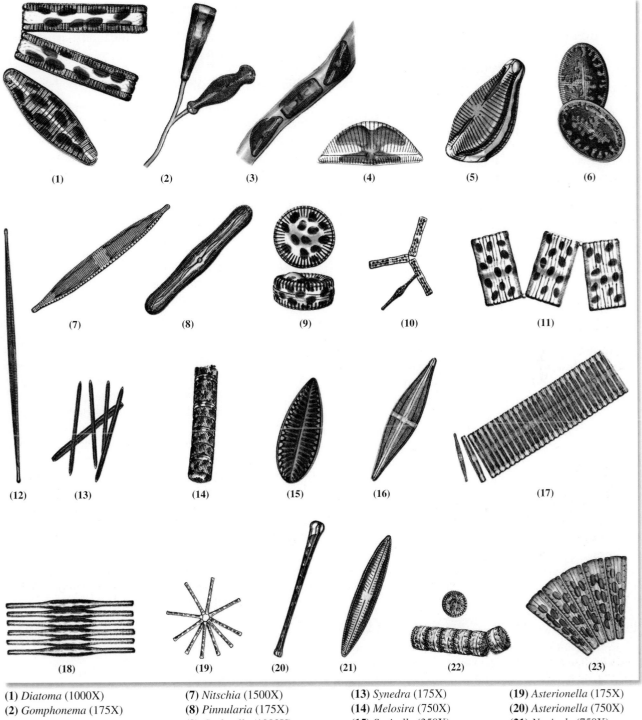

FIGURE 5.5 **Diatoms** (Courtesy of the U.S. Environmental Protection Agency, Office of Research & Development, Cincinnati, OH 45268)

(1) *Diatoma* (1000X)
(2) *Gomphonema* (175X)
(3) *Cymbella* (175X)
(4) *Cymbella* (1000X)
(5) *Gomphonema* (2000X)
(6) *Cocconeis* (750X)

(7) *Nitschia* (1500X)
(8) *Pinnularia* (175X)
(9) *Cyclotella* (1000X)
(10) *Tabellaria* (175X)
(11) *Tabellaria* (1000X)
(12) *Synedra* (350X)

(13) *Synedra* (175X)
(14) *Melosira* (750X)
(15) *Surirella* (350X)
(16) *Stauroneis* (350X)
(17) *Fragillaria* (750X)
(18) *Fragillaria* (750X)

(19) *Asterionella* (175X)
(20) *Asterionella* (750X)
(21) *Navicula* (750X)
(22) *Stephanodiscus* (750X)
(23) *Meridion* (750X)

called the **diatoms.** The diatoms are unique in that they have hard cell walls of pectin, cellulose, or silicon oxide that are constructed in two halves. The two halves fit together like lid and box.

Skeletons of dead diatoms accumulate on the ocean bottom to form *diatomite*, or "diatomaceous earth," which is commercially available as an excellent polishing compound. It is postulated by some that much of our petroleum reserves may have been formulated by the accumulation of oil from dead diatoms over millions of years.

Division 4 Phaeophycophyta

(Brown Algae)

With the exception of three freshwater species, all algal protists of this division exist in salt water (marine); thus, it is unlikely that you will encounter any phaeophycophytes in this laboratory experience. These algae have essentially the same pigments seen in the chrysophycophytes, but they appear brown because of the masking effect of the greater amount of fucoxanthin. Food storage in the brown algae is in the form of **laminarin,** a polysaccharide, and **mannitol,** a sugar alcohol. All species of brown algae are multicellular and sessile. Most seaweeds are brown algae.

Division 5 Pyrrophycophyta

(Fire Algae)

The principal members of this division are the **dinoflagellates.** Since the majority of these protists are marine, only two freshwater forms are shown in figure 5.2: *Peridinium* and *Ceratium* (illustrations 17 and 18). Most of these protists possess cellulose walls of interlocking armor plates, as in *Ceratium.* Two flagella are present: one is directed backward when swimming and the other moves within a transverse groove. Many marine dinoflagellates are bioluminescent. Some species of marine *Gymnodinium,* when present in large numbers, produce the **red tides** that cause water discoloration and unpleasant odors along our coastal shores.

These algae have **chlorophylls a** and **c** and several xanthophylls. Foods are variously stored in the form of **starch, fats,** and **oils.**

THE PROKARYOTES

The prokaryotes differ from the protists in that they are considerably smaller, lack distinct nuclei with nu-

clear membranes, and are enclosed in rigid cell walls. Since all members of this group are bacteria, the three-domain system of classification puts them in Domain Bacteria.

Division Cyanobacteria

Division Cyanobacteria in the Domain Bacteria includes microorganisms that were once classified as the blue-green algae. Their initial classification as algae was due in part to the fact that they have **chlorophyll a** and evolve oxygen as a result of photosynthetic metabolism. However, they are more properly classified with the Bacteria because they possess a typical prokaryotic cell structure and contain peptidoglycan in their cell walls. They differ from the green sulfur and the purple sulfur bacteria in that the latter use *bacteriochlorophyll* for photosynthesis and do not evolve oxygen because their photosynthetic metabolism is carried out under anaerobic conditions.

Over a thousand species of cyanobacteria have been reported. They are present in almost all moist environments from the tropics to the poles, including both freshwater and marine. Figure 5.6 illustrates only a random few that are frequently seen.

The designation of these bacteria as "blue-green" is somewhat misleading in that many are black, purple, red, and various shades of green, including blue-green. The varying colors are due to the varying proportions of photosynthetic pigments present, which include chlorophyll a, carotenoids, and phycobiliproteins. The latter pigments consist of allophycocyanin, phycocyanin, and phycoerythrin, which are in a complex with protein molecules to form **phycobilisomes.** These structures are also found in the red algae.

Cyanobacteria do not contain chloroplasts. Their photosynthic apparatus is contained in parallel arrays of membranes called thylakoids. The phycobilisomes are attached to the surface of the thylakoids in irregular arrays.

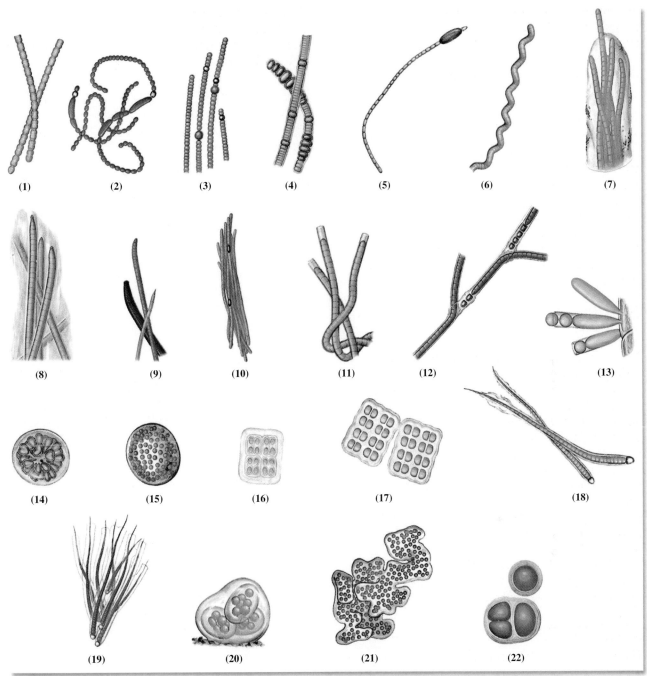

(1) (2) (3) (4) (5) (6) (7)

(8) (9) (10) (11) (12) (13)

(14) (15) (16) (17) (18)

(19) (20) (21) (22)

(1) *Anabaena* (350X)
(2) *Anabaena* (350X)
(3) *Anabaena* (175X)
(4) *Nodularia* (350X)
(5) *Cylindrospermum* (175X)
(6) *Arthrospira* (700X)

(7) *Microcoleus* (350X)
(8) *Phormidium* (350X)
(9) *Oscillatoria* (175X)
(10) *Aphanizomenon* (175X)
(11) *Lyngbya* (700X)
(12) *Tolypothrix* (350X)

(13) *Entophysalis* (1000X)
(14) *Gomphosphaeria* (1000X)
(15) *Gomphosphaeria* (350X)
(16) *Agmenellum* (700X)
(17) *Agmenellum* (175X)
(18) *Calothrix* (350X)

(19) *Rivularia* (175X)
(20) *Anacystis* (700X)
(21) *Anacystis* (175X)
(22) *Anacystis* (700X)

FIGURE 5.6 **Cyanobacteria** (Courtesy of the U.S. Environmental Protection Agency, Office of Research & Development, Cincinnati, OH 45268)

LABORATORY REPORT

Student: _____

Date: _____   Section: _____

EXERCISE 5   Protozoans, Algae, and Cyanobacteria

A. Tabulation of Observations

In this study of freshwater microorganisms, record your observations in the following tables. The number of organisms to be identified will depend on the availability of time and materials. Your instructor will indicate the number of each type that should be recorded.

Record the genus of each identifiable type. Also, indicate the phylum or division to which the organism belongs. Microorganisms that you are unable to identify should be sketched in the space provided. It is not necessary to draw those that are identified.

PROTOZOANS

GENUS	PHYLUM	BOTTLE NO.	SKETCHES OF UNIDENTIFIED

ALGAE

GENUS	DIVISION	BOTTLE NO.	SKETCHES OF UNIDENTIFIED

CYANOBACTERIA

GENUS	BOTTLE NO.	SKETCHES OF UNIDENTIFIED

B. General Questions

Record the answers to the following questions in the answer column. It may be necessary to consult your text or library references for one or two of the answers.

1. Give the kingdom in which each of the following groups of organisms is found:
   a. protozoans
   b. algae
   c. cyanobacteria
   d. bacteria
   e. fungi
   f. microscopic invertebrates

2. Four kingdoms are represented by the organisms in question 1. Name the fifth kingdom.

3. What is the most significant characteristic seen in eukaryotes that is lacking in prokaryotes?

4. What characteristic in the microscopic invertebrates distinguishes them from protozoans?

5. Which protozoan phylum was not found in pond samples because phylum members are all parasitic?

6. Indicate whether the following are *present* or *absent* in the algae:
   a. cilia
   b. flagella
   c. chloroplasts

ANSWERS

1a. _____

b. _____

c. _____

d. _____

e. _____

f. _____

2. _____

3. _____

4. _____

5. _____

6a. _____

b. _____

c. _____

7. Indicate whether the following are *present* or *absent* in the proto-zoans:
   a. cilia
   b. chloroplasts
   c. mitochondria
   d. mitosis

8. Which photosynthetic pigment is common to all algae and cyanobacteria?

9. Name two photosynthetic pigments that are found in the cyanobac-teria but not in the algae.

10. What photosynthetic pigment is found in bacteria but is lacking in all other photosynthetic organisms?

11. What type of movement is exhibited by the diatoms?

C. **Protozoan Characterization**

Select the protozoan groups in the right-hand column that have the following characteristics:

1. move with flagella
2. move with cilia
3. move with pseudopodia
4. have nuclear membranes
5. lack nuclear membranes
6. all species are parasitic
7. produce resistant cysts

1. Sarcodina
2. Mastigophora
3. Ciliophora
4. Sporozoa
5. all of above
6. none of above

ANSWERS

General Questions

7a. _____

b. _____

c. _____

d. _____

8. _____

9a. _____

b. _____

10. _____

11. _____

Protozoans

1. _____

2. _____

3. _____

4. _____

5. _____

6. _____

7. _____

D. Characterization of Algae and Cyanobacteria

Select the groups in the right-hand column that have the following characteristics:

PIGMENTS

1. chlorophyll a
2. chlorophyll b
3. chlorophyll c
4. fucoxanthin
5. c-phycocyanin
6. c-phycoerythrin

FOOD STORAGE

7. fats
8. oils
9. starches
10. laminarin
11. leucosin
12. paramylum
13. mannitol

OTHER STRUCTURES

14. pellicle, no cell wall
15. cell walls, box and lid
16. chloroplasts
17. phycobilisomes
18. thylakoids

1. Euglenophycophyta
2. Chlorophycophyta
3. Chrysophycophyta
4. Phaeophycophyta
5. Pyrrophycophyta
6. Cyanobacteria
7. all of above
8. none of above

ANSWERS

Algae

1. _____
2. _____
3. _____
4. _____
5. _____
6. _____
7. _____
8. _____
9. _____
10. _____
11. _____
12. _____
13. _____
14. _____
15. _____
16. _____
17. _____
18. _____

Ubiquity of Bacteria

Bacteria are the most widely distributed organisms in the biosphere. They occur as part of the normal flora of humans and animals, they are disease-causing parasites on many organisms, they abound in the soil and in water systems, and they are found deep within the earth's crust. We depend on them to degrade our sewage and solid wastes in landfills and to drive geochemical cycles involving nitrogen, carbon, and sulfur. Even though they are small in size and simple in their cell structure, they are considerably diverse in their metabolism and activities.

They are defined primarily by their cellular structure and small size. Bacteria have a simple cell structure that lacks a defined nucleus surrounded by nuclear membrane. Their nuclear genetic material is primarily supercoiled, circular DNA molecules that reside in the cell cytoplasm. Bacteria lack cellular organelles such as mitochondria and chloroplasts, but through modifications in their cell membrane, they carry out respiration and photosynthesis. Their ribosomes for synthesizing proteins are structurally different than higher cells and they are inhibited by many broad-spectrum antibiotics. For most bacteria, the cell is surrounded by a cell wall composed of a unique molecule called **peptidoglycan.** It is not found in any other kind of cell in the biological world. Most bacteria are small in size, averaging from 0.5 microns to 10 microns. Cyanobacteria tend to be larger, and one recently discovered bacterium is almost large enough to be seen without the aid of a microscope.

Figure 6.1 illustrates most of the common shapes of bacteria. They can be grouped into three morphological types: **rods** (bacilli); **cocci** (spherical); and **spirals** or curved rods. The rods or bacilli can have rounded, flat, or tapered ends and they can be motile or nonmotile. Cocci may occur singly, in chains, in a tetrad (packet of four cells), or in irregular masses. Most cocci are nonmotile because they lack flagella, the organelles of motility. The spiral bacteria can exist as slender spirochaetes, as a spirillum, or as a comma-shaped curved rod (vibrio).

During this laboratory period, you will be provided with three kinds of sterile bacteriological media that you will expose to the environment in various

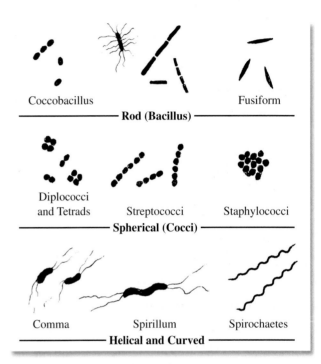

FIGURE 6.1   Bacterial morphology

ways. To ensure that these exposures cover as wide a spectrum as possible, specific assignments will be made for each student. In some instances, a moistened swab will be used to remove bacteria from some object; in other instances a petri plate of medium will be exposed to the air or a cough. You will be issued a number that will enable you to determine your specific assignment from the chart on page 46.

MATERIALS

per student:
- 1 tube of nutrient broth
- 1 petri plate of trypticase soy agar (TSA)
- 1 sterile cotton swab
- Sharpie marking pen

per two or more students:
- 1 petri plate of blood agar

1. Scrub down your desktop with a disinfectant (see Exercise 8, Aseptic Technique).

2. Expose your TSA plate according to your assignment in the table below. *Label the bottom* of your plate with your initials, your assignment number, and the date.

3. Moisten a sterile swab by immersing it into a tube of nutrient broth and expressing most of the broth out of it by pressing the swab against the inside wall of the tube.

4. Rub the moistened swab over a part of your body such as a finger or ear, or some object such as a doorknob or telephone mouthpiece, and return the swab to the tube of broth. It may be necessary to break off the stick end of the swab so that you can replace the cap on the tube.

5. Label the tube with your initials and the source of the bacteria.

6. Expose the blood agar plate by coughing onto it. Label the bottom of the plate with the initials of the individuals that cough onto it. Be sure to date the plate also.

7. Incubate the plates and tube at 37°C for 48 hours.

EVALUATION

After 48 hours incubation, examine the tube of nutrient broth and two plates. Shake the tube vigorously without wetting the cap. Is it cloudy or clear? Compare it with an uncontaminated tube of broth. What is the significance of cloudiness? Do you see any colonies growing on the blood agar plate? Are the colonies all the same size and color? If not, what does this indicate? Group together a set of TSA plates representing all nine types of exposure. Record your results on the Laboratory Report.

Your instructor will indicate whether these tubes and plates are to be used for making slides in Exercise 11 (Simple Staining). If the plates and tubes are to be saved, containers will be provided for their storage in the refrigerator. Place the plates and tubes in the designated containers.

LABORATORY REPORT

Record your results in Laboratory Report 6.

TABLE 6.1

Exposure Method for TSA Plate	Student Number
1. To the air in laboratory for 30 minutes	1, 10, 19, 28
2. To the air in room other than laboratory for 30 minutes	2, 11, 20, 29
3. To the air outside of building for 30 minutes	3, 12, 21, 30
4. Blow dust onto exposed medium	4, 13, 22, 31
5. Moist lips pressed against medium	5, 14, 23, 32
6. Fingertips pressed lightly on medium	6, 15, 24, 33
7. Several coins pressed temporarily on medium	7, 16, 25, 34
8. Hair combed over exposed medium (10 strokes)	8, 17, 26, 35
9. Optional: Any method not listed above	9, 18, 27, 36

LABORATORY REPORT

Student: _____

Date: _____   Section: _____

EXERCISE 6   The Bacteria

A. Questions

After you have tabulated your results on the table provided in part B, answer the following questions:

1. Using the number of colonies as an indicator, which habitat sampled by the class appears to be the most contaminated one? _____

2. Why do you suppose this habitat contains such a high microbial count? _____
   _____

3. a. Were any plates completely lacking in colonies? _____

   b. Do you think that the habitat sampled was really sterile? _____

   c. If your answer to *b* is *no,* then how can you account for the lack of growth on the plate?
      _____

   d. If your answer to *b* is *yes,* defend it: _____
      _____

4. In a few words, describe some differences in the macroscopic appearance of bacteria and mold colonies:
   _____
   _____
   _____

B. Tabulation

After examining your TSA and blood agar plates, record your results in the following table and on a similar table that your instructor has drawn on the chalkboard. With respect to the plates, we are concerned with a quantitative evaluation of the degree of contamination and differentiation as to whether the organisms are bacteria or molds. Quantify your recording as follows:

0 no growth	+ + + 51 to 100 colonies
+ 1 to 10 colonies	+ + + + over 100 colonies
+ + 11 to 50 colonies	

After shaking the tube of broth to disperse the organisms, look for cloudiness (turbidity). If the broth is clear, no bacterial growth occurred. Record no growth as 0. If tube is turbid, record + in the last column.

STUDENT INITIALS	PLATE EXPOSURE METHOD		COLONY COUNTS		BROTH	
	TSA	Blood Agar	Bacteria	Mold	Source	Result

The Fungi:
Yeasts and Molds

The fungi comprise a large group of eukaryotic non-photosynthetic organisms that include such diverse forms as slime molds, water molds, mushrooms, puffballs, bracket fungi, yeasts, and molds. Fungi belong to Kingdom **Myceteae.** The study of fungi is called **mycology.**

Myceteae consist of three divisions: Gymnomycota (slime molds), Mastigomycota (water molds and others), and Amastigomycota (yeasts, molds, bracket fungi, and others). It is the last division that we will study in this exercise.

Fungi may be saprophytic or parasitic and unicellular or filamentous. Some organisms, such as the slime molds (Exercise 21), are borderline between fungi and protozoa in that amoeboid characteristics are present and fungi-like spores are produced.

The distinguishing characteristics of the group as a whole are that they (1) are eukaryotic, (2) are non-photosynthetic, (3) lack tissue differentiation, (4) have cell walls of chitin or other polysaccharides, and (5) propagate by spores (sexual and/or asexual).

In this study, we will examine prepared stained slides and slides made from living cultures of yeasts and molds. Molds that are normally present in the air will be cultured and studied macroscopically and microscopically. In addition, an attempt will be made to identify the various types that are cultured.

Before attempting to identify the various molds, familiarize yourself with the basic differences between molds and yeasts. Note in figure 7.1 that yeasts are essentially unicellular and molds are multicellular.

MOLD AND YEAST DIFFERENCES

Species within the Amastigomycota may have cottony (moldlike) appearance or moist (yeasty) characteristics that set them apart. As pronounced as these differences are, we do not classify the various fungi in this group on the basis of their being mold or yeast. The reason that this type of division doesn't work is that some species exist as molds under certain conditions and as yeasts under other conditions. Such species are said to be **dimorphic,** or **biphasic.**

The principal differences between molds and yeasts are as follows:

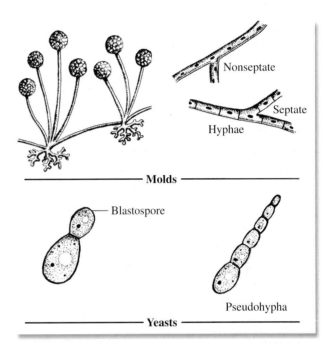

FIGURE 7.1   Structural differences between molds and yeasts

Molds

Hyphae   Molds have microscopic filaments called *hyphae* (hypha, singular). As shown in figure 7.1, if the filament has crosswalls, it is referred to as having **septate hyphae.** If no crosswalls are present, the coenocytic filament is said to be **nonseptate,** or **aseptate.** Actually, most of the fungi that are classified as being septate are incompletely septate since the septae have central openings that allow the streaming of cytoplasm from one compartment to the next. A mass of intermeshed hyphae, as seen macroscopically, is a *mycelium.*

Asexual Spores   Two kinds of asexual spores are seen in molds: sporangiospores and conidia. **Sporangiospores** are spores that form within a sac called a *sporangium.* The sporangia are attached to stalks called *sporangiophores.* See illustration 1, figure 7.2.

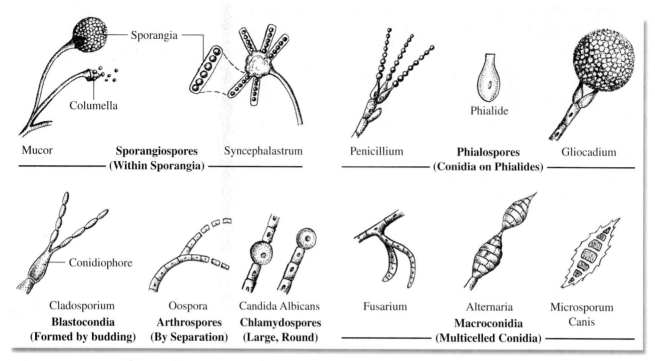

FIGURE 7.2 Types of asexual spores seen in fungi

Conidia are asexual spores that form on specialized hyphae called *conidiophores*. If the conidia are small they are called *microconidia;* large multicellular conidia are known as *macroconidia*. The following four types of conidia are shown in figure 7.2.

- **Phialospores:** Conidia of this type are produced by vase-shaped cells called *phialides*. Note in figure 7.2 that *Penicillium* and *Gliocadium* produce this type.
- **Blastoconidia:** Conidia of this type are produced by budding from cells of preexisting conidia, as in *Cladosporium,* which typically has lemon-shaped spores.

- **Arthrospores:** This type of conidia forms by separation from preexisting hyphal cells. Example: *Oospora.*
- **Chlamydospores:** These spores are large, thick-walled, round, or irregular structures formed within or on the ends of a hypha. Common to most fungi, they generally form on old cultures. Example: *Candida albicans.*

Sexual Spores   Three kinds of sexual spores are seen in molds: zygospores, ascospores, and basidiospores. Figure 7.3 illustrates the three types.

FIGURE 7.3 Types of sexual spores seen in the Amastigomycota

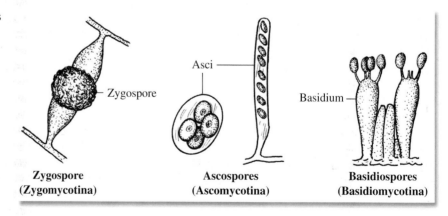

Zygospores are formed by the union of nuclear material from the hyphae of two different strains. **Ascospores,** on the other hand, are sexual spores produced in enclosures, which may be oval sacs or elongated tubes. **Basidiospores** are sexually produced on club-shaped bodies called *basidia*. A basidium is considered by some to be a modified type of ascus.

Yeasts

Hyphae Unlike molds, yeasts do not have true hyphae. Instead, they form multicellular structures called **pseudohyphae.** See figure 7.1.

Asexual Spores The only asexual spore produced by yeasts is called a **blastospore,** or **bud.** These spores form as an outpouching of a cell by a budding process. It is easily differentiated from the parent cell by its small size. It may separate from the original cell or remain attached. If successive buds remain attached in the budding process, the result is the formation of a **pseudohypha.**

SUBDIVISIONS OF THE AMASTIGOMYCOTA

Division Amastigomycota consists of four subdivisions: Zygomycotina, Ascomycotina, Basidiomycotina, and Deuteromycotina. They are separated on the basis of the type of sexual reproductive spores as follows:

Zygomycotina

These fungi have nonseptate hyphae and produce zygospores. They also produce sporangiospores. *Rhizopus, Mucor*, and *Syncephalastrum* are representative genera of this subdivision.

Ascomycotina

Since all the fungi in this subdivision produce ascospores, they are grouped into one class, the *Ascomycetes.* They are commonly referred to as the "ascomycetes" and are also called "sac fungi." All of them have septate hyphae and most of them have chitinous walls.

Fungi in this group that produce a single ascus are called *ascomycetes yeasts.* Other ascomycetes produce numerous asci in complex flask-shaped fruiting bodies called **perithecia** or **pseudothecia,** in cup-shaped structures, or in hollow spherical bodies, as in powdery mildews, *Eupenicillium* or *Talaromyces* (the sexual stages for *Penicillium*).

Basidiomycotina

All fungi in this subdivision belong to one class, the *Basidiomycetes.* Puffballs, mushrooms, smuts, rust, and shelf fungi on tree branches are also basidiomycetes. The sexual spores of this class are **basidiospores.**

Deuteromycotina

This fourth division of the Amastigomycota is an artificial group that was created to include any fungus that has not been shown to have some means of sexual reproduction. Often, species that are relegated to this division remain here for only a short period of time: as soon as the right conditions have been provided for sexual spores to form, they are reclassified into one of the first three subdivisions. Sometimes, however, the asexual and sexual stages of a fungus are discovered and named separately by different mycologists, with the result that a single species acquires two different names. Although generally there is a switch over to the sexual-stage name, not all mycologists conform to this practice.

Members of this group are commonly referred to as the *fungi imperfecti* or *deuteromycetes.* It is a large group, containing over 15,000 species.

LABORATORY PROCEDURES

Several options are provided here for the study of molds and yeasts. The procedures to be followed will be outlined by your instructor.

Yeast Study

The organism *Saccharomyces cerevisiae,* which is used in bread making and alcohol fermentation, will be used for this study. Either prepared slides or living organisms may be used.

MATERIALS

- prepared slides of *Saccharomyces cerevisiae*
- broth cultures of *Saccharomyces cerevisiae*
- methylene blue stain
- microscope slides and cover glasses

Prepared Slides If prepared slides are used, they may be examined under high-dry or oil immersion. One should look for typical **blastospores** and **ascospores.** Space is provided on the Laboratory Report for drawing the organisms.

Living Material If broth cultures of *Saccharomyces cerevisiae* are available, they should be examined on a wet mount slide with phase-contrast or brightfield optics. Two or three loopfuls of the organisms should be placed on the slide with a drop of methylene blue stain. Oil immersion will reveal the greatest amount of detail. Look for the **nucleus** and **vacuole**. The nucleus is the smaller body. Draw a few cells on the Laboratory Report.

Mold Study

Examine a petri plate of Sabouraud's agar that has been exposed to the air for about an hour and incubated at room temperature for 3–5 days. This medium has a low pH, which makes it selective for molds. A good plate will have many different-colored colonies. Note the characteristic "cottony" nature of the colonies. Also, look at the bottom of the plate and observe how the colonies differ in color here. The identification of molds is based on surface color, backside color, hyphal structure, and types of spores.

Figure 7.4 reveals how some common molds appear when grown on Sabouraud's agar. Keep in mind when using figure 7.4 that the appearance of a mold colony can change appreciably as it gets older. The photographs in figure 7.4 are of colonies that are 10 to 21 days old.

Conclusive identification cannot be made unless a microscope slide is made to determine the type of hyphae and spores that are present. Figure 7.5 reveals, diagrammatically, the microscopic differences that one looks for when identifying mold genera.

Two Options In making slides from mold colonies, one can make either wet mounts directly from the colonies by the procedure outlined here or make cultured slides as outlined in Exercise 22. The following steps should be used for making stained slides directly from the colonies. Your instructor will indicate the number of identifications that are to be made.

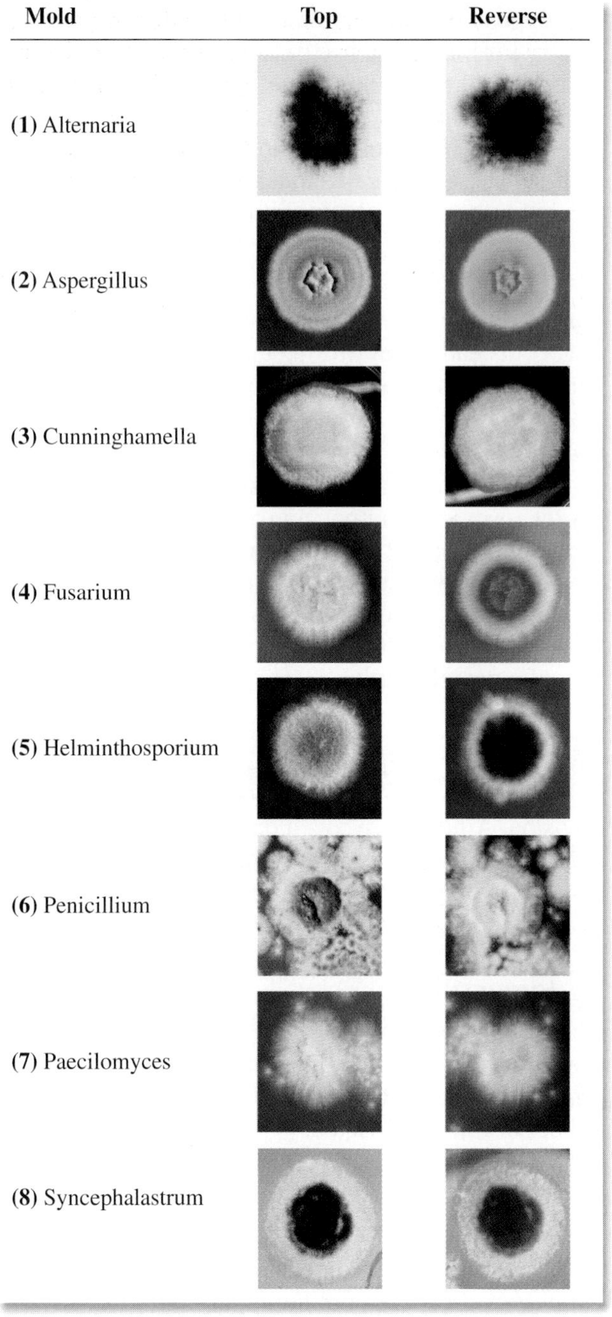

MATERIALS

- mold cultures on Sabouraud's agar
- microscope slides and cover glasses
- lactophenol cotton blue stain
- sharp-pointed scalpels or dissecting needles

1. Place an uncovered plate on a dissecting microscope and examine the colony. Look for hyphal structures and spore arrangement. Increase the magnification if necessary to more clearly see spores. Ignore white colonies as they are usually young colonies that have not begun the sporulation process.

FIGURE 7.4 Colony characteristics of some of the more common molds

CAUTION Avoid leaving the cover off the mold culture plates or disturbing the colonies very much. Dispersal of mold spores to the air must be kept to a minimum.

2. Consult figures 7.4 and 7.5 to make a preliminary identification based on colony characteristics and low-power magnification of hyphae and spores.

3. Make a wet mount slide by transferring a small amount of culture with a sharp scalpel or dissecting needle to a drop of lactophenol cotton blue on a microscope slide. Gently try to tease apart the mycelium with a dissecting needle. Cover the specimen with a coverslip and examine with the low-power objective. Look for hyphae that have spore structures. Go to the high-dry objective to discern more detail about the spores. Compare your specimen to figure 7.5 and see if you can identify the culture based on microscopic morphology.

4. Repeat the above procedure for each different colony.

LABORATORY REPORT

After recording your results on the Laboratory Report, answer all the questions.

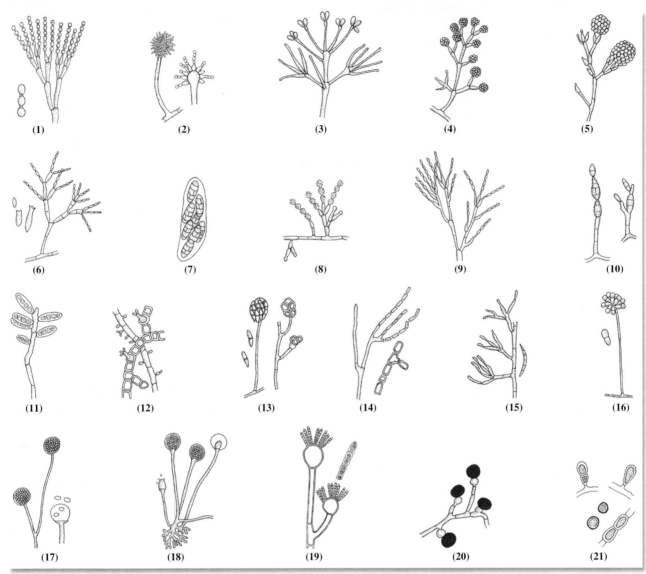

(1)   (2)   (3)   (4)   (5)

(6)   (7)   (8)   (9)   (10)

(11)   (12)   (13)   (14)   (15)   (16)

(17)   (18)   (19)   (20)   (21)

(1) *Penicillium*– bluish-green; brush arrangement of phialospores.

(2) *Aspergillus*– bluish-green with sulfur-yellow areas on the surface. *Aspergillus niger* is black.

(3) *Verticillium*– pinkish-brown, elliptical microconidia.

(4) *Trichoderma*– green, resemble *Penicillium* macroscopically.

(5) *Gliocadium*– dark-green; conidia (phialospores) borne on phialides, similar to *Penicillium*; grows faster than *Penicillium*.

(6) *Cladosporium (Hormodendrum)*– light green to grayish surface; gray to black back surface; blastoconidia.

(7) *Pleospora*– tan to green surface with brown to black back; ascospores shown are produced in sacs borne within brown, flask-shaped fruiting bodies called pseudothecia.

(8) *Scopulariopsis*– light-brown; rough-walled microconidia.

(9) *Paecilomyces*– yellowish-brown, elliptical microconidia.

(10) *Alternaria*– dark greenish-black surface with gray periphery; black on reverse side; chains of macroconidia.

(11) *Bipolaris*– black surface with grayish periphery; macroconidia shown.

(12) *Pullularia*– black, shiny, leathery surface; thick-walled; budding spores.

(13) *Diplosporium*– buff-colored wooly surface; reverse side has red center surrounded by brown.

(14) *Oospora (Geotrichum)*– buff-colored surface; hyphae break up into thin-walled rectangular arthrospores.

(15) *Fusarium*– variants, of yellow, orange, red, and purple colonies; sickle-shaped macroconidia.

(16) *Trichothecium*– white to pink surface; two-celled conidia.

(17) *Mucor*– a zygomycete; sporangia with a slimy texture; spores with dark pigment.

(18) *Rhizopus*– a zygomycete; spores with dark pigment.

(19) *Syncephalastrum*– a zygomycete; sporangiophores bear rod-shaped sporangioles, each containing a row of spherical spores.

(20) *Nigrospora*– conidia black, globose, one-celled, borne on a flattened, colorless vesicle at the end of a conidiophore.

(21) *Montospora*– dark gray center with light gray periphery; yellow-brown conidia.

FIGURE 7.5   Microscopic appearance of some of the more common molds

LABORATORY REPORT

EXERCISE 7   The Fungi: Yeasts and Molds

A. Yeast Study

Draw a few representative cells of *Saccharomyces cerevisiae* in the appropriate circles below. Blastospores (buds) and ascospores, if seen, should be shown and labeled.

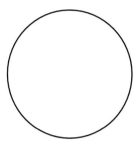

Prepared Slide

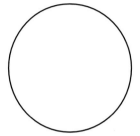

Living Cells

B. Mold Study

In the following table, list the genera of molds identified in this exercise. Under colony description, give the approximate diameter of the colony, its topside color and backside (bottom) color. For microscopic appearance, make a sketch of the organism as it appears on slide preparation.

GENUS	COLONY DESCRIPTION	MICROSCOPIC APPEARANCE (DRAWING)

Yeasts and Molds (continued)

C. Questions

Record the answers for the following questions in the answer column.

1. The science that is concerned with the study of fungi is called _____.

2. The kingdom to which the fungi belong is _____.

3. Microscopic filaments of molds are called _____.

4. A filamentlike structure formed by a yeast from a chain of blastospores is called a _____.

5. A mass of mold filaments, as observed by the naked eye, is called a _____.

6. Most molds have _____ hyphae (*septate* or *non-septate*).

7. List three kinds of sexual spores that are the basis for classifying the molds.

8. What is the name of the rootlike structure that is seen in *Rhizopus?*

9. What type of hypha is seen in *Mucor* and *Rhizopus*?

10. What kind of asexual spores are seen in *Mucor* and *Rhizopus?*

11. What kind of asexual spores are seen in *Penicillium?*

12. What kind of asexual spores are seen in *Alternaria?*

13. Which subdivision of the Amastigomycota contains individuals that lack sexual spores?

14. What division of Myceteae consists of slime molds?

15. Fungi that exist both as yeasts and molds are said to be _____.

ANSWERS

1. _____
2. _____
3. _____
4. _____
5. _____
6. _____
7a. _____
b. _____
c. _____
8. _____
9. _____
10. _____
11. _____
12. _____
13. _____
14. _____
15. _____

MANIPULATION OF MICROORGANISMS

One of the most critical techniques that any beginning student in microbiology must learn is aseptic technique. This technique insures that an aseptic environment is maintained when handling microorganisms. This means two things:

1. no contaminating microorganisms are introduced into cultures or culture materials and
2. the microbiologist is not contaminated by cultures that are being manipulated.

Aseptic technique is crucial in characterizing an unknown organism. Oftentimes, multiple transfers must be made from a stock culture to various test media. It is imperative that only the desired organism is transferred each time and that no foreign bacteria are introduced during the transfer. Aseptic technique is also obligatory in isolating and purifying bacteria from a mixed source of organisms. The streak-plate and pour-plate techniques provide a means to isolate an individual species. And once an organism is in pure culture and stored as a stock culture, aseptic technique insures that the culture remains pure when it is necessary to retrieve the organism.

Individuals who work with pathogenic bacteria must be sure that any pathogen that is being handled is not accidently released causing harm to themselves or to coworkers. Failure to observe aseptic technique can obviously pose a serious threat to many.

In the following exercises, you will learn the techniques that allow you to handle and manipulate cultures of microorganisms. Once you have mastered these procedures, you will be able to make transfers of microorganisms from one kind of medium to another with confidence. You will also be able to isolate an organism from a mixed culture to obtain a pure isolate. It is imperative that you have a good grasp of these procedures, as they will be required over and over in the exercises in this manual.

Aseptic Technique

The use of aseptic technique insures that no contaminating organisms are introduced into culture materials when the latter are inoculated or handled in some manner. It also insures that organisms that are being handled do not contaminate the handler or others who may be present. And its use means that no contamination remains after you have worked with cultures.

As you work with these procedures, with time, they will become routine and second nature to you. You will automatically know that a set of procedures outlined below will be used when dealing with cultures of microorganisms. This may involve the transfer of a broth culture to a plate for streaking, or inoculating an isolated colony from a plate onto a slant culture to prepare a stock culture. It may also involve inoculating many tubes of media and agar plates from a stock culture in order to characterize and identify an unknown bacterium. Making sure that only the desired organism is transferred in each inoculation is paramount to being successful in the identification process. The general procedure for aseptic technique follows.

Work Area Disinfection The work area is first treated with a disinfectant to kill any microorganisms that may be present. This process destroys vegetative cells and viruses but may not destroy endospores.

Loops and Needles The transfer of cultures will be achieved using inoculating loops and needles. These implements must be sterilized before transferring any culture. A loop or needle is sterilized by inserting it into a Bunsen burner flame until it is red hot. This will incinerate any contaminating organisms that may be present.

Culture Tube Flaming and Inoculation Prior to inserting a cooled loop or needle into a culture tube, the cap is removed and the mouth of the tube is flamed. If the tube is a broth tube, the loop is inserted into the tube and twisted several times to ensure that the organisms on the loop are delivered to the liquid. If the tube is an agar slant, the surface of the slant is inoculated by drawing the loop up the surface of the slant from the bottom of the slant to its top. For stab cultures, a needle is inserted into the agar medium by stabbing it into the agar. After the culture is inocu-

lated, the mouth of the tube is reflamed and the tube is recapped.

Final Flaming of the Loop or Needle After the inoculation is complete, the loop or needle is flamed in the Bunsen burner to destroy any organisms that remain on these implements. The loop or needle is then returned to its receptacle for storage. It should never be placed on the desk surface.

Petri Plate Inoculations Loops are used to inoculate or streak petri plates. The plate cover is raised and held diagonally over the plate to protect the surface from any contamination in the air. The loop containing the inoculum is then streaked gently over the surface of the agar. It is important not to gouge or disturb the surface of the agar with the loop. The cover is replaced and the loop is flamed in a Bunsen burner.

Final Disinfection of the Work Area When all work for the day is complete, the work area is treated with disinfectant to insure that any organism that might have been deposited during any of the procedures is killed.

To gain some practice in aseptic transfer of bacterial cultures, three simple transfers will be performed in this exercise:
1. broth culture to broth tube
2. agar slant culture to an agar slant and
3. agar plate to an agar slant.

TRANSFER FROM BROTH CULTURE TO ANOTHER BROTH

Do a broth tube to broth tube inoculation, using the following technique. Figure 8.1 illustrates the procedure for removing organisms from a culture, and figure 8.2 shows how to inoculate a tube of sterile broth.

MATERIALS

- broth culture of *Escherichia coli*
- tubes of sterile nutrient broth
- inoculating loop
- Bunsen burner
- disinfectant for desktop and sponge
- Sharpie marking pen

1. Prepare your desktop by swabbing down its surface with a disinfectant. Use a sponge.
2. With a marking pen, label a tube of sterile nutrient broth with your initials and *E. coli.*
3. Sterilize your inoculating loop by holding it over the flame of a Bunsen burner **until it becomes bright red.** The entire wire must be heated. See illustration 1, figure 8.1.
4. Using your free hand, gently shake the tube to disperse the culture (illustration 2, figure 8.1).
5. Grasp the tube cap with the little finger of your hand holding the inoculating loop and remove it from the tube. Flame the mouth of the tube as shown in illustration 3, figure 8.1.
6. Insert the inoculating loop into the culture (illustration 4, figure 8.1).
7. Remove the loop containing the culture, flame the mouth of the tube again (illustration 5, figure 8.1), and recap the tube (illustration 6). Place the culture tube back on the test-tube rack.

(1) Inoculating loop is heated until it is red-hot.

(2) Organisms in culture are dispersed by shaking tube.

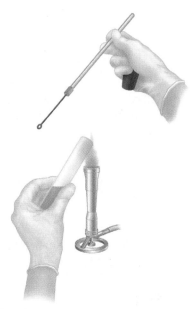

(3) Tube enclosure is removed and mouth of tube is flamed.

(4) A loopful of organisms is removed from tube.

(5) Loop is removed from culture and tube mouth is flamed.

(6) Tube enclosure is returned to tube.

FIGURE 8.1 Procedure for removing organisms from a broth culture with inoculating loop

8. Grasp a tube of sterile nutrient broth with your free hand, carefully remove the cap with your little finger, and flame the mouth of this tube (illustration 1, figure 8.2).

9. Without flaming the loop, insert it into the sterile broth, inoculating it (illustration 2, figure 8.2). To disperse the organisms into the medium, move the loop back and forth in the tube.

10. Remove the loop from the tube and flame the mouth (illustration 3, figure 8.2). Replace the cap on the tube (illustration 4, figure 8.2).

11. Sterilize the loop by flaming it (illustration 5, figure 8.2). Return the loop to its container.

12. Incubate the culture you just inoculated at 37°C for 24–48 hours.

  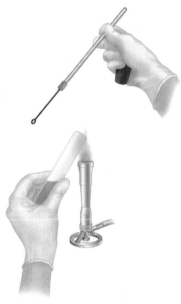

(**1**) Cap is removed from sterile broth and tube mouth is flamed.

(**2**) Unheated loop is inserted into tube of sterile broth.

(**3**) Loop is removed from broth and tube mouth is flamed.

(**4**) Tube enclosure is returned to tube.

(**5**) Loop is flamed and returned to receptacle.

FIGURE 8.2 Procedure for inoculating a nutrient broth

TRANSFER OF BACTERIA FROM A SLANT

To inoculate a sterile nutrient agar slant from an agar slant culture, use the following procedure. Figure 8.3 illustrates the entire process.

MATERIALS

- agar slant culture of *E. coli*
- sterile nutrient agar slant
- inoculating loop
- Bunsen burner
- Sharpie marking pen

1. If you have not already done so, prepare your desktop by swabbing down its surface with a disinfectant.
2. With a marking pen label a tube of nutrient agar slant with your initials and *E. coli*.
3. Sterilize your inoculating loop by holding it over the flame of a Bunsen burner **until it becomes bright red** (illustration 1, figure 8.3). The entire wire must be heated. Allow the loop to cool completely.
4. Using your free hand, pick up the slant culture of *E. coli* and remove the cap using the little finger of the hand that is holding the loop (illustration 2, figure 8.3).
5. Flame the mouth of the tube and insert the cooled loop into the tube. Pick up some of the culture on the loop (illustration 3, figure 8.3) and remove the loop from the tube.
6. Flame the mouth of the tube (illustrations 4 and 5, figure 8.3) and replace the cap, being careful not to burn your hand. Return tube to rack.
7. Pick up a sterile nutrient agar slant with your free hand, remove the cap with your little finger as before, and flame the mouth of the tube (illustration 6, figure 8.3).
8. Without flaming the loop containing the culture, insert the loop into the tube and gently inoculate the surface of the slant by moving the loop back and forth over the agar surface, while moving up the surface of the slant (illustration 7, figure 8.3). This should involve a type of serpentine motion.
9. Remove the loop, flame the mouth of the tube, and recap the tube (illustration 8, figure 8.3). Replace the tube in the rack.
10. Flame the loop, heating the entire wire to red-hot (illustration 9, figure 8.3), allow to cool, and place the loop in its container.
11. Incubate the inoculated agar slant at 30°C for 24–48 hours.

WORKING WITH AGAR PLATES

(Inoculating a slant from a petri plate)

The transfer of organisms from colonies on agar plates to slants or broth tubes is very similar to the

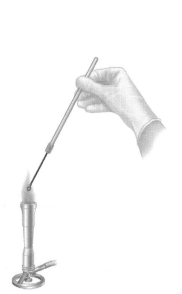

(1) Inoculating loop is heated until it is red-hot.

(2) Cap is removed from slant culture and tube mouth is heated.

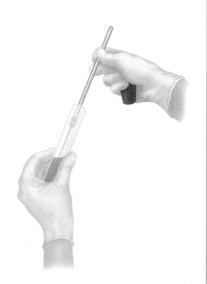

(3) Organism is picked up from slant with inoculating loop.

FIGURE 8.3   Procedure for inoculating a nutrient agar slant from a slant culture

continued

(4) Mouth of tube is flamed. Inoculating loop is not flamed.

(5) Slant culture is re-capped and returned to test tube rack.

(6) Tube of sterile agar slant is uncapped and mouth is flamed.

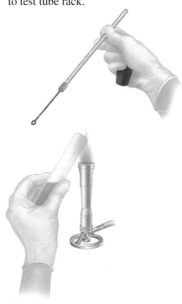

(7) Slant surface is streaked with unflamed loop in serpentine manner.

(8) Tube mouth is flamed, recapped and incubated.

(9) Loop is flamed red-hot and returned to receptacle.

FIGURE 8.3 *(continued)*

procedures used in the last two transfers (broth to broth and slant to slant). The following rules should be observed.

Loops and Needles　Loops are routinely used when streaking agar plates and slants. When used properly, a loop will not gouge or tear the agar surface. Needles are used in transfers involving stab cultures.

Plate Handling　Media in plates must always be protected against contamination. To prevent exposure

to air contamination, covers should always be left closed. When organisms are removed from a plate culture, the cover should be only partially opened as shown in illustration 2, figure 8.4.

Flaming Procedures　Inoculating loops or needles must be flamed in the same manner that you used when working with previous tubes. One difference when working with plates is that plates are never flamed!

(1) Inoculating loop is heated until it is red-hot.

(2) With free hand, raise the lid of the Petri plate just enough to access a colony to pick up a loopful of organisms.

(3) After flaming the mouth of a sterile slant, streak its surface.

(4) Flame the mouth of the tube and re-cap the tube.

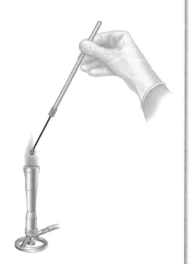

(5) Flame the inoculating loop and return it to receptacle.

FIGURE 8.4 **Procedure for inoculating a nutrient agar slant from an agar plate**

Plate Labeling and Incubation   Petri plates containing inoculated media are labeled on the bottom of the plate. Inoculated plates are almost always incubated upside down. This prevents moisture from condensing on the agar surface and spreading the inoculated organisms.

To transfer organisms from a petri plate to an agar slant, use the following procedure:

MATERIALS

- nutrient agar plate with bacterial colonies
- sterile nutrient agar slant
- inoculating loop
- Sharpie marking pen

1. If you have not done so, swab your work area with disinfectant. Allow area to dry.
2. Label a sterile nutrient agar slant with your name and organism to be transferred.
3. Flame an inoculating loop until it is red-hot (illustration 1, figure 8.4). Allow the loop to cool.
4. As shown in illustration 2, figure 8.4, raise the lid of a petri plate sufficiently to access a colony with your sterile loop.

   Do not gouge into the agar with your loop as you pick up organisms, and do not completely remove the lid, exposing the surface to the air. Close the lid once you have picked up the organisms.

5. With your free hand, pick up the sterile nutrient agar slant tube. Remove the cap by grasping the cap with the little finger of the hand that is holding the loop.

6. Flame the mouth of the tube and insert the loop into the tube to inoculate the surface of the slant, using a serpentine motion (illustration 3, figure 8.4). Avoid disrupting the agar surface with the loop.

7. Remove the loop from the tube and flame the mouth of the tube. Replace the cap on the tube (illustration 4, figure 8.4).

8. Flame the loop (illustration 5, figure 8.4) and place it in its container.

9. Incubate the nutrient agar slant at 37° C for 24–48 hours.

SECOND PERIOD

Examine all three tubes and record your results on Laboratory Report 8.

Pure Culture Techniques

When we try to study the bacterial flora of the body, soil, water, or just about any environment, we realize quickly that bacteria exist in natural environments as mixed populations. It is only in very rare instances that they occur as a single species. Robert Koch, the father of medical microbiology, was one of the first to recognize that if he was going to prove that a particular bacterium causes a specific disease, it would be necessary to isolate the agent from all other bacteria and characterize the pathogen. From his studies on pathogenic bacteria, his laboratory contributed many techniques to the science of microbiology, including the method for obtaining **pure cultures** of bacteria. With a pure culture, we are able to study the cultural, morphological, and physiological characteristics of an individual organism.

Several methods for obtaining pure cultures are available to the microbiologist. Two commonly used procedures are the **streak plate** and the **pour plate.** Both procedures involve diluting the bacterial cells in a sample to an end point where a single cell divides giving rise to single **pure colony.** The colony is therefore assumed to be the identical progeny of the original cell and can be picked and used for further study of the bacterium.

In this exercise, you will use both the streak plate and pour plate methods to separate a mixed culture of bacteria. The bacteria may be differentiated by the characteristics of the colony, such as color, shape, and other colony characteristics. Isolated colonies can then be subcultured and stains prepared to check for purity.

STREAK PLATE METHOD

For economy of materials and time, the streak plate method is best. It requires a certain amount of skill, however, which is forthcoming with experience. A properly executed streak plate will give as good an isolation as is desired for most work. Figure 9.1 illustrates how colonies of a mixed culture should be spread out on

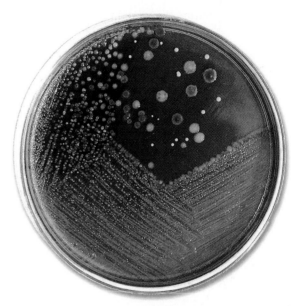

FIGURE 9.1   If your streak plate reveals well-isolated colonies of three colors (red, white, and yellow), you have a plate suitable for subculturing.

a properly made streak plate. The important thing is to produce good spacing between colonies.

MATERIALS

- electric hot plate (or tripod and wire gauze)
- Bunsen burner and beaker of water
- wire loop, thermometer, and Sharpie marking pen
- nutrient agar pour and 1 sterile petri plate
- mixed culture of *Serratia marcescens* or *Escherichia coli* and *Micrococcus luteus* or *Chromobacterium violaceum*

1. Prepare your tabletop by disinfecting its surface with the disinfectant that is available in the laboratory (Roccal, Zephiran, Betadine, etc.). Use a sponge or paper towels to scrub it clean.

2. Label the bottom surface of a sterile petri plate with your name and date. Use a marking pen such as a Sharpie.

3. Liquefy a tube of nutrient agar, cool to 50° C, and pour the medium into the bottom of the plate, following the procedure illustrated in figure 9.2. Be sure to flame the neck of the tube prior to pouring to destroy any bacteria around the end of the tube.

After pouring the medium into the plate, gently rotate the plate so that it becomes evenly distributed, but do not splash any medium up over the sides.

Agar-agar, the solidifying agent in this medium, becomes liquid when boiled and resolidifies at around 42° C. Failure to cool it prior to pouring into the plate will result in condensation of moisture on the cover. Any moisture on the cover is undesirable because if it drops down on the colonies, the organisms of one colony can spread to other colonies, defeating the entire isolation technique.

(1) Liquefy a nutrient agar pour by boiling for 5 minutes.

(2) Cool down the nutrient agar pour to 50° C by pouring off some of the hot water and adding cold water to the beaker. Hold at 50° C for 5 minutes.

(3) Remove the cap from the tube and flame the open end of the tube.

(4) Pour the contents of the tube into the bottom of the Petri plate and allow it to solidify.

FIGURE 9.2   Procedure for pouring an agar plate for streaking

4. Streak the plate by one of the methods shown in figure 9.4. Your instructor will indicate which technique you should use.

Caution: Be sure to follow the routine in figure 9.3 for getting the organism out of culture.

5. Incubate the plate in an *inverted position* at 25° C for 24–48 hours. By incubating plates upside down, the problem of moisture on the cover is minimized.

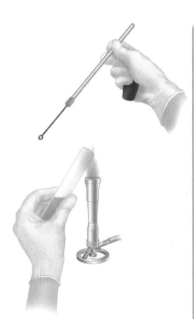

(**1**) Shake the culture tube from side to side to suspend organisms. Do not moisten cap on tube.

(**2**) Heat loop and wire to red-hot. Flame the handle slightly also.

(**3**) Remove the cap and flame the neck of the tube. Do not place the cap down on the table.

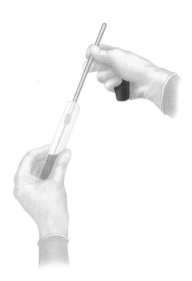

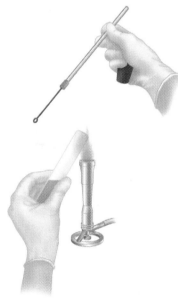

(**4**) After allowing the loop to cool for at least 5 seconds, remove a loopful of organisms. Avoid touching the side of the tube.

(**5**) Flame the mouth of the culture tube again.

(**6**) Return the cap to the tube and place the tube in a test-tube rack.

continued

FIGURE 9.3   Routine for inoculating a petri plate

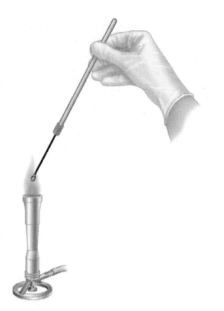

(7) Streak the plate, holding it as shown. Do not gouge into the medium with the loop.

(8) Flame the loop before placing it down.

FIGURE 9.3 *(continued)*

Quadrant Streak
(Method A)

(1) Streak one loopful of organisms over Area 1 near edge of the plate. Apply the loop lightly. Don't gouge into the medium.
(2) Flame the loop, cool 5 seconds, and make 5 or 6 streaks from Area 1 through Area 2. Momentarily touching the loop to a sterile area of the medium before streaking insures a cool loop.
(3) Flame the loop again, cool it, and make 6 or 7 streaks from Area 2 through Area 3.
(4) Flame the loop again, and make as many streaks as possible from Area 3 through Area 4, using up the remainder of the plate surface.
(5) Flame the loop before putting it aside.

Quadrant Streak
(Method B)

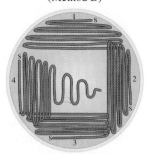

(1) Streak one loopful of organisms back and forth over Area 1, starting at point designated by "s". Apply loop lightly. Don't gouge into the medium.
(2) Flame the loop, cool 5 seconds, and touch the medium in sterile area momentarily to insure coolness.
(3) Rotate dish 90 degrees while keeping the dish closed. Streak Area 2 with several back and forth strokes, hitting the original streak a few times.
(4) Flame the loop again. Rotate the dish and streak Area 3 several times, hitting last area several times.
(5) Flame the loop, cool it, and rotate the dish 90 degrees again. Streak Area 4, contacting Area 3 several times and drag out the culture as illustrated.
(6) Flame the loop before putting it aside.

Radiant Streak

(1) Spread a loopful of organisms in small area near the edge of the plate in Area 1. Apply the loop lightly. Don't gouge medium.
(2) Flame the loop and allow it to cool for 5 seconds. Touching a sterile area will insure coolness.
(3) **From the edge** of Area 1 make 7 or 8 straight streaks to the opposite side of the plate.
(4) Flame the loop again, cool it sufficiently, and cross streak over the last streaks, **starting near Area 1**.
(5) Flame the loop before putting it aside.

Continuous Streak

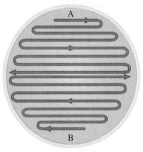

(1) Starting at the edge of the plate (Area A) with a loopful of organisms, spread the organisms in a single continuous movement to the center of the plate. Use light pressure and avoid gouging the medium.
(2) Rotate the plate 180 degrees so that the uninoculated portion of the plate is away from you.
(3) Rotate dish 90 degrees while keeping the dish closed. Streak Area 2 with several back and forth strokes, hitting the original streak a few times.
(4) Without flaming the loop again, and using the same face of the loop, continue streaking the other half of the plate by starting at Area B and working toward the center.
(5) Flame the loop before putting it aside.

FIGURE 9.4   Four different streak techniques

POUR PLATE METHOD

(Loop Dilution)

This method of separating one species of bacteria from another consists of diluting out one loopful of organisms with three tubes of liquefied nutrient agar in such a manner that one of the plates poured will have an optimum number of organisms to provide good isolation. Figure 9.5 illustrates the general procedure. One advantage of this method is that it requires somewhat less skill than that required for a good streak plate; a disadvantage, however, is that it requires more media, tubes, and plates. Proceed as follows to make three dilution pour plates, using the same mixed culture you used for your streak plate.

MATERIALS

- mixed culture of bacteria
- 3 nutrient agar pours
- 3 sterile petri plates
- electric hot plate
- beaker of water
- thermometer
- inoculating loop and Sharpie marking pen

1. Label the three nutrient agar pours **I, II,** and **III** with a marking pen and place them in a beaker of water on an electric hot plate to be liquefied. To save time, start with hot tap water if it is available.

2. While the tubes of media are being heated, label the bottoms of the three Petri plates **I, II,** and **III.**
3. Cool down the tubes of media to 50° C, using the same method that was used for the streak plate.
4. Following the routine in figure 9.5, inoculate tube I with one loopful of organisms from the mixed culture. Note the sequence and manner of handling the tubes in figure 9.6.
5. Inoculate tube II with one loopful from tube I after thoroughly mixing the organisms in tube I by shaking the tube from side to side or by rolling the tube vigorously between the palms of both hands. ***Do not splash any of the medium up onto the tube closure***. Return tube I to the water bath.
6. Agitate tube II to completely disperse the organisms and inoculate tube III with one loopful from tube II. Return tube II to the water bath.
7. Agitate tube III, flame its neck, and pour its contents into plate III.
8. Flame the necks of tubes I and II and pour their contents into their respective plates.
9. After the medium has completely solidified, incubate the *inverted* plates at 25° C for 24–48 hours.

EVALUATION OF THE TWO METHODS

After 24 to 48 hours of incubation examine all four Petri plates. Look for colonies that are well isolated from the others. Note how crowded the colonies ap-

FIGURE 9.5 Three steps in the loop dilution technique for separating out organisms

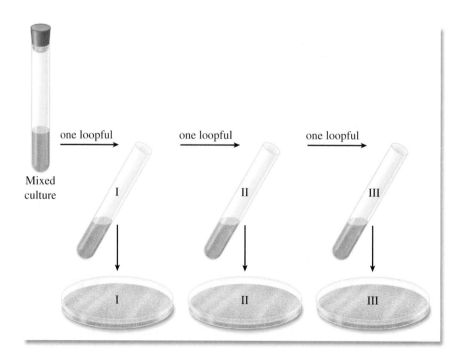

Mixed culture   one loopful → I   one loopful → II   one loopful → III

I   II   III

pear on plate I as compared with plates II and III. Plate I will be unusable. Either plate II or III will have the most favorable isolation of colonies. Can you pick out well-isolated colonies on your best pour plate that are distinct from one another?

Draw the appearance of your streak plate and pour plates on the Laboratory Report.

SUBCULTURING TECHNIQUES

The next step in the development of a pure culture is the transfer of an isolated colony from the petri plate to a tube of nutrient broth or a slant of nutrient agar. Use your loop to carefully pick an isolated colony and aseptically transfer a colony to the broth tube or slant.

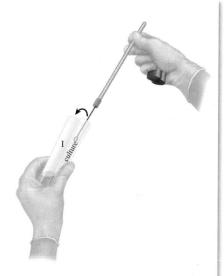

(1) Liquefy three nutrient agar pours, cool to 50° C, and let stand for 10 minutes.

(2) After shaking the culture to disperse the organisms, flame the loop and necks of the tubes.

(3) Transfer one loopful of the culture to tube I.

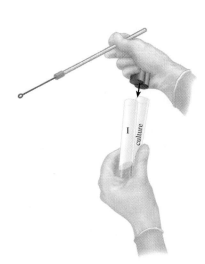

(4) Flame the loop and the necks of both tubes.

(5) Replace the caps on the tubes and return culture to the test-tube rack.

(6) Disperse the organisms in tube I by shaking the tube or rolling it between the palms.

continued

FIGURE 9.6   Tube-handling procedure in making inoculations for pour plates

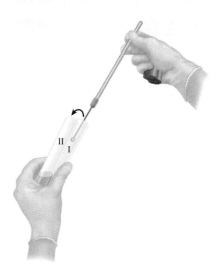

(7) Transfer one loopful from tube I to tube II. Return tube I to the water bath.

(8) After shaking tube II and transferring one loopful to tube III, flame the necks of each tube.

(9) Pour the inoculated pours into their respective Petri plates.

FIGURE 9.6 *(continued)*

To insure that the broth is inoculated, rotate the loop in the broth several times before withdrawing it from the broth tube. For the slant, make an "S" motion by drawing the loop from the bottom of the tube up the surface of the slant. Use the following routine to sub-culture the different organisms that you have isolated.

MATERIALS

- nutrient agar slants
- inoculating loops
- Bunsen burners

1. Label one tube *Staphylococcus aureus* and a second *Serratia marcescens*.
2. Select a well-isolated red colony on either the streak plate or the pour plate. Use your inoculating loop to pick a well-isolated colony and transfer it to the tube labeled *S. marcescens*.
3. Repeat this procedure for a yellow/cream-colored colony and transfer the colony to the tube labeled *S. aureus*.
4. Incubate the tubes at 25° C for 24–48 hours.

EVALUATION OF SLANTS

After incubation, examine the slants. Is the *S. marcescens* culture red? If the culture was incubated at a temperature higher than 25° C it may not be red because higher temperatures inhibit the formation of the organism's red pigment. Draw the appearance of the slant. What color is the *S. aureus* culture? Draw the slant.

You cannot be sure that your cultures are pure until you have made a microscopic examination of the respective cultures. It is entirely possible that the *S. marcescens* culture harbors some contaminating *S. aureus* and vice versa. Prepare smears for each culture and Gram stain the smears. *S. marcescens* is a gram-negative short rod while *S. aureus* is a gram-positive coccus. Draw the Gram stained smears on the Laboratory Report.

LABORATORY REPORT

Complete the Laboratory Report for this exercise.

5. Why should a petri plate be discarded if media is splashed up the side to the top?

   _____

   _____

6. Give two reasons why it is important to invert plates during incubation:

   _____

   _____

7. Why is it important not to dig into the agar with the loop? _____

   _____

   _____

8. Why must the loop be flamed before entering a culture? _____

   Why must it be flamed after making an inoculation? _____

STAINING AND OBSERVATION OF MICROORGANISMS

The eight exercises in this unit include the procedures for ten slide techniques that one might employ in morphological studies of bacteria. A culture method in Exercise 17 also is included as a substitute for slide techniques when pathogens are encountered.

These exercises are intended to serve two equally important functions: (1) to help you to develop the necessary skills in making slides and (2) to introduce you to the morphology of bacteria. Although the title of each exercise pertains to a specific technique, the organisms chosen for each method have been carefully selected so that you can learn to recognize certain morphological features. For example, in the exercise on simple staining (Exercise 11), the organisms selected exhibit metachromatic granules, pleomorphism, and palisade arrangement of cells. In Exercise 14 (Gram Staining), you will observe the differences between cocci and bacilli, as well as learn how to execute the staining routine.

The importance of the mastery of these techniques cannot be overemphasized. Although one is seldom able to make species identification on the basis of morphological characteristics alone, it is a very significant starting point. This fact will become increasingly clear with subsequent experiments.

Although the steps in the various staining procedures may seem relatively simple, student success is often quite unpredictable. Unless your instructor suggests a variation in the procedure, try to follow the procedures exactly as stated, without improvisation. Photomicrographs in color have been provided for many of the techniques; use them as a guide to evaluate the slides you have prepared. Once you have mastered a specific technique, feel free to experiment.

Smear Preparation

The success for most staining procedures depends upon the preparation of a good **smear.** There are several goals in preparing a smear. The first goal is to cause the cells to adhere to the microscope slide so that they are not washed off during subsequent staining and washing procedures. Second, it is important to insure that shrinkage of cells does not occur during staining, otherwise distortion and artifacts can result. A third goal is to prepare thin smears because the thickness of the smear will determine if you can visualize individual cells, their arrangement, or details regarding microstructures associated with cells. Thick smears of cells with large clumps obscure details about individual cells and, furthermore, the smear can entrap stain keeping it from being removed by washing or destaining, leading to erroneous results. The procedure for making a smear is illustrated in figure 10.1.

The first step in preparing a bacteriological smear differs according to the source of the organisms. If the bacteria are growing in a liquid medium (broths, milk, saliva, urine, etc.), one starts by placing two or more loopfuls of the liquid medium directly on the slide.

From solid media such as nutrient agar, blood agar, or some part of the body, one starts by placing one or two loopfuls of water on the slide and then uses an inoculating loop to disperse the organisms in the water. Bacteria growing on solid media tend to cling to each other and must be dispersed sufficiently by dilution in water; unless this is done, the smear will be too thick. *The most difficult concept for students to understand about making slides from solid media is that it takes only a very small amount of material to make a good smear.* When your instructor demonstrates this step, pay very careful attention to the amount of material that is placed on the slide.

The organisms to be used for your first slides may be from several different sources. If the plates from Exercise 6 were saved, some slides may be made from them. If they were discarded, the first slides may be made for Exercise 11, which pertains to simple staining. Your instructor will indicate which cultures to use.

FROM LIQUID MEDIA

(Broths, saliva, milk, etc.)

If you are preparing a bacterial smear from liquid media, follow this routine, which is depicted on the left side of figure 10.1.

MATERIALS

- microscope slides
- Bunsen burner
- wire loop
- Sharpie marking pen
- slide holder (clothespin)

1. Wash a slide with soap or Bon Ami and hot water, removing all dirt and grease. Handle the clean slide by its edges.
2. Write the initials of the organism or organisms on the left-hand side of the slide with a marking pen.
3. To provide a target on which to place the organisms, make a $\frac{1}{2}''$ circle on the *bottom* side of the slide, centrally located, with a marking pen. Later on, when you become more skilled, you may wish to omit the use of this "target circle."
4. Shake the culture vigorously and transfer two loopfuls of organisms to the center of the slide over the target circle. Follow the routine for inoculations shown in figure 10.2. *Be sure to flame the loop after it has touched the slide.*

> **CAUTION:** Be sure to cool the loop completely before inserting it into a medium. A loop that is too hot will spatter the medium and move bacteria into the air.

5. Spread the organisms over the area of the target circle.
6. Allow the slide to dry by normal evaporation of the water. Don't apply heat.

7. After the smear has become completely dry, place the slide in a clothespin and pass it several times through the flame of a Bunsen burner. Avoid prolonged heating of the slide as it can shatter from excessive exposure to heat. The underside of the slide should feel warm to the touch.

Note that in this step one has the option of using or not using a clothespin to hold the slide. *Use the option preferred by your instructor.*

FROM SOLID MEDIA

When preparing a bacterial smear from solid media, such as nutrient agar or a part of the body, follow this routine, which is depicted on the right side of figure 10.1.

MATERIALS

- microscope slides
- inoculating needle and loop

FIGURE 10.1   Procedure for making a bacterial smear

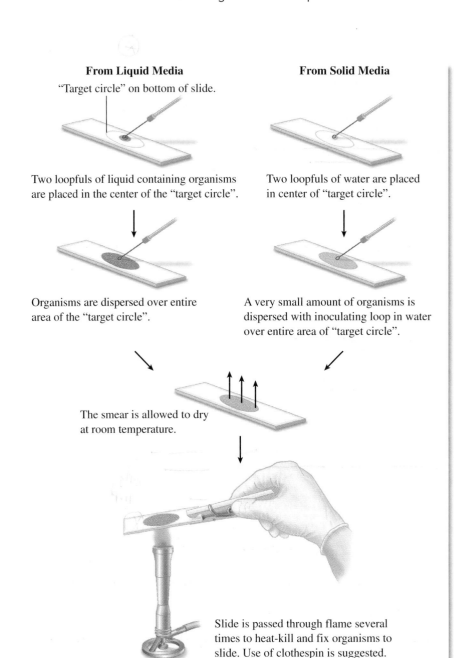

From Liquid Media

"Target circle" on bottom of slide.

Two loopfuls of liquid containing organisms are placed in the center of the "target circle".

Organisms are dispersed over entire area of the "target circle".

The smear is allowed to dry at room temperature.

From Solid Media

Two loopfuls of water are placed in center of "target circle".

A very small amount of organisms is dispersed with inoculating loop in water over entire area of "target circle".

Slide is passed through flame several times to heat-kill and fix organisms to slide. Use of clothespin is suggested.

- Sharpie marking pen
- slide holder (clothespin)
- Bunsen burner

1. Wash a slide with soap or Bon Ami and hot water, removing all dirt and grease. Handle the clean slide by its edges.

2. Write the initials of the organism or organisms on the left-hand side of the slide with a marking pen.

3. Mark a "target circle" on the bottom side of the slide with a marking pen. (See comments in step 3 on page 83.)

(1) Shake the culture tube from side to side to suspend organisms. Do not moisten cap on tube.

(2) Heat loop and wire to red-hot. Flame the handle slightly also.

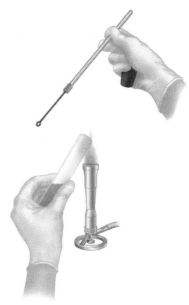

(3) Remove the cap and flame the neck of the tube. Do not place the cap down on the table.

(4) After allowing the loop to cool for at least 5 seconds, remove a loopful of organisms. Avoid touching the side of the tube.

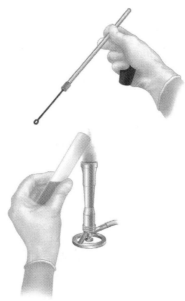

(5) Flame the mouth of the culture tube again.

(6) Return the cap to the tube and place the tube in a test-tube rack.

continued

FIGURE 10.2   Aseptic procedure for organism removal

(7) Place the loopful of organisms in the center of the target circle on the slide.

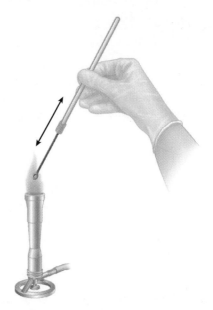

(8) Flame the loop again before removing another loopful from the culture or setting the inoculating loop aside.

FIGURE 10.2 *(continued)*

4. Flame an inoculating loop, let it cool, and transfer two loopfuls of water to the center of the target circle.
5. Flame an inoculating needle then let it cool. Pick up *a very small amount of the organisms,* and mix it into the water on the slide. Disperse the mixture over the area of the target circle. Be certain that the organisms have been well emulsified in the liquid. *Be sure to flame the inoculating loop before placing it aside.*
6. Allow the slide to dry by normal evaporation of the water. Don't apply heat.

7. After the slide has become completely air dry, place it in a clothespin and pass it several times through the flame of a Bunsen burner. Avoid prolonged heating of the slide as it can shatter from excessive exposure to heat. The underside of the slide should feel warm to the touch.

LABORATORY REPORT

Answer the question on Laboratory Report 10–13 that relate to this exercise.

Simple Staining

The use of a single stain to color a bacterial organism is commonly referred to as **simple staining.** Some of the most commonly used dyes for simple staining are methylene blue, basic fuchsin, and crystal violet. All of these dyes work well on bacteria because they have color-bearing ions (*chromophores*) that are positively charged (cationic).

The fact that bacteria are slightly negatively charged produces a pronounced attraction between these cationic chromophores and the organism. Such dyes are classified as **basic dyes.** The basic dye methylene blue (methylene$^+$ chloride$^-$) will be used in this exercise. Those dyes that have anionic chromophores are called **acidic dyes.** Eosin (sodium$^+$ eosinate$^-$) is such a dye. The anionic chromophore, eosinate$^-$, will not stain bacteria because of the electrostatic repelling forces that are involved.

The staining times for most simple stains are relatively short, usually from 30 seconds to 2 minutes, depending on the affinity of the dye. After a smear has been stained for the required time, it is washed off gently, blotted dry, and examined directly under oil immersion. Such a slide is useful in determining basic morphology and the presence or absence of certain kinds of granules.

An avirulent strain of *Corynebacterium diphtheriae* will be used here for simple staining. In its pathogenic form, this organism is the cause of diphtheria, a very serious disease. One of the steps in identifying this pathogen is to do a simple stain of it to demonstrate the following unique characteristics: pleomorphism, metachromatic granules, and palisade arrangement of cells.

Pleomorphism pertains to irregularity of form: that is, demonstrating several different shapes. While *C. diphtheriae* is basically rod-shaped, it also appears club-shaped, spermlike, or needle-shaped. *Bergey's Manual* uses the terms "pleomorphic" and "irregular" interchangeably.

Metachromatic granules are distinct reddish-purple granules within cells that show up when the organisms are stained with methylene blue. These granules are masses of *volutin*, a polymetaphosphate.

Palisade arrangement pertains to parallel arrangement of rod-shaped cells. This characteristic, also called "picket fence" arrangement, is common to many corynebacteria.

PROCEDURE

Prepare a slide of *C. diphtheriae,* using the procedure outlined in figure 11.1. It will be necessary to refer back to Exercise 10 for the smear preparation procedure.

MATERIALS

- slant culture of avirulent strain of *Corynebacterium diphtheriae*
- methylene blue (Loeffler's)
- wash bottle
- bibulous paper

After examining the slide, compare it with the photomicrograph in illustration 1, figure 14.4. Record your observations on Laboratory Report 10–13.

(1) A bacterial smear is stained with methylene blue for one minute.

(2) Stain is briefly washed off slide with water.

(3) Water drops are carefully blotted off slide with bibulous paper.

FIGURE 11.1   Procedure for simple staining

Negative Staining

Another stain that can be used to study the morphology of bacterial cells is the **negative stain.** Negative stains such as nigrosin and india ink do not penetrate the bacterial cell but rather cause the background area around a cell to be opaque or dark. Cells appear as transparent objects against the dark background. The negative stain reveals the shape of the cell and extracellular features such as capsules. The method consists of mixing the organism with a small amount of stain and spreading a very thin film over the surface of the slide. For the negative stain, cells are not heat fixed prior to the application of the negative stain.

The method can be useful for determining cell morphology and size of the cells. Because no heat fixation was performed, no shrinkage of the cells has occurred and size determinations are more accurate than those determined on fixed material. Avoiding heating is also important if the capsule surrounding bacterial cells is to be observed because heating severely shrinks this structure. The negative stain is also useful for observing spirochaetes that do not stain readily with ordinary dyes.

THREE METHODS

Negative staining can be done by one of three different methods. Figure 12.1 illustrates the more commonly used method in which the organisms are mixed in a drop of nigrosine and spread over the slide with another slide. The goal is to produce a smear that is thick at one end and feather-thin at the other end. Somewhere between the too thick and too thin areas will be an ideal spot to study the organisms.

Figure 12.2 illustrates a second method, in which organisms are mixed in only a loopful of nigrosine instead of a full drop. In this method, the organisms are spread over a smaller area in the center of the slide with an inoculating needle. No spreader slide is used in this method.

The third procedure (Woeste-Demchick's method), which is not illustrated here, involves applying ink to a conventional smear with a black felt-tip marking pen. If this method is used, it should be done on a smear prepared in the manner described in exercise 13. Simply put, the technique involves applying a *single coat* of marking-pen ink over a smear.

(1) Organisms are dispersed into a small drop of nigrosine or india ink. Drop should not exceed ⅛" diameter and should be near the end of the slide.

(2) Spreader slide is moved toward drop of suspension until it contacts the drop causing the liquid to be spread along it's spreading edge.

(3) Once spreader slide contacts the drop on the bottom slide, the suspension will spread out along the spreading edge as shown.

(4) Spreader slide is pushed to the left, dragging the suspension over the bottom slide. After the slide has air-dried, it may be examined under oil immersion.

FIGURE 12.1   Negative staining technique, using a spreader slide

(1) A loopful of nigrosine or india ink is placed in the center of a clean microscope slide.

(2) A sterile inoculating wire is used to transfer the organisms to the liquid and mix the organisms into the stain.

(3) Suspension of bacteria is spread evenly over an area of one or two centimeters with the straight wire.

(4) Once the preparation has completely air-dried, it can be examined under oil immersion. No heat should be used to hasten drying.

FIGURE 12.2   A second method for negative staining

Note in the procedure below that slides may be made from organisms between your teeth or from specific bacterial cultures. Your instructor will indicate which method or methods you should use and demonstrate some basic aseptic techniques. Various options are provided here to ensure success.

MATERIALS

- microscope slides (with polished edges)
- nigrosine solution or india ink
- slant cultures of *S. aureus* and *B. megaterium*
- inoculating loop
- sterile toothpicks
- Bunsen burner
- Sharpie marking pen
- felt-tip marking pen (see Instructor's Handbook)

1. Swab down your tabletop with disinfectant in preparation for making slides.
2. Clean two or three microscope slides with Bon Ami to rid them of all dirt and grease.
3. By referring to figure 12.1 or 12.2, place the proper amount of stain on the slide.

4. **Oral Organisms:** Remove a small amount of material from between your teeth with a sterile straight toothpick and mix it into the stain on the slide. Be sure to break up any clumps of organisms with the toothpick.

> **CAUTION** If you use a toothpick, discard it into a beaker of disinfectant.

5. **From Cultures:** With a *sterile* loop, transfer a very small amount of bacteria from the slant to the center of the stain on the slide.
6. Spread the mixture over the slide according to the procedure used in figure 12.1 or 12.2.
7. Allow the slide to air-dry and examine with an oil immersion objective.

LABORATORY REPORT

Draw a few representative types of organisms on Laboratory Report 10–13. If slide is of oral organisms, look for yeasts and hyphae as well as bacteria. Spirochaetes may also be present.

Capsular Staining

Some bacterial cells are surrounded by an extracellular slime layer called a **capsule** or **glycocalyx.** This structure can play a protective role for certain pathogenic bacteria such as *Streptococcus pneumoniae.* The capsule prevents phagocytic white blood cells from engulfing and destroying this bacterial pathogen, enabling the organism to invade the lungs and cause pneumonia. The capsule is also a means for many bacteria to attach to solid surfaces in the environment. For example, *Streptococcus mutans* can attach to the surface of a tooth by its capsular material resulting in the formation of dental plaque, which contributes to the process of tooth decay in humans. Most capsules are usually composed of polysaccharides but in some cases a capsule can consist of polypeptides with unique amino acids. Evidence supports the view that probably all bacterial cells have some amount of slime layer, but in most cases the amount is not enough to be readily discernible.

Staining of the bacterial capsule cannot be accomplished by ordinary staining procedures. If smears are heat fixed prior to staining, the capsule shrinks or is destroyed and cannot be seen in stains. If the cells are not heat fixed to the slide, they can wash off during washing procedures. However, the capsule can be demonstrated by combining the methods for the simple stain and the negative stain as shown in figure 13.1. You will use this method to stain the capsule of *Klebsiella pneumoniae.*

MATERIALS

- 36–48 hour milk culture of *Klebsiella pneumoniae*
- india ink
- crystal violet

Observation   Examine the slide under oil immersion and compare your slide with illustration 2, figure 14.4 on page 97. Record your results on Laboratory Report 10–13.

(1) Two loopfuls of the organism are mixed in a small drop of india ink.

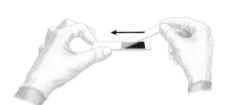

(2) The ink suspension of bacteria is spread over slide and air-dried.

(3) The slide is *gently* heat-dried to fix the organisms to the slide.

(4) Smear is stained with crystal violet for one minute.

(5) Crystal violet is *gently* washed off with water.

(6) Slide is blotted dry with bibulous paper, and examined with oil immersion objective.

FIGURE 13.1   Procedure for demonstration of a capsule

Gram staining: Bacterial cells from eukaryotic nuclei

Gram staining: gram positive and gram-negative →Red
violed.

↓
Retain crystal
violet iodine

First gram-positive and gram-negative are stained with the
primary stain crystal violet.

* second step; the gram's iodine is added to smear.

Gram Staining

In 1884, the Danish physician Christian Gram was trying to develop a staining procedure that would differentiate bacterial cells from eukaryotic nuclei in stained tissue samples. Although Gram was not completely successful in developing a tissue stain, what resulted from his work is the most important stain in bacteriology, the Gram stain. This technique separates bacteria into two groups: gram-positive and gram-negative bacteria. The procedure is based on the fact that gram-positive bacteria retain a crystal violet-iodine complex through decolorization with alcohol or acetone. Gram-positive bacteria appear as purple when viewed by microscopy. In contrast, alcohol or acetone removes the crystal violet-iodine complex from gram-negative bacteria. These bacteria must, therefore, be counterstained with a red dye, safranin, after the decolorization step in order to be visualized by microscopy. Hence, gram-negative bacteria appear as red cells when viewed by microscopy.

Figure 14.1 illustrates the appearance of cells after each step in the Gram-stain procedure. Note that initially both gram-positive and gram-negative cells are stained by the **primary stain,** crystal violet. In the second step of the procedure, Gram's iodine is added to the smear. Iodine is a **mordant** that complexes with the crystal violet and forms an insoluble complex in gram-positive cells. At this point, both types of cells will still appear as purple. The dye-mordant complex is not removed from gram-positive bacteria but is leached from gram-negative cells by the alcohol or acetone in the **decolorization** step. After decoloriza-

*Most imp
Stain: Gram
Stain.*

Reagent	Gram positive	Gram negative
None (Heat-fixed cells)		
Crystal Violet (20 seconds)		
Gram's Iodine (1 minute)		
Ethyl Alcohol (10–20 seconds)		
Safranin (1 minute)		

FIGURE 14.1 Color changes that occur at each step in the gram-staining process

tion, gram-positive cells are purple but gram-negative cells are colorless. In the final step, a **counterstain,** safranin, is applied, which stains the colorless gram-negative cells. The appearance of the gram-positive cells is unchanged because the crystal violet is a much more intense stain than safranin.

The mechanism for how the Gram stain works is related to chemical differences in the cell walls of gram-positive and gram-negative bacteria (figure 14.2). When viewed by electron microscopy, gram-positive cells have a thick layer of **peptidoglycan** that comprises the cell wall of these organisms. In contrast, the cell wall in gram-negative cells consists of

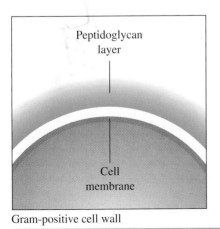

Peptidoglycan layer

Cell membrane

Gram-positive cell wall

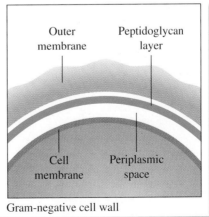

Outer membrane

Peptidoglycan layer

Cell membrane

Periplasmic space

Gram-negative cell wall

FIGURE 14.2 Comparison of gram-positive and gram-negative cell walls

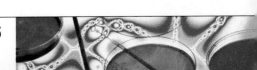

an outer membrane that covers a much thinner layer of peptidoglycan. It is these significant differences in structure that probably determines whether the dye-mordant complex is removed from the gram-negative cell or remains associated with the gram-positive cell.

Of all the staining techniques you will use in microbiology, the Gram stain is one of the most important. It will be critical in identifying your unknown bacteria and you will use it routinely in many exercises in this manual. Although this technique seems quite simple, performing it with a high degree of reliability requires some practice and experience. Several factors can affect the outcome of the procedure:

1. It is important to use cultures that are 16–18 hours old. Gram-positive cultures older than this can convert to gram-variable or gram-negative and give erroneous results. (It is important to note that gram-negative bacteria never convert to gram-positive.)
2. It is critical to prepare thin smears. Thin smears allow the observation of individual cells and any arrangement in which the cells occur. Furthermore, the thickness of your smears can affect decolorization. Thick smears can entrap the primary

stain, which is not removed by alcohol or acetone. Cells that occur in the entrapped stain can appear gram-positive leading to erroneous results.
3. Decolorization is the most critical step in the Gram-stain procedure. If the destain is overapplied, the dye-mordant complex can eventually be removed from gram-positive cells, converting them to gram-negative cells.

During this laboratory period, you will be provided an opportunity to stain several different kinds of bacteria to see if you can achieve the degree of success that is required. Remember, if you don't master this technique now, you will have difficulty with your unknowns later.

STAINING PROCEDURE

MATERIALS

- slides with heat-fixed smears
- Gram-staining kit and wash bottle
- bibulous paper

1. Cover the smear with **crystal violet** and let stand for *20 seconds* (see figure 14.3).

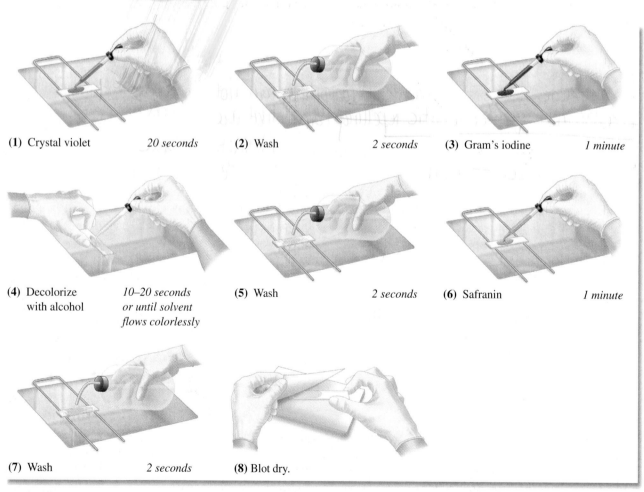

(1) Crystal violet   *20 seconds*   (2) Wash   *2 seconds*   (3) Gram's iodine   *1 minute*

(4) Decolorize   *10–20 seconds*   (5) Wash   *2 seconds*   (6) Safranin   *1 minute*
     with alcohol   *or until solvent*
               flows colorlessly

(7) Wash   *2 seconds*   (8) Blot dry.

FIGURE 14.3 The gram-staining procedure

2. Briefly wash off the stain, using a wash bottle of distilled water. Drain off excess water.
3. Cover the smear with **Gram's iodine** solution and let it stand for *one minute*. (Your instructor may prefer only 30 seconds for this step.)
4. Wash off the Gram's iodine. Hold the slide at a 45-degree angle and allow the 95% alcohol to flow down the surface of the slide. Do this until the alcohol is colorless as it flows from the smear down the surface of the slide. *This should take no more than 20 seconds for properly prepared smears.* Note: thick smears can take longer than 20 seconds for decolorization.
5. Stop decolorization by washing the slide with a gentle stream of water.
6. Cover the smear with **safranin** for 1 minute.
7. Wash gently for a few seconds, blot dry with bibulous paper, and air-dry.
8. Examine the slide under oil immersion.

ASSIGNMENTS

The organisms that will be used here for Gram staining represent a diversity of form and staining characteristics. Some of the rods and cocci are gram-positive; others are gram-negative. One rod-shaped organism is a spore-former and another is acid-fast. The challenge here is to make Gram-stained slides of various combinations that reveal their differences.

MATERIALS

- broth cultures of *Staphylococcus aureus, Pseudomonas aeruginosa,* and *Moraxella (Branhamella) catarrhalis*
- nutrient agar slant cultures of *Bacillus megaterium* and *Mycobacterium smegmatis*

Mixed Organisms I (Triple Smear Practice Slides) Prepare three slides with three smears on each slide. On the left portion of each slide make a smear of *Staphylococcus aureus*. On the right portion of each slide make a smear of *Pseudomonas aeruginosa*. In the middle of the slide make a smear that is a mixture of both organisms, using two loopfuls of each organism. ***Be sure to flame the loop sufficiently to avoid contaminating cultures.***

Gram stain one slide first, saving the other two for later. Examine the center smear. If done properly, you should see purple cocci and pink rods as shown in illustration 3, figure 14.4.

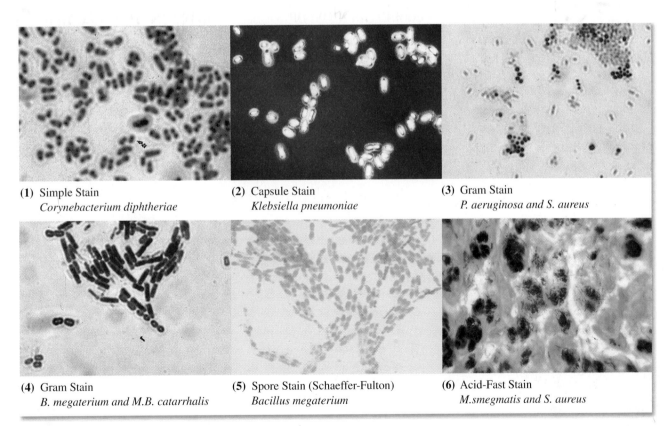

(1) Simple Stain
Corynebacterium diphtheriae

(2) Capsule Stain
Klebsiella pneumoniae

(3) Gram Stain
P. aeruginosa and S. aureus

(4) Gram Stain
B. megaterium and M.B. catarrhalis

(5) Spore Stain (Schaeffer-Fulton)
Bacillus megaterium

(6) Acid-Fast Stain
M.smegmatis and S. aureus

FIGURE 14.4 **Photomicrographs of representative staining techniques (8000x)**
(2) © Science VU/Visuals Unlimited (5) Courtesy of Lansing Prescott (6) © Science VU/Visuals Unlimited

Call your instructor over to evaluate your slide. If the slide is improperly stained, the instructor will be able to tell what went wrong by examining all three smears. He or she will inform you how to correct your technique when you stain the next triple smear reserve slide.

Record your results on Laboratory Report 14–16 by drawing a few cells in the appropriate circle.

Mixed Organisms II Make a Gram-stained slide of a mixture of *Bacillus megaterium* and *Moraxella (Branhamella) catarrhalis*.

This mixture differs from the previous slide in that the rods (*B. megaterium*) will be purple and the cocci (*M.B. catarrhalis*) will be large pink diplococci. See illustration 4, figure 14.4.

As you examine this slide look for clear areas on the rods, which represent endospores. Since endospores are refractile and impermeable to crystal violet, they will appear as transparent holes in the cells.

Draw a few cells in the appropriate circle on your Laboratory Report sheet.

Acid-Fast Bacteria To see how acid-fast mycobacteria react to Gram's stain, make a Gram-stained slide of *Mycobacterium smegmatis*. If your staining technique is correct, the organisms should appear gram-positive.

Draw a few cells in the appropriate circle on your Laboratory Report sheet.

Spore Staining: two methods.

Thick protein coat → which is an exosporium that forms a protective barrier around the spore.

When species of bacteria belonging to the genera *Bacillus* and *Clostridia* exhaust essential nutrients, they undergo a complex developmental cycle that produces resting stages called **endospores.** Endospores allow these bacteria to survive environmental conditions that are not favorable for growth. If nutrients once again become available, the endospore can go through the process of germination to form a new vegetative cell and growth will resume. Endospores are very dehydrated structures that are not actively metabolizing. Furthermore, they are resistant to heat, radiation, acids, and many chemicals, such as disinfectants, that normally harm or kill vegetative cells. Their resistance is due in part to the fact that they have a thick protein coat, or **exosporium,** that forms a protective barrier around the spore. Heat resistance is associated with the water content of endospores. The higher the water content of an endospore, the less heat resistant the endospore will be. During sporulation, the water content of the endospore is reduced to 10–30% of the vegetative cell. This results because calcium ions complex with spore-specific proteins and a chemical, dipicolinic acid. The latter compound is not found in vegetative cells. This complex forms a gel that controls the amount of water that can enter the endospore, thus maintaining its dehydrated state.

Since endospores are not easily destroyed by heat or chemicals, they define the conditions necessary to establish sterility. For example, to destroy endospores by heating, they must be exposed for 15 to 20 minutes to steam under pressure, which generates temperatures of 121° C. Such conditions are produced in an **autoclave.**

The resistant properties of endospores also mean that they are not easily penetrated by stains. For example in exercise 14, you observed that endospores did not readily Gram stain. If endospore-containing cells are stained by basic stains such as crystal violet, the spores appear as unstained areas in the vegetative cell. However, if heat is applied while staining with malachite green, the stain penetrates the endospore and becomes entrapped in the endospore. The malachite green is not removed by subsequent washing with decolorizing agents or water. In this instance, heat is acting as mordant to facilitate the uptake of the stain.

SCHAEFFER-FULTON METHOD

The Schaeffer-Fulton method, which is depicted in figure 15.1, utilizes malachite green to stain the endospore and safranin to stain the vegetative portion of the cell. Utilizing this technique, a properly stained spore-former will have a green endospore contained in a pink sporangium. Illustration 5, figure 14.4, on page 97 reveals what such a slide looks like under oil immersion.

After preparing a smear of *Bacillus megaterium,* follow the steps outlined in figure 15.1 to stain the spores.

Rachna Sachdev

MATERIALS

- 24–36 hour nutrient agar slant culture of *Bacillus megaterium*
- electric hot plate and small beaker (25 ml)
- spore-staining kit consisting of a bottle each of 5% malachite green and safranin

DORNER METHOD

The Dorner method for staining endospores produces a red spore within a colorless sporangium. Nigrosine is used to provide a dark background for contrast. The six steps involved in this technique are shown in figure 15.2. Although both the sporangium and endospore are stained during boiling in step 3, the sporangium is decolorized by the diffusion of safranin molecules into the nigrosine.

Prepare a slide of *Bacillus megaterium* that utilizes the Dorner method. Follow the steps in figure 15.2.

(1) Cover smear with small piece of paper toweling and saturate it with malachite green. Steam over boiling water for *5 minutes*. Add additional stain if stain boils off.

(2) After the slide has cooled sufficiently, remove the paper toweling and rinse with water for 30 seconds.

(3) Counterstain with safranin for about *20 seconds*.

(4) Rinse briefly with water to remove safranin.

(5) Blot dry with bibulous paper, and examine slide under oil immersion.

FIGURE 15.1   The Schaeffer-Fulton spore stain method

MATERIALS

- nigrosine
- electric hot plate and small beaker (25 ml)
- small test tube (10 × 75 mm size)
- test-tube holder
- 24–36 hour nutrient agar slant culture of *Bacillus megaterium*

QUICK SPORE STAIN

A variation on the Schaeffer-Fulton method is a quick method that uses the same stains.

MATERIALS

- *Bacillus subtilis* slant cultures, older than 36 hours
- malachite green stain
- safranin stain
- staining racks
- clothespins

PROCEDURE

1. Prepare a smear of the organism and allow it to air-dry.
2. Grasp the slide with the air-dried smear with a clothespin and pass it through a Bunsen burner flame 10 times. Be careful not to overdo the heating as the slide can break.
3. Immediately flood the smear with malachite green and allow to stand for 5 minutes.
4. Wash the smear with a gentle stream of water.
5. Stain with safranin for 45 seconds. Spores will be green and the vegetative cell will be red.

LABORATORY REPORT

After examining the organisms under oil immersion, draw a few cells in the appropriate circles on Laboratory Report 14–16.

(1) Make a heavy suspension of bacteria by dispersing several loopfuls of bacteria in 5 drops of sterile water.

(2) Add 5 drops of carbolfuchsin to the bacterial suspension.

(3) Heat the carbolfuchsin suspension of bacteria in a beaker of boiling water for *10 minutes*.

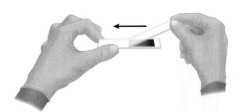

(4) Mix several loopfuls of bacteria in a drop of nigrosine on the slide.

(5) Spread the nigrosine-bacteria mixture on the slide in the same manner as in Exercise 11 (Negative Staining).

(6) Allow the smear to air-dry. Examine the slide under oil immersion.

FIGURE 15.2   The Dorner spore stain method

Acid-Fast Staining:
Ziehl-Neelsen Method

Bacteria in the genus *Mycobacterium* and some in the genus *Nocardia* contain a waxy material in their cell walls called **mycolic acid.** This material significantly affects the staining properties of these organisms and prevents them from being stained by many of the stains used in microbiology. However, if they are stained with carbolfuchsin and heat is applied during the staining procedure, the carbolfuchsin is able to penetrate the cell and it is not removed by subsequent washing with acid-alcohol. Such bacteria are said to be **acid-fast** and appear pink or red in stained smears. This property sets them apart from most other bacteria, which are decolorized by the acid-alcohol and must be counterstained with methylene blue to be seen. In the acid-fast stain, heat is acting as a mordant to soften the mycolic acid so the stain can penetrate the cell.

The acid-fast stain is an important diagnostic tool in the identification of *Mycobacterium tuberculosis* the causative agent of tuberculosis, and *Mycobacterium leprae,* the bacterium that causes leprosy in humans. When the stain is used in the diagnosis of these diseases, the mycobacteria appear as red rods whereas tissue cells and non-acid-fast bacteria are stained blue. An example of an acid-fast stain is seen in photo 6 in figure 14.4.

In the following exercise, you will prepare an acid-fast stain of a mixture of *Mycobacterium smegmatis* and *Staphylococcus aureus*. *M. smegmatis* is a nonpathogenic acid-fast rod that occurs in soil and on the external genitalia of humans. *S. aureus* is a non-acid-fast rod coccus that can also be part of the normal flora of humans as well as a potential pathogen.

MATERIALS

- nutrient agar slant culture of *Mycobacterium smegmatis* (48-hour culture)
- nutrient broth culture of *S. aureus*
- electric hot plate and small beaker
- acid-fast staining kit (carbolfuchsin, acid-alcohol, and methylene blue)

Smear Preparation Prepare a mixed culture smear by placing two loopfuls of *S. aureus* on a slide and transferring a small amount of *M. smegmatis* to the broth on the slide with an inoculating loop. Since the smegma bacilli are waxy and tend to cling to each other in clumps, break up the masses of organisms with the inoculating loop. After air-drying the smear, heat-fix it.

Staining Follow the staining procedure outlined in figure 16.1.

Examination Examine under oil immersion and compare your slide with illustration 6, figure 14.4.

Laboratory Report Record your results on Laboratory Report 14–16.

(1) Cover smear with carbolfuchsin. Steam over boiling water *5 minutes*. Add additional stain if stain boils off.

(2) After slide has cooled, decolorize with acid-alcohol for *15–20 seconds*.

(3) Stop decolorization action of acid-alcohol by rinsing *briefly* with water.

(4) Counterstain with methylene blue for *30 seconds*.

(5) Rinse *briefly* with water to remove excess methylene blue.

(6) Blot dry with bibulous paper. Examine directly under oil immersion.

FIGURE 16.1 Ziehl-Neelsen acid-fast staining procedure

Motility Determination

The major organelles of motility in bacteria are **flagella.** Flagella allow cells to move toward nutrients in the environment or move away from harmful substances, such as acids, in a complicated process called **chemotaxis.** The flagellum is a rigid helical structure that extends as much as 10 microns out from the cell. However, flagella are very thin structures, less than 0.2 microns, and, therefore, they are below the resolution of the light microscope. For flagella to be observed by light microscopy, they must be stained by special techniques. Flagella cause a bacterial cell to move because they rotate, in a way similar to a screw on a boat engine that rotates to propel a boat through the water.

Motility and the arrangement of flagella around the cell (figure 17.1) are important taxonomic characteristics that are useful in characterizing bacteria. Motility can be determined by several methods. It can be determined microscopically by observing cells in a **wet mount.** In this procedure, a drop of viable cells is placed on a microscope slide and covered with a cover glass. The slide is then observed with a phase-contrast microscope. The rapid swimming movement of cells in the microscopic field confirms motility. However, wet mounts can easily dry out by evaporation, which is especially troublesome if observations need to made for prolonged periods of time. Drying can be delayed by using the **hanging drop technique,** shown in figure 17.2. In this procedure, a drop of cells is placed on a cover glass, which is then placed over a special slide that has a concave depression in its center. The glass is held in place with petroleum jelly,

thus forming an enclosed glass chamber that prevents drying.

For the beginner, true swimming motility under the microscope must be differentiated from **Brownian motion** of cells or movement caused by currents under the cover glass. Brownian motion is movement due to molecular bombardment of cells causing cells to shake or "jiggle about" but not move in any vectorial way. Cells can also appear to move because currents can be created under the cover glass when pressure is exerted by focusing the oil immersion lens or by the wet mount drying out. This causes cells to "sweep" across the field.

Another method for determining motility involves inoculating semisoft agar medium. This medium has an agar concentration of 0.4%, which does not inhibit bacteria from "swimming" through the medium. In this procedure, the organism is inoculated by stabbing the semisolid medium with an inoculating needle. If the organism is motile, it will swim away from the line of inoculation into the uninoculated surrounding medium. Nonmotile bacteria will be found only along the line of inoculation. For pathogenic bacteria, such as the typhoid bacillus, the use of semisoft agar medium to determine motility is often preferred over microscope techniques because of the potential for infection posed by pathogens in making wet mounts.

In the following exercise, you will use both microscopic and culture media procedures to determine motility of bacterial cultures.

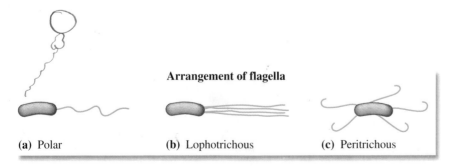

Arrangement of flagella

(a) Polar **(b)** Lophotrichous **(c)** Peritrichous

FIGURE 17.1 Arrangement of flagella

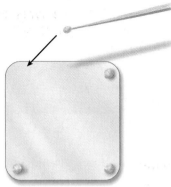

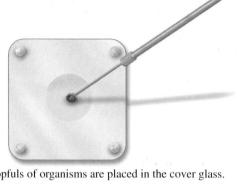

(1) A small amount of Vaseline is placed near each corner of the cover glass with a toothpick.

(2) Two loopfuls of organisms are placed in the cover glass.

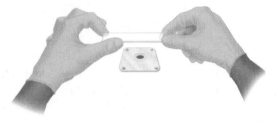

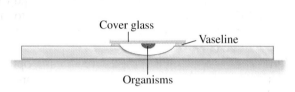

(3) Depression slide is pressed against Vaseline on cover glass and quickly inverted.

(4) The completed preparation can be examined under oil immersion.

FIGURE 17.2 The hanging drop slide

FIRST PERIOD

During the first period, you will make wet mount and hanging drop slides of two organisms: *Proteus vulgaris* and *Micrococcus luteus*. Tube media (semisolid medium or SIM medium) and a soft agar plate will also be inoculated. The media inoculations will have to be incubated to be studied in the next period. Proceed as follows:

MATERIALS

- microscope slides and cover glasses
- depression slide
- 2 tubes of semisolid or SIM medium
- 1 petri plate of soft nutrient agar (20–25 ml of soft agar per plate)
- nutrient broth cultures of *Micrococcus luteus* and *Proteus vulgaris* (young cultures)
- inoculating loop and needle
- Bunsen burner

Wet Mounts   Prepare wet mount slides of each of the organisms, using several loopfuls of the organism on the slides. Examine under an oil immersion objective. Observe the following guidelines:

- Use only scratch-free, clean slides and cover glasses. This is particularly important when using phase-contrast optics.

- Label each slide with the name of the organism.
- By manipulating the diaphragm and voltage control, reduce the lighting sufficiently to make the organisms visible. Unstained bacteria are very transparent and difficult to see.
- For proof of true motility, look for directional movement that is several times the long dimension of the bacterium. The movement will also occur in different directions in the same field.
- Ignore Brownian movement. *Brownian movement* is vibrational movement caused by invisible molecules bombarding bacterial cells. If the only movement you see is vibrational and not directional, the organism is nonmotile.
- If you see only a few cells exhibiting motility, consider the organism to be motile. Characteristically, only a few of the cells will be motile at a given moment.
- Don't confuse water current movements with true motility. Water currents are due to capillary action caused by temperature changes and drying out. All objects move in a straight line in one direction.
- And, finally, always *examine a wet mount immediately,* once it has been prepared, because motility decreases with time after preparation.

Hanging Drop Slides   By referring to figure 17.2, prepare hanging drop slides of each organism. Be sure to use clean cover glasses and label each slide with a marking pen. When placing loopfuls of organisms on the cover glass, be sure to flame the loop between applications. Once the slide is placed on the microscope stage, do as follows:

1. Examine the slide first with the low-power objective. If your microscope is equipped with an automatic stop, avoid using the stop; instead, use the coarse adjustment knob for bringing the image into focus. The greater thickness of the depression slide prevents one from being able to focus at the stop point.
2. Once the image is visible under low power, swing the high-dry objective into position and readjust the lighting. Since most bacteria are drawn to the edge of the drop by surface tension, **focus near the edge of the drop.**
3. If your microscope has phase-contrast optics, switch to high-dry phase. Although a hanging drop does not provide the shallow field desired for phase-contrast, you may find that it works fairly well.
4. If you wish to use oil immersion, simply rotate the high-dry objective out of position, add immersion oil to the cover glass, and swing the oil immersion lens into position.
5. Avoid delay in using this setup. Water condensation may develop to decrease clarity and the organisms become less motile with time.
6. Review all the characteristics of bacterial motility that are stated on page 108 under wetmounts.

Tube Method   Inoculate tubes of semisolid or SIM media with each organism according to the following instructions:

1. Label the tubes of semisolid (or SIM) media with the names of the organisms. Place your initials on the tubes also.
2. Flame and cool the inoculating needle, and insert it into the culture after flaming the neck of the tube.
3. Remove the cap from the tube of medium, flame the neck, and stab it two-thirds of the way down to the bottom, as shown in figure 17.3. Flame the neck of the tube again before returning the cap to the tube.
4. Repeat steps 2 and 3 for the other culture.
5. Incubate the tubes at room temperature for 24 to 48 hours.

Plate Method   Mark the bottom of a plate of soft agar with two one-half inch circles about one inch apart. Label one circle ML and the other PV. These circles will be targets for your culture stabs. Put your initials on the plate also.

Using proper aseptic techniques, stab the medium in the center of the ML circle with *M. luteus* and the center of the other circle with *P. vulgaris*. Incubate the plate for 24 to 48 hours at room temperature.

SECOND PERIOD

Assemble the following materials that were inoculated during the last period and incubated.

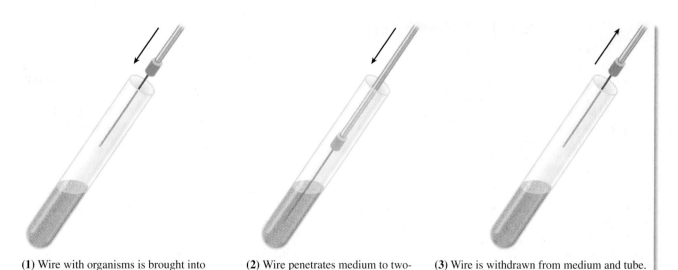

(**1**) Wire with organisms is brought into tube without touching walls of tube.

(**2**) Wire penetrates medium to two-thirds of its depth.

(**3**) Wire is withdrawn from medium and tube. Neck of tube is flamed and plugged.

FIGURE 17.3   Stab technique for motility test

MATERIALS

- culture tubes of motility medium that have been incubated
- inoculated petri plate that has been incubated

Compare the two tubes that were inoculated with *M. luteus* and *P. vulgaris*. Look for cloudiness as evidence of motility. *Proteus* should exhibit motility. Does it? Record your results on the Laboratory Report.

Compare the appearance of the two stabs in the soft agar. Describe the differences that exist in the two stabs.

Does the plate method provide any better differentiation of results than the tube method?

LABORATORY REPORT

Complete the Laboratory Report for this exercise.

CULTURE METHODS

All nutritional types are represented among the protists. This diversity requires a multiplicity of culture methods. This unit presents those techniques that have proven most successful for the culture of heterotrophic bacteria, molds, and slime molds.

The first three exercises (18, 19, and 20) pertain to basic techniques applicable to heterotrophs.

This unit culminates the basic techniques phase of this course. A thorough understanding of microscopy, slide techniques, and culture methods provides a substantial foundation for the remainder of the exercises in this manual. If independent study projects are to be pursued as a part of this course, the completion of this unit will round out the background knowledge and skills for such work.

The cultivation of microorganisms on an artificial growth medium requires that the medium supply all the nutritional and energy requirements necessary for growth. However, in some cases, we may not know what the specific nutrient requirements are for a certain organism to grow. In order to cultivate such an organism, we construct a medium using rich extracts of meat or plants that would supply all the amino acids, nucleotide bases, vitamins, and other growth factors required by our organism. Such a medium is called a **complex medium** because the exact composition and amounts of the individual amino acids, vitamins, growth factors, and other components that make up the medium are not exactly known. Many of the media used in microbiology are complex, such as nutrient agar, which is used to cultivate a variety of bacteria, especially those used in exercises in this manual. For some organisms we know what the specific nutritional requirements are for growth. For example, for *Escherichia coli,* we prepare a medium composed of specific components and amounts, such as a glucose and various salts. This medium is called a **defined medium** because the specific chemical composition is known and the individual components are weighed out exactly to make up the medium.

NUTRITIONAL REQUIREMENTS OF BACTERIA

Any medium, be it complex or defined, must supply certain basic nutritional requirements that are necessary for all cells to grow. These include a carbon and energy source, nitrogen, minerals, vitamins and growth factors, and water.

Carbon Source Organisms can be divided into two groups based on their carbon requirements. **Heterotrophs** obtain their carbon from organic compounds such as polysaccharides, carbohydrates, amino acids, peptides, and proteins. Meat and plant extracts are added to complex media to supply these nutrients. In contrast, **autotrophs** derive their carbon requirements from fixing carbon dioxide. From the latter, they must synthesize all the complex molecules that comprise the bacterial cell.

Energy Bacterial cells require energy to carry out biosynthetic processes that lead to growth. These include synthesizing nucleic acids, proteins, and structural elements such as cell walls. **Chemoorganotrophs** derive their energy needs from the breakdown of organic molecules by fermentation or respiration. Most bacteria belong to this metabolic group. **Chemolithotrophs** oxidize inorganic ions such as nitrate or iron to obtain energy. Examples of the chemolithotrophs are the nitrifying and iron bacteria. **Photoautotrophs** contain photosynthetic pigments such as chlorophyll or bacteriochlorophyll that convert solar energy into chemical energy by the process of photosynthesis. Energy derived from photosynthesis can then be used by the cell to fix carbon dioxide and synthesize the various cellular materials necessary for growth. For these organisms, no energy source is supplied in the medium but rather energy needs to be supplied in the form of illumination. The cyanobacteria and the green and purple sulfur bacteria are examples of photoautotrophs that carry out photosynthesis. A few photosynthetic bacteria are **photoheterotrophs.**

These organisms also derive their energy requirements from photosynthesis but their carbon needs come from growth on organic molecules such as succinate or glutamate. Some of the purple nonsulfur bacteria are found in this category.

Nitrogen Nitrogen is an essential element in biological molecules such as amino acids, nucleotide bases, and vitamins. Some bacteria can synthesize these compounds using carbon intermediates and inorganic forms of nitrogen (e.g., ammonia and nitrate). Others lack this capability and must gain their nitrogen from organic molecules such as proteins, peptides, or amino acids. Beef extract and peptones are incorporated into complex media to provide a source of nitrogen for these bacteria. Some bacteria are even capable of fixing atmospheric nitrogen into inorganic nitrogen, which can then be used for biosynthesis of amino acids. Bacteria such as *Rhizobium* and *Azotobacter* are examples of nitrogen-fixing bacteria.

Minerals Metals are essential in bacterial metabolism because they are cofactors in enzymatic reactions and are integral parts of molecules such as cytochromes, bacteriochlorophyll, and vitamins. Metals required for growth include sodium, potassium, calcium, magnesium, manganese, iron, zinc, copper, cobalt, and phosphorus. Most are required in catalytic or very small amounts.

Vitamins and Growth Factors Vitamins serve as coenzymes in metabolism. For example, the vitamin niacin is a part of the coenzyme NAD, and flavin is a component of FAD. Some bacteria, like the streptococci and lactobacilli, require vitamins because they are unable to synthesize them. Other bacteria (e.g., *Escherichia coli*) can synthesize vitamins and hence do not require them in media in order to grow. However, sometimes, even in addition to supplying all the normal components, it is necessary to add growth factors for ample growth of certain bacteria. Many pathogens are fastidious and grow better if blood or serum components are incorporated into their media. Blood and serum may provide additional metabolic factors not found in the normal components.

Water The cell consists of 70% to 80% water. Cells require an aqueous environment because enzymatic reactions and transport only occur in its presence. Furthermore, water maintains the various components of the cytoplasm in solution. When preparing media, it is essential to always use **distilled water.** Tap water can contain minerals such as calcium, phosphorus, and magnesium ions that could react with peptones and meat extracts to cause unwanted precipitates and cloudiness.

In addition to having the right components, it is important to make sure that the pH of the medium is adjusted to optimal values so that growth is not inhibited. Most bacteria grow best at a neutral pH value around 7. Fungi prefer pH values around 5 for best growth. Most commercial media do not require adjusting the pH, but it would probably be necessary to adjust the pH of a defined synthetic medium. This can be done with acids such as HCl or bases such as sodium hydroxide.

DIFFERENTIAL AND SELECTIVE MEDIA

Media can be made with components that will allow certain bacteria to grow but will inhibit others from growing. Such a medium is a **selective medium.** Antibiotics, dyes, and various inhibitory compounds are often incorporated into media to create selective conditions for growing specific organisms. For example, the dyes eosin and methylene blue when incorporated into EMB media do not affect the growth of gram-negative bacteria but do inhibit the growth of gram-positive bacteria. Incorporation of sodium chloride into mannitol-salt agar selects for *Staphylococcus aureus* but inhibits the growth of other bacteria that cannot tolerate the salt concentration.

A **differential medium** contains substances that cause some bacteria to take on an appearance that distinguishes them from other bacteria. When *Staphylococcus aureus* grows on mannitol-salt agar, it ferments mannitol, changing a pH indicator from red to yellow around colonies. Other staphylococci cannot ferment mannitol and their growth on this medium results in no change in the indicator. EMB is also a differential medium. Gram-negative bacteria that ferment lactose in this medium form colonies with a metallic-green sheen. Non-lactose-fermenting bacteria do not form colonies with the characteristic metallic sheen.

Media can be prepared in liquid or solid form depending on the application of its use. Liquid broth cultures are used to grow large volumes of bacteria. Fermentation studies, indole utilization, and the methyl red and Voges-Proskauer tests are done in broth cultures. Streaking of bacteria and selection of isolated colonies are done on solid media. **Agar,** a complex polysaccharide isolated from seaweed, is added in a concentration of 1.5% to solidify liquid media. The use of agar in bacteriological media was first introduced in Robert Koch's laboratory. Agar has unique properties that make it ideal for use in microbiology. First, it melts at 100° C but does not solidify until it cools to 45° C. Bacteria can be inoculated into agar media at this temperature (e.g., pour plates) without killing the cells. Second, agar is not a nutrient for most bacteria (the exceptions are a few bacteria found in marine environments). Sometimes, agar is added at lower concentrations (e.g., 0.4%) to make semisoft media. This type of media is used in motility studies.

Prior to 1930, it was necessary for laboratory workers to prepare media using various raw materials. This often involved boiling plant material or meat to prepare extracts that would be used in the preparation of media. Today, commercial companies prepare and sell media components, which are used for most routine bacteriological media. It is only necessary to weigh out a measured amount of a specific medium, such as nutrient agar, and dissolve it in water. In some cases, it may be necessary to adjust the pH prior to sterilizing the medium.

Before any medium can be used to grow bacteria or other microorganisms, it must be **sterilized,** that is, any contaminating bacteria introduced during preparation must be killed or removed. Most media can be **autoclaved** to achieve sterilization. This involves heating the media to **121° C** for at least **15 minutes** at **15 psi** of steam pressure. These conditions are sufficient to kill cells and any endospores present. Sometimes a medium may require a component that is heat sensitive and it cannot be subjected to autoclave temperatures. The component can be filter sterilized by passing a solution through a bacteriological filter of 0.45 microns. This filter will retain any cells and endospores that may be present. After filtration, the component can be added to the sterilized medium.

MEDIA PREPARATION ASSIGNMENT

In this laboratory period, you will work with your laboratory partner to prepare tubes of media that will be used in future laboratory experiments (figure 18.1). Your instructor will indicate which media you are to prepare. Record in the space below the number of tubes of specific media that have been assigned to you and your partner.

nutrient broth   _____
nutrient agar pours   _____
nutrient agar slants   _____
other   _____

Several different sizes of test tubes are used for media, but the two sizes most generally used are either 16 mm or 20 mm diameter by 15 cm long. Select the correct size tubes first, according to these guidelines:

Large tubes (20 mm dia): Use these test tubes for *all pours* (i.e., nutrient agar, Sabouraud's agar, EMB agar, etc.). Pours are used for filling petri plates.

Small tubes (16 mm dia): Use these tubes for all *broths, deeps,* and *slants.*

If the tubes are clean and have been protected from dust or other contamination, they can be used without cleaning. If they need cleaning, scrub out the insides with warm water and detergent, using a test-tube brush. Rinse twice, first with tap water and finally with distilled water to rid them of all traces of detergent. Place them in a wire basket or rack, inverted, so that they can drain. Do not dry with a towel.

Measurement and Mixing

The amount of medium you make for a batch should be determined as precisely as possible to avoid shortage or excess.

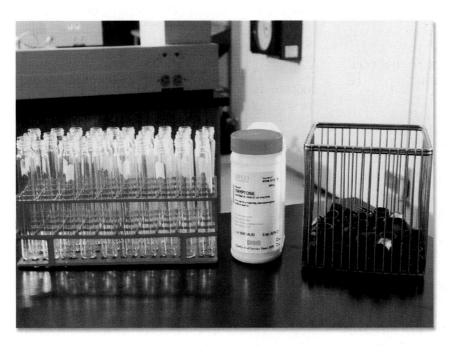

FIGURE 18.1 Basic supplies for making up a batch of medium
©The McGraw-Hill Companies/Auburn University Photographic Service

- graduate, beaker, glass stirring rod
- bottles of dehydrated media
- Bunsen burner and tripod, or hot plate

1. Measure the correct amount of water needed to make up your batch. The following volumes required per tube must be taken into consideration:

pours . . . . 12 ml
deeps . . . . 6 ml
slants . . . . 4 ml
broths . . . . 5 ml
broths with fermentation tubes . . . . 5–7 ml

2. Consult the label on the bottle to determine how much powder is needed for 1,000 ml and then determine by proportionate methods how much you need for the amount of water you are using. Weigh this amount on a balance (figure 18.2) and add it to the beaker of water. If the medium does not contain agar, the mixture usually goes into solution without heating (figure 18.3).

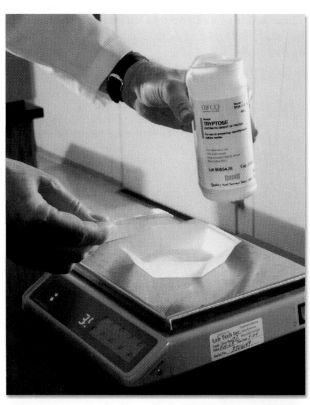

FIGURE 18.2 Correct amount of dehydrated medium is carefully weighed on a balance
©The McGraw-Hill Companies/Auburn University Photographic Service

3. **If the medium contains agar,** heat the mixture on a stirring hot plate (figure 18.4) or on an electric hot plate until it comes to a boil. To safeguard against water loss, *before heating, mark the level of the top of the medium on the side of the beaker with a marking pen.* As soon as it "froths up," turn off the heat. If an electric hot plate is used, the medium must be removed from the hot plate or it will boil over the sides of the container.

   Caution: Be sure to keep stirring from the bottom with a glass stirring rod so that the medium does not char on the bottom of the beaker.

4. Check the level of the medium with the mark on the beaker to note if any water has been lost. Add sufficient distilled water as indicated. Keep the temperature of the medium at about 60° C to avoid solidification. The medium will solidify at around 40° C.

Adjusting the pH

Although dehydrated media contain buffering agents to keep the pH of the medium in a desired range, the pH of a batch of medium may differ from that stated on the label of the bottle. Before the medium is tubed, therefore, one should check the pH and make any necessary adjustments.

If a pH meter (figure 18.5) is available and already standardized, use it to check the pH of your medium. If the medium needs adjustment, use the bottles of HCl and NaOH to correct the pH. If no meter is available, pH papers will work about as well. Make pH adjustment as follows:

- beaker of medium
- acid and base kits (dropping bottles of 1N and 0.1N HCl and NaOH)
- glass stirring rod
- pH papers
- pH meter (optional)

1. Dip a piece of pH test paper into the medium to determine the pH of the medium.
2. **If the pH is too high,** add a drop or two of HCl to lower the pH. For large batches use 1N HCl. If the pH difference is slight, use the 0.1N HCl. Use a glass stirring rod to mix the solution as the drops are added.

FIGURE 18.3 Dehydrated medium is dissolved in a measured amount of water
©The McGraw-Hill Companies/Auburn University Photographic Service

FIGURE 18.4 If the medium contains agar, it must be heated to dissolve the agar
©The McGraw-Hill Companies/Auburn University Photographic Service

FIGURE 18.5 The pH of the medium is adjusted by adding acid or base as per recommendations
©The McGraw-Hill Companies/Auburn University Photographic Service

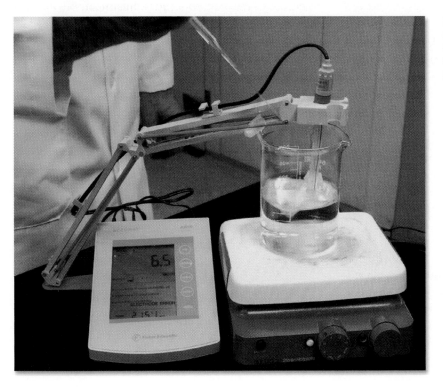

FIGURE 18.6 An automatic pipetting machine will deliver precise amounts of media to test tubes
©The McGraw-Hill Companies/Auburn University Photographic Service

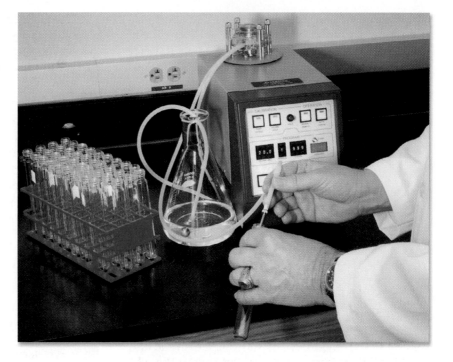

3. **If the pH is too low,** add NaOH, one drop at a time, to raise the pH. For slight pH differences, use 0.1N NaOH; for large differences use 1N NaOH. Use a glass stirring rod to mix the solution as the drops are added.

FILLING THE TEST TUBES

Once the pH of the medium is adjusted, it must be dispensed into test tubes. If an automatic pipetting machine is to be used, as shown in figure 18.6, it will have to be set up for you by your instructor. These machines can be adjusted to deliver any amount of medium at any desired speed. When large numbers of tubes are to be filled, the automatic pipetting machine should be used.

MATERIALS

• automatic pipetters

1. Follow the instructions provided by your instructor for setting up the automatic pipetter. This will involve adjusting the desired amount of medium to be delivered to each test tube and possibly other settings. If you are using an automatic pipette aid, you will need to repeatedly draw up medium in a pipette and deliver the desired amount by pressing the release button on the pipette aid (figure 18.7).

FIGURE 18.7 Small batches of media can be delivered with hand-held automatic pipetters
©The McGraw-Hill Companies/Auburn University Photographic Service

2. Place the supply tube into the medium and proceed to fill each test tube according to the type of delivery system you are using. Your instructor will help you with this step.

3. If you are delivering agar medium, keep the beaker of medium on a stirring hot plate to maintain the agar in solution. A magnetic stirring bar placed in the medium will aid in constantly stirring the solution.

4. If the medium is to be used for fermentation, add a Durham tube to each tube before filling the test tube. This should be placed in the tube *with the open end of the Durham tube down.* When medium is placed in the test tube, the Durham tube may float on the top of the medium but it will submerge during autoclaving.

Capping the Tubes

The last step before sterilization is to provide a closure for each tube. Plastic (polypropylene) caps are suitable in most cases. All caps that slip over the tube end have inside ridges that grip the side of the tube and provide an air gap to allow steam to escape during sterilization (figure 18.8). If you are using tubes with plastic screw-caps, *the caps should not be screwed tightly before sterilization; instead, each one must be left partly unscrewed.*

If no slip-on caps of the correct size are available, it may be necessary to make up some cotton plugs. A properly made cotton plug should hold firmly in the tube so that it is not easily dislodged.

Sterilization

As soon as the tubes of media have been stoppered, they must be sterilized. Organisms on the walls of the tubes, in the distilled water, and in the dehydrated medium will begin to grow within a short period of time at room temperature, destroying the medium.

Prior to sterilization, the tubes of media should be placed in a wire basket with a label taped on the outside of the basket. The label should indicate the type of medium, the date, and your name.

Sterilization must be done in an autoclave (figure 18.9). The following considerations are important in using an autoclave:

- Check with your instructor on the procedure to be used with your particular type of autoclave. Complete sterilization occurs at 250° F (121.6° C). To achieve this temperature the autoclave has to develop 15 pounds per square inch (psi) of steam pressure. To reach the correct temperature, there must be some provision in the chamber for the escape of air. On some of the older units it is necessary to allow the steam to force air out through the door before closing it.

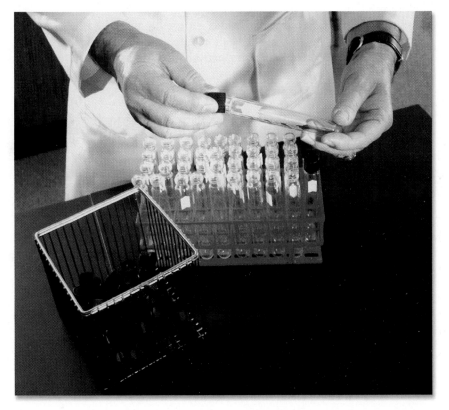

FIGURE 18.8 Once the medium has been dispensed, tubes are capped prior to autoclaving
©The McGraw-Hill Companies/Auburn University Photographic Service

FIGURE 18.9 Media is sterilized in an autoclave for 15–20 minutes at 15 psi steam pressure
©The McGraw-Hill Companies/Auburn University Photographic Service

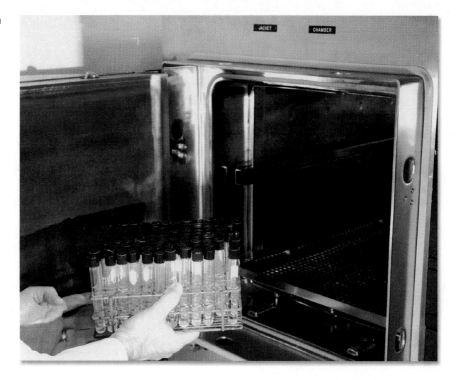

- *Don't overload the chamber.* One should not attempt to see how much media can be packed into it. Provide ample space between baskets of media to allow for circulation of steam.
- *Adjust the time of sterilization to the size of load.* Small loads may take only 10 to 15 minutes. An autoclave full of media may require 30 minutes for complete sterilization.

After Sterilization

Slants   If you have a basket of tubes that are to be converted to slants, it is necessary to lay the tubes down in a near-horizontal manner as soon as they are removed from the autoclave. The easiest way to do this is to use a piece of rubber tubing (1/2″ dia) to support the capped end of the tube as it rests on the countertop. Solidification should occur in about 30–60 minutes.

Other Media   Tubes of broth, agar deeps, nutrient gelatin, etc., should be allowed to cool to room temperature after removal from the autoclave. Once they have cooled down, place them in a refrigerator or cold-storage room.

Storage   If tubes of media are not to be used immediately, they should be stored in a cool place. When stored for long periods of time at room temperature media tend to lose moisture. At refrigerated temperatures media will keep for months.

LABORATORY REPORT

Complete the Laboratory Report for exercise 18.

Cultivation of Anaerobes

Bacteria can be classified into five groups based upon their requirement for air that contains 20% oxygen.

Obligate (Strict) Aerobes   These organism require oxygen for growth. They primarily carry out respiratory metabolism in which oxygen acts as a terminal electron acceptor. Examples are *Bacillus* and *Pseudomonas*.

Obligate (Strict) Anaerobes   Bacteria belonging to this group cannot tolerate oxygen and must be cultivated under growth conditions in which oxygen is removed. Anaerobes carry out fermentation or employ **anaerobic respiration** for growth. Their sensitivity is not to oxygen, but rather to by-products of oxygen such as peroxides and superoxide. The latter are a strong oxidant and a free radical compound, respectively, that damage and destroy biological molecules in the cell, causing death. Aerobic bacteria have catalase and superoxide dismutase, which are enzymes that convert highly reactive hydrogen peroxide and superoxide to harmless compounds, such as water and oxygen. Anaerobes lack these enzyme systems to deal with peroxide and superoxide. Because of their sensitivity to oxygen, obligate anaerobes require specialized conditions for growth where the reactive by-products of oxygen are not present. Anaerobic jars or media that contain chemicals that react with oxygen are used to grow these bacteria. Examples of obligate anaerobes are *Clostridium* and *Bacteroides*.

Facultative Anaerobes   The facultative anaerobes are bacteria that can grow by respiratory means if oxygen is present, but if oxygen is absent they can also grow using other modes of metabolism such as fermentation. *Escherichia coli* is a facultative anaerobe.

Microaerophiles   These organisms require oxygen in amounts lower than the atmosphere (5–10% vs. 20%). Higher concentrations of oxygen are toxic to these bacteria. It is thought that their sensitivity to oxygen is due to their limited ability to carry out respiration or because they have an oxygen-sensitive enzyme. Growth of microaerophiles can be enhanced by cultivation in a candle jar. Cultures are set up in a jar in which a lighted candle is placed. A tight lid is

placed on the jar after which the candle is extinguished because the oxygen is consumed by combustion. The oxygen concentration decreases and the carbon dioxide increases to about 3.5% in the jar. *Heliobacter pylori*, the bacterium that causes stomach ulcers, is a microaerophile.

Aerotolerant Organisms   These bacteria can grow in oxygen but they cannot use it for metabolism. Hence, they derive no metabolic benefit from oxygen. They are said to be **indifferent** to oxygen. They tend to grow better in conditions where carbon dioxide and water-vapor levels are higher. They contain superoxide dismutase but lack catalase. Members of the genus *Streptococcus* are aerotolerant organisms.

Figure 19.1 illustrates where the various classes of bacteria grow in a tube in relation to the degree of oxygen tension in the medium. In this experiment, you will inoculate various media with several organisms that have different oxygen requirements. The media you will use are fluid thioglycollate (FTM), tryptone glucose yeast extract agar (TGYA), and Brewer's anaerobic agar. In this experiment you will

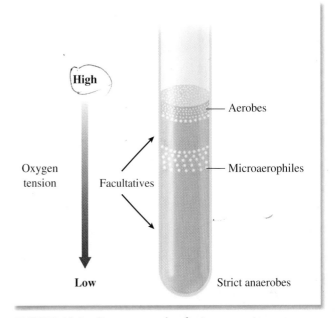

FIGURE 19.1   Oxygen needs of microorganisms

inoculate one liquid medium and two solid media with several organisms that have different oxygen requirements. The media are fluid thyioglycollate medium (FTM), tryptone glucose yeast extract (TGYA), and Brewer's anaerobic agar. Each medium will serve a different purpose. A description for each medium follows:

TGYA Shake This solid medium will be used to prepare "shake tubes." The medium is not primarily for the cultivation of anaerobes but will be used to determine the oxygen requirements of different bacteria. It will be inoculated in the liquefied state, shaken to mix the organisms throughout the medium, and allowed to solidify. After incubation one determines the oxygen requirements on the basis of where growth occurs in the shake tube: top, middle, or bottom.

FTM Fluid thioglycollate medium is a rich liquid medium that supports the growth of both aerobic and anaerobic bacteria. It contains glucose, cystine, and sodium thioglycollate to reduce its oxidation-reduction (O/R) potential. It also contains the dye resazurin, which is an indicator for the presence of oxygen. In the presence of oxygen the dye becomes pink. Since the oxygen tension is always higher near the surface of the medium, the medium will be pink at the top and colorless in the middle and bottom. The medium also contains a small amount of agar, which helps to localize the organisms and favors anaerobiasis in the bottom of the tube.

Brewer's Anaerobic Agar This solid medium is excellent for culturing anaerobic bacteria in petri dishes. It contains thioglycollate as a reducing agent and resazurin as an O/R indicator. For strict anaerobic growth, it is essential that plates be incubated in an oxygen-free environment.

To provide an oxygen-free incubation environment for the petri plates of anaerobic agar, we will use the **GasPak anaerobic jar.** Note in figure 19.2 that hydrogen is generated in the jar, which removes the oxygen by forming water. Palladium pellets catalyze the reaction at room temperature. The generation of hydrogen is achieved by adding water to a plastic envelope of chemicals. Note also that CO_2 is produced, which is a requirement for the growth of many fastidious bacteria. To make certain that anaerobic conditions actually exist in the jar, an indicator strip of methylene blue becomes colorless in the total absence of oxygen. If the strip is not reduced (decolorized) within 2 hours, the jar has not been sealed properly, or the chemical reaction has failed to occur.

In addition to doing a study of the oxygen requirements of six organisms in this experiment, you will have an opportunity during the second period to do a microscopic study of the types of endospores formed by three spore-formers used in the inoculations. Proceed as follows:

FIRST PERIOD
(Inoculations and Incubation)

Since six microorganisms and three kinds of media are involved in this experiment, it will be necessary for economy of time and materials to have each student work with only three organisms. The materials list for this period indicates how the organisms will be distributed.

During this period, each student will inoculate three tubes of medium and only one petri plate of Brewer's anaerobic agar. The tubes and all of the plates will be placed in a GasPak jar to be incubated in a 37° C incubator. Students will share results.

FIGURE 19.2 The GasPak anaerobic jar

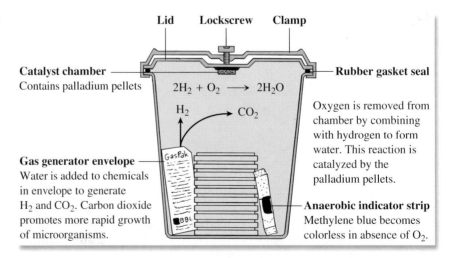

Lid　Lockscrew　Clamp

Catalyst chamber
Contains palladium pellets

$$2H_2 + O_2 \longrightarrow 2H_2O$$

H_2 → CO_2

Rubber gasket seal

Oxygen is removed from chamber by combining with hydrogen to form water. This reaction is catalyzed by the palladium pellets.

Gas generator envelope
Water is added to chemicals in envelope to generate H_2 and CO_2. Carbon dioxide promotes more rapid growth of microorganisms.

Anaerobic indicator strip
Methylene blue becomes colorless in absence of O_2.

MATERIALS

per student:
- 3 tubes of fluid thioglycollate medium
- 3 TGYA shake tubes (liquefied)
- 1 petri plate of Brewer's anaerobic agar

broth cultures for **odd-numbered students:**
- *Staphylococcus aureus, Streptococcus faecalis,* and *Clostridium sporogenes*

broth cultures for **even-numbered students:**
- *Bacillus subtilis, Escherichia coli,* and *Clostridium rubrum*
- GasPak anaerobic jar, 3 GasPak generator envelopes, 1 GasPak anaerobic generator strip, scissors, and one 10 ml pipette water baths at student stations (electric hot plate, beaker of water, and thermometer)

1. Set up a 45° C water bath at your station in which you can keep your tubes of TGYA shakes from solidifying. One water bath for you and your laboratory partner will suffice. (Note in the materials list that the agar shakes have been liquefied for you prior to lab time.)

2. Label the six tubes with the organisms assigned to you (one organism per tube), you initials, and assignment number.

   Note: *Handle the tubes gently to avoid taking on any unwanted oxygen into the media. If the tubes of FTM are pink in the upper 30%, they must be boiled a few minutes to drive off the oxygen, then cooled to inoculate.*

3. Heavily inoculate each of the TGYA shake tubes with several loopfuls of the appropriate organism for that tube. To get good dispersion of the organisms in the medium, roll each tube gently between the palms as shown in figure 19.3. To prevent oxygen uptake, do not overly agitate the medium. Allow these tubes to solidify at room temperature.

4. Inoculate each of the FTM tubes with the appropriate organisms.

5. Streak your three organisms on the plate of anaerobic agar in the manner shown in figure 19.4. Note that only three straight-line streaks, well separated, are made. Place the petri plate (inverted) in a cannister with the plates of other students that is to go into the GasPak jar.

6. Once all the students' plates are in cannisters, place the cannisters and tubes into the jar.

7. To activate and seal the GasPak jar, proceed as follows:

   a. Peel apart the foil at one end of a GasPak indicator strip and pull it halfway down. The indicator will turn blue on exposure to the air. Place the indicator strip in the jar so that the wick is visible.

   b. Cut off the corner of each of three GasPak gas generator envelopes with a pair of scissors. Place them in the jar in an upright position.

   c. Pipette 10 ml of tap or distilled water into the open corner of each envelope. Avoid forcing the pipette into the envelope.

   d. Place the inner section of the lid on the jar, making certain it is centered on top of the jar. Do not use grease or other sealant on the rim of the jar since the O-ring gasket provides an effective seal when pressed down on a clean surface.

   e. Unscrew the thumbscrew of the outer lid until the exposed end is completely withdrawn into the threaded hole. Unless this is done, it will be impossible to engage the lugs of the jar with the outer lid.

FIGURE 19.3 Organisms are dispersed in medium by rolling tube gently between palms

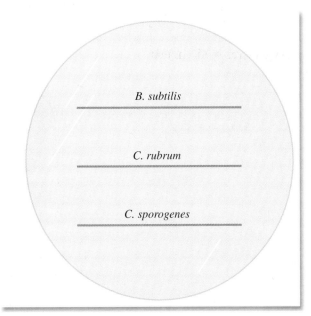

B. subtilis

C. rubrum

C. sporogenes

FIGURE 19.4 Three organisms are streaked on agar plate as straight-line streaks

f. Place the outer lid on the jar directly over the inner lid and rotate the lid slightly to allow it to drop in place. Now rotate the lid firmly to engage the lugs. The lid may be rotated in either direction.

g. Tighten the thumbscrew by turning clockwise. If the outer lid raises up, the lugs are not properly engaged.

8. Place the jar in a 37° C incubator. After 2 or 3 hours, check the jar to note if the indicator strip has lost its blue color. If decolorization has not occurred, replace the palladium pellets and repeat the entire process.

9. Incubate the tubes and plates for 24 to 48 hours.

SECOND PERIOD

(Culture Evaluations and Spore Staining)

Remove the lid from the GasPak jar. If vacuum holds the inner lid firmly in place, break the vacuum by sliding the lid to the edge. When transporting the plates and tubes to your desk *take care not to agitate the FTM tubes*. The position of growth in the medium can be easily changed if handled carelessly.

MATERIALS

- tubes of FTM
- shake tubes of TGYA
- 2 Brewer's anaerobic agar plates
- spore-staining kits and slides

1. Compare the six FTM and TGYA shake tubes that you and your laboratory partner share with figure 19.5 to evaluate the oxygen needs of the six organisms.

2. Compare the growths (or lack of growth) on your petri plate and the plate of your laboratory partner.

3. Record your results on Laboratory Report 19.

4. If time permits, make a combined slide with three separate smears of the three spore-formers, using either one of the two spore-staining methods in Exercise 15. Draw the organisms in the circles provided in Laboratory Report 19.

LABORATORY REPORT

Complete Laboratory Report 19.

FIGURE 19.5 Growth patterns for different types of bacteria

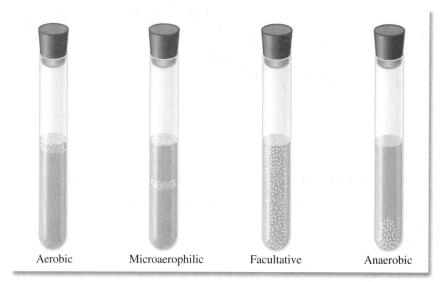

Aerobic   Microaerophilic   Facultative   Anaerobic

Enumeration of Bacteria:
The Standard Plate Count

It is often essential to determine the numbers of bacteria in a sample. For example, the grading of milk is based on the numbers of bacteria present. Whether a patient has a bladder infection is dependent on a certain threshold level of bacteria present in a urine sample. Sometimes it is just important to know how many bacteria are present in food or water to determine purity. Several different methods can be used to determine the number of bacterial cells and each method has its own advantages and disadvantages. The use of one method over another will be dictated by the purpose of the study. The following are some of the methods for determining bacterial numbers:

Microscopic Counts   A sample can be diluted and the cells in the sample can be counted with the aid of a microscope. Special slides, such as the Petroff-Hauser chamber, facilitate counting because the slide has a grid pattern and the amount of sample delivered to the grid is known. Milk samples can be counted by microscopic means with a great deal of reliability and confidence.

Most Probable Number (MPN)   The number of bacteria in a sample can be determined by the relationship of some growth parameter to statistical probability. The safety of drinking water is dependent on there being no sewage contamination of potable water, and this is tested using the MPN method. Indicator bacteria called **coliforms**, which are found in the intestines of humans and warm-blooded animals, ferment lactose to produce acid and gas. The presence of these bacteria in a water sample suggests the potential for disease. A series of tubes with lactose is inoculated with water samples and the pattern of tubes showing acid and gas is compared to statistical tables that give the probable numbers of coliforms present. You will use this procedure in Exercise 47 to test water for the presence of coliforms. The MPN method is limited to testing where statistical tables have been set up for a particular growth parameter.

Standard Plate Count (SPC)   The standard plate, or viable count, is one of the most common methods for determining bacterial numbers in a sample. A sample is diluted in a series of dilution blanks as shown in figure 20.1. Aliquots of the dilutions are then plated

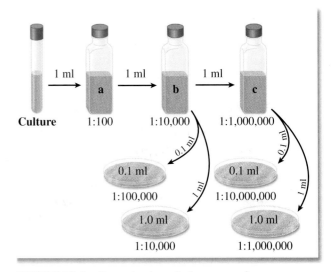

FIGURE 20.1   Quantitative plating procedure

onto media and the numbers of colonies are counted after incubation for 24–48 hours. It is assumed that the bacterial cells are diluted to an end point where a single cell divides giving rise to a visible colony on a plate. The number of bacteria in the original sample is determined by multiplying the number of colonies by the dilution factor. However, the assumption that a colony represents a single cell is not always correct because cells in a chain, such as *Streptococcus,* will also give rise to a colony on a plate. Because of the uncertainty in how many actual cells form a colony, counts by the SPC are reported as **colony forming units** (cfus). Only numbers between 30 and 300 cfus are considered statistically valid. If the cfus are greater than 300, there is a probability that overcrowding on the plate could have inhibited some cells from growing. Less than 30 cfus could involve a sampling error and an underestimate of numbers. The SPC method determines only viable cells, whereas a microscopic count determines both living and dead cells. Also, the SPC method is biased because specific conditions and media are used and these factors may exclude certain bacteria in the counts. For example, the SPC would severely underestimate the numbers of bacteria in a soil sample because the conditions and the medium used for the count probably favor heterotrophs that grow aerobically at neutral pH values. These conditions do not allow for the growth of anaerobes, chemolithotrophs, or bacteria that may grow at extremes of pH.

Indirect Methods Sometimes one only wants to know if cells are growing and, therefore, increasing in number. Growth can be related to some parameter that increases with cell division. Growing cells increase their protein, nucleic acid content, and mass because cells are dividing. Thus, measurements of protein, DNA, and dry weight can be used to monitor growth. Likewise, a culture will become more turbid as cells divide, and the **turbidity** of a culture can be determined and related to growth. Cell turbidity can be measured in a spectrophotometer, which measures the **absorbance** or **optical density** of a culture. Oftentimes, a standard curve is constructed that relates optical density to actual numbers of bacteria determined by a SPC. However, one must bear in mind that both living and dead cells will contribute to the culture turbidity, which is also a disadvantage of this method.

In the following exercise, you will use the SPC to determine the numbers of bacteria in a culture. You will also measure the turbidity of a culture and plot the optical density values of diluted samples.

QUANTITATIVE PLATING METHOD
(Standard Plate Count)

In determining the number of organisms present in water, milk, and food, the **standard plate count** (SPC) is universally used. It is relatively easy to perform and gives excellent results. We can also use this basic technique to calculate the number of organisms in a bacterial culture. It is in this respect that this assignment is set up.

The procedure consists of diluting the organisms with a series of sterile water blanks as illustrated in figure 20.1. Generally, only three bottles are needed, but more could be used if necessary. By using the dilution procedure indicated here, a final dilution of 1:1,000,000 occurs in blank C. From blanks B and C, measured amounts of the diluted organisms are transferred into empty petri plates. Nutrient agar, cooled to 50°C, is then poured into each plate. After the nutrient agar has solidified, the plates are incubated for 24 to 48 hours and examined. A plate that has between 30 and 300 colonies is selected for counting. From the count it is a simple matter to calculate the number of organisms per milliliter of the original culture. It should be pointed out that greater accuracy can be achieved by pouring two plates for each dilution and

averaging the counts. Duplicate plating, however, has been avoided for economic reasons.

Pipette Handling

Success in this experiment depends considerably on proper pipetting techniques. Pipettes may be available to you in metal cannisters or in individual envelopes; they may be disposable or reusable. In the distant past, pipetting by mouth was routine practice. However, the hazards are obvious, and today it must be avoided. Your instructor will indicate the techniques that will prevail in this laboratory. If this is the first time that you have used sterile pipettes, consult figure 20.2, keeping the following points in mind:

- When removing a sterile pipette from a cannister, do so without contaminating the ends of the other pipettes with your fingers. This can be accomplished by *gently* moving the cannister from side to side in an attempt to isolate one pipette from the rest.
- After removing your pipette, replace the cover on the cannister to maintain sterility of the remaining pipettes.
- Don't touch the body of the pipette with your fingers or lay the pipette down on the table before or after you use it. **Keep that pipette sterile** until you have used it, and don't contaminate the table or yourself with it after you have used it.
- Always use a mechanical pipetting device such as the one in illustration 3, figure 20.2. For safety reasons, deliveries by mouth are not acceptable in this laboratory.
- Remove and use only one pipette at a time; if you need 3 pipettes for the whole experiment and remove all 3 of them at once, there is no way that you will be able to keep 2 of them sterile while you are using the first one.
- When finished with a pipette, place it in the *discard cannister*. The discard cannister will have a disinfectant in it. At the end of the period, reusable pipettes will be washed and sterilized by the laboratory assistant. Disposable pipettes will be discarded. Students have been known to absentmindedly return used pipettes to the original sterile cannister, and, occasionally, even toss them into the wastebasket. We are certain that no one in this laboratory would *ever* do that!

DILUTING AND PLATING PROCEDURE

Proceed as follows to dilute out a culture of *E. coli* and pour four plates, as illustrated in figure 20.1.

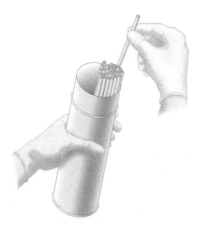

(1) Reusable pipettes may be available in disposable envelopes or metal cannisters. When using pipettes from cannisters be sure to cap them after removing a pipette.

(2) Never touch the tip or barrel of a pipette with your fingers. Contaminating the pipette will contaminate your work.

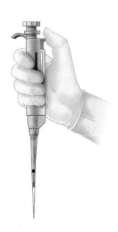

(3) Use a mechanical pipetter for all pipetting in this laboratory. Pipetting by mouth is too hazardous.

(4) After using a pipette place it in the discard cannister. Even "disposable" pipettes must be placed here.

FIGURE 20.2 Pipette handling techniques

MATERIALS

per 4 students:
- 1 bottle (40 ml) broth culture of *E. coli*

per student:
- 1 bottle (80 ml) nutrient agar
- 4 petri plates
- 1.1 ml pipettes
- 3 sterile 99 ml water blanks
- cannister for discarded pipettes

1. Liquefy a bottle of nutrient agar. While it is being heated, label three 99 ml sterile water blanks **A, B,** and **C.** Also, label the four petri plates **1:10,000, 1:100,000, 1:1,000,000,** and **1:10,000,000.** In addition, indicate with la-

bels the amount to be pipetted into each plate (**0.1 ml** or **1.0 ml**).

2. Shake the culture of *E. coli* and transfer 1 ml of the organisms to blank A, using a sterile 1.1 ml pipette. After using the pipette, place it in the discard cannister.

3. Shake blank A 25 times in an arc of 1 foot for 7 seconds with your elbow on the table as shown in figure 20.3. Forceful shaking not only brings about good distribution, but it also breaks up clumps of bacteria.

4. With a different 1.1 ml pipette, transfer 1 ml from blank A to blank B.

5. Shake water blank B 25 times in same manner.

FIGURE 20.3 Standard procedure for shaking water blanks requires elbow to remain fixed on table

6. With another sterile pipette, transfer 0.1 ml from blank B to the 1:100,000 plate and 1.0 ml to the 1:10,000 plate. With the same pipette, transfer 1.0 ml to blank C.
7. Shake blank C 25 times.
8. With another sterile pipette, transfer from blank C 0.1 ml to the 1:10,000,000 plate and 1.0 ml to the 1:1,000,000 plate.
9. After the bottle of nutrient agar has boiled for 8 minutes, cool it down in a water bath at 50° C for **at least 10 minutes.**
10. Pour one-fourth of the nutrient agar (20 ml) into each of 4 plates. Rotate the plates **gently** to get adequate mixing of medium and organisms. **This step is critical!** Too little action will result in poor dispersion and too much action may slop inoculated medium over the edge.
11. After the medium has cooled completely, incubate at 35° C for 48 hours, inverted.

COUNTING AND CALCULATIONS
MATERIALS

• 4 culture plates
• Quebec colony counter
• mechanical hand counter
• felt-tip pen (optional)

1. Lay out the plates on the table in order of dilution and compare them. *Select the plates that have no fewer than 30 nor more than 300 colonies for your count.* Plates with less than 30 or more than 300 colonies are statistically unreliable.

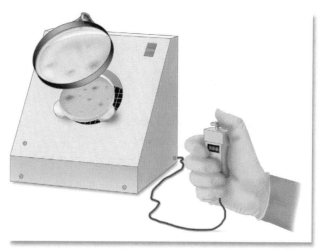

FIGURE 20.4 Colony counts are made on a Quebec counter, using a mechanical hand tally

2. Place the plate on the Quebec colony counter with the lid removed. See figure 20.4. Start counting at the top of the plate, using the grid lines to prevent counting the same colony twice. Use a mechanical hand counter. Count every colony, regardless of how small or insignificant. Record counts on the table in section A of Laboratory Report 20.
   Alternative Counting Method: Another way to do the count is to remove the lid and place the plate upside down on the colony counter. Instead of using the grid to keep track, use a felt-tip pen to mark off each colony as you do the count.
3. Calculate the number of bacteria per ml of undiluted culture using the data recorded in section A of Laboratory Report 20. Multiply the number of colonies counted by the dilution factor (the reciprocal of the dilution).
   Example: If you counted 220 colonies on the plate that received 1.0 ml of the 1:1,000,000 dilution: $220 \times 1,000,000$ (or 2.2×10^8) bacteria per ml. If 220 colonies were counted on the plate that received 0.1 ml of the 1:1,000,000 dilution, then the above results would be multiplied by 10 to convert from number of bacteria per 0.1 ml to number of bacteria per 1.0 ml (2,200,000,000, or 2.2×10^9).
   Use only two significant figures. If the number of bacteria per ml was calculated to be 227,000,000, it should be recorded as 230,000,000, or 2.3×10^8.

DETERMINATION OF GROWTH BY OPTICAL DENSITY

Turbidity can give a quick indication that a culture is growing but it does not give actual cell numbers. For an actively growing culture, turbidity will increase

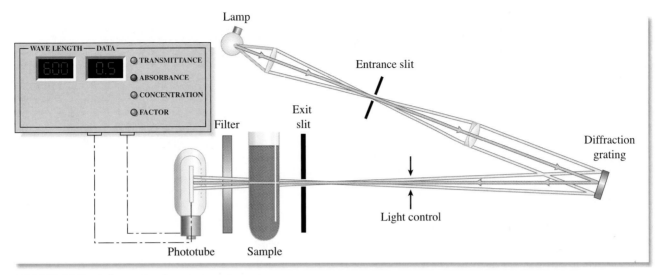

FIGURE 20.5 Schematic of a spectrophotometer

with time. The turbidity of a culture can be measured in a spectrophotometer because a bacterial culture acts like a colloidal suspension. As light passes through the culture, it will be absorbed by the bacterial cells and the light emerging from the culture will be proportionally decreased by the number of cells present. Therefore, within certain defined limits, the amount of light absorbed is proportional to the amount of cells present.

Figure 20.5 illustrates the path of light through a spectrophotometer. A beam of white light passes through a series of lenses and a slit and onto a diffraction grading where the light is separated into different wavelengths of the visible spectrum. A specific wavelength of monochromatic light can be selected from the diffraction grading by an exit slit. Adjusting the wavelength control on the instrument will reorient the diffraction grading so that a different wavelength can be selected by the exit slit. The monochromatic light then passes through a sample and activates a photomutiplier tube that measures the **absorbance (optical density, O.D.)** on a galvanometer. The higher the absorbance, the greater the concentration of bacterial cells. In the following exercise, you will demonstrate the relationship between O.D. and cell turbidity by measuring the optical density values for various dilutions of a culture.

There should be a direct proportional relationship between the concentration of bacterial cells and the absorbance (optical density, O.D.) of the culture. To demonstrate this principle, you will measure the O.D. of various dilutions of the culture provided to you. These values will be plotted on a graph as a function of culture dilution. You may find that there is a linear relationship between concentration of cells and O.D. only up to a certain O.D. At higher O.D. values, the

relationship may not be linear. That is, for a doubling in cell concentration, there may be less than a doubling in O.D.

MATERIALS

- broth culture of *E. coli* (same one as used for plate count)
- spectrophotometer cuvettes (2 per student)
- 4 small test tubes and test-tube rack
- 5 ml pipettes
- bottle of sterile nutrient broth (20 ml per student)

1. Calibrate the spectrophotometer, using the procedure described in figure 20.6. These instructions are specifically for the Bausch and Lomb Spectronic 20. In handling the cuvettes, keep the following points in mind:
   a. Rinse the cuvette several times with distilled water to get it clean before using.
   b. Keep the lower part of the cuvette spotlessly clean by keeping it free of liquids, smudges, and fingerprints. Wipe it clean with Kimwipes or some other lint-free tissue. Don't wipe the cuvettes with towels or handkerchiefs.
   c. Insert the cuvette into the sample holder with its index line registered with the index line on the holder.
   d. After the cuvette is seated, line up the index lines exactly.
   e. Handle these tubes with great care. They are expensive.
2. Label a cuvette 1:1 (near top of tube) and four test tubes 1:2, 1:4, 1:8, and 1:16. These tubes will be used for the serial dilutions shown in figure 20.7.

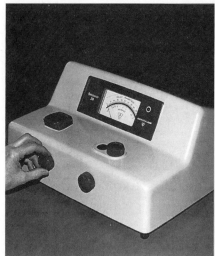

1) Turn on instrument by rotating zero control knob clockwise. Do this 20 minutes before measurements are to be made. Also, set wavelength knob (top of instrument) at 686 nanometers wavelength. Adjust the meter needle to zero by rotating zero control knob.

2) Insert a cuvette containing 3 ml of sterile nutrient broth into sample holder. The cover must be closed. Keep the index line of cuvette in line with index line on sample holder. Refer to instructions 1a through 1e on page 147 concerning care of cuvette.

3) Adjust the meter to read 100% transmittance by rotating light–control knob. Remove cuvette of nutrient broth and close lid. If needle does not return to zero, readjust accordingly. Reinsert nutrient broth again to see if 100% transmittance still registers. If it has changed, re-adjust with light–control knob. Once meter is adjusted for 0 and 100%,transmittance, turbidity measurements can be made. Recheck calibration from time to time to make certain instrument is set properly.

FIGURE 20.6   Calibration procedure for the B&L Spectronic 2 on page 147

3. With a 5 ml pipette, dispense 4 ml of sterile nutrient broth into tubes 1:2, 1:4, 1:8, and 1:16.

4. Shake the culture of *E. coli* vigorously to suspend the organisms, and with the same 5 ml pipette, transfer 4 ml to the 1:1 cuvette and 4 ml to the 1:2 test tube.

5. Mix the contents in the 1:2 tube by drawing the mixture up into the pipette and discharging it into the tube three times.

6. Transfer 4 ml from the 1:2 tube to the 1:4 tube, mix three times, and go on to the other tubes in a similar manner. Tube 1:16 will have 8 ml of diluted organisms.

7. Measure the optical density of each of the five tubes, starting with the 1:16 tube first. The contents of each of the test tubes must be transferred to a cuvette for measurement. Be sure to close the lid on the sample holder when making measurements. A single cuvette can be used for all the measurements.

8. Record the O.D. values in the table of Laboratory Report 20.

9. Plot the O.D. values on the graph of Laboratory Report 20.

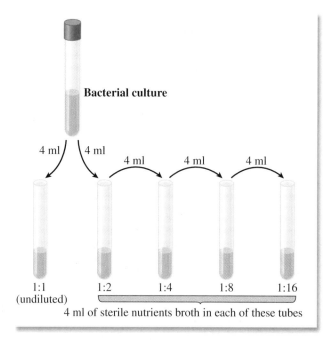

FIGURE 20.7   Dilution procedure for cuvettes

Slime Mold Culture

The classification system proposed by Alexopoulos and Mims places the slime molds in Division Gymnomycota of the Kingdom Myceteae. These heterotrophic microorganisms exist in cool, shady, moist places in the woods—on decaying logs, dead leaves, and other organic matter. Unlike the holophytic bacteria and other Myceteae, they ingest their food in a manner similar to the amoebas; that is, they are phagotrophic. In the vegetative stages, these microorganisms are unlike the other Myceteae in that the cells lack cell walls; when fruiting bodies are formed, however, cell walls are present.

The categorization of slime molds as protozoans or as fungi has always been problematical. Certainly, they are intermediate in that they have characteristics of both groups.

Figure 21.1 illustrates the life cycle of one type of slime mold, the plasmodial type. The genus *Physarum* is the one to be studied in this experiment. The assimilative stage of this organism is the **plasmodium.** This multinucleate structure is slimy in appearance and moves slowly by flowing its cytoplasm in amoeboid fashion over surfaces on which it feeds. Most species feed on bacteria and possibly on other small organisms that they encounter.

Plasmodial growth continues as long as adequate food supply and moisture are available. Eventually, however, environmental changes may result in the formation of sclerotia or sporangia. A **sclerotium** is a hardened mass of irregular shape that forms from the plasmodium when moisture and temperature conditions become less than ideal. When conditions improve, the sclerotium reverts back to a plasmodium. Figure 21.2 is a photograph of two sclerotia that formed on a laboratory culture. **Sporangia** are fructifications that form under conditions similar to those required for sclerotia. Exactly why sporangia form instead of sclerotia is still not clearly understood. Sporangia form by the separation of the plasmodium into many rounded mounds of protoplasm that extend upward on stalks. The nuclei within the sporangia undergo meiosis to become haploid spores with tough cell walls. The sclerotia and sporangia of figures 21.2 and 21.3 were photographed on the same culture of laboratory-grown *Physarum.*

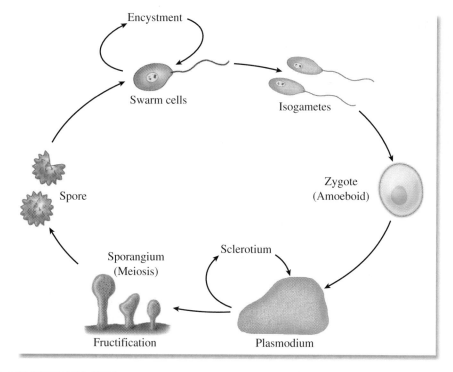

FIGURE 21.1 Life cycle of *Physarum polycephalum*

Encystment

Swarm cells

Isogametes

Zygote (Amoeboid)

Spore

Sclerotium

Sporangium (Meiosis)

Fructification

Plasmodium

FIGURE 21.2 Sclerotia of *Physarum polycephalum* (3X)

FIGURE 21.3 Sporangium of *Physarum polycephalum* (actual size is 3mm tall) Image originally published in Henry C. Aldrich and Johm W. Daniel, Cell Biology of Physarum and Didymium. New York: Academic Press, 1982. Used with permission

Both sclerotia and spores may survive adverse environmental conditions for long periods of time. Once environmental conditions improve, the spores germinate to produce flagellated pear-shaped **swarm cells.**

These swarm cells may do one of three things: (1) they may encyst if conditions suddenly become adverse, (2) they may divide one or more times to form isogametes, or (3) they may act as isogametes and unite directly to form a **zygote.** Once a zygote is formed, it takes on an amoeboid form and undergoes a series of mitotic divisions to produce a plasmodium. This completes the life cycle.

Three procedures will be described here for the study of *Physarum polycephalum:* (1) moist chamber culture, (2) agar culture method, and (3) spore germination technique. The techniques used will be determined by the availability of time and materials.

MOIST CHAMBER CULTURE

To grow large numbers of plasmodia, sclerotia, and sporangia that can be used for an entire class, one needs to create a rather large moisture chamber. Any covered glass or plastic container that is 10 to 12 inches square or round is suitable.

MATERIALS

- sclerotia of *Physarum polycephalum*
- container for culture (10 $\frac{1}{2}$" dia Pyrex casserole dish with cover or 10–12" square plastic box with cover)
- glass petri dish cover
- sharp scalpel
- rolled oat flakes (long-cooking type)
- 10" dia filter paper or paper toweling

1. In the center of the container place a petri dish cover, open end down. Lay a large piece of filter paper or paper toweling over the petri dish and saturate with distilled water. The petri dish provides a raised area above any excess water that may make the paper too wet.
2. With a sharp scalpel transfer a small fragment of sclerotium from the *Physarum* culture to the filter paper. A sclerotium may vary from dark orange to brown in color. See figure 21.2. Moisten the sclerotium with a drop of distilled water.
3. After a few hours the organism will be awakened to activity and begin to seek food. At this point, place a flake of rolled oats near the edge of the spreading growth for it to feed on.
4. Incubate the moist chamber in a dark place at room temperature. Add moisture (distilled water) and oat flakes periodically as needed. It is better to add a few fresh flakes daily than to overfeed by applying all flakes at once. Such a culture should keep for several weeks. To promote the formation of sclerotia, allow some of the water to evaporate

by leaving the lid partially open for a while. To bring about sporangia formation, withhol⌐ while keeping the culture m⌐

AGAR ⌐ RE ⌐

(Plasmodia ⌐

An actively metabolizing plasmodium is dark yellow and streaked with vessels. The streaming of protoplasm in these vessels is best observed under the microscope. To be able to study this unique structure, it is best to culture the organism on nonnutrient agar. Make such a culture as follows:

MATERIALS

- rolled oat flakes
- scalpel
- petri plate with 15 ml of nonsterile, nonnutrient agar

1. Lift some occupied oat flakes from the filter paper in the moist chamber and transfer to a plate of nonsterile, nonnutrient 1.5% agar. Maintain this culture by adding fresh oat flakes periodically, but don't add water.
2. After a well-developed plasmodium has formed, study the streaming protoplasm under low power of the microscope. Observation is made by transmitted light through the agar on the microscope stage. Look for periodical reversal of direction of flow.
3. Cut one of the vessels through in which the flow is active and observe the effect.
4. Transfer a piece of plasmodium to another part of the medium and watch it reconstitute itself.
5. Leave the cover slightly open on the petri dish for several days and note any changes that might occur as time goes by.

⌐ORE GERMINATION

T⌐ observation of spore germination can be achieved with a hanging drop slide. Once sporangia are in abundance, one can make such a slide as follows:

MATERIALS

- depression slides (sterile)
- plain microscope slides (sterile)
- cover glasses (sterile)
- Vaseline
- toothpicks
- sporangia of *Physarum polycephalum*
- Bunsen burner
- 70% alcohol

1. With a toothpick, place a small amount of Vaseline near each corner of the cover glass. (See figure 18.1, page 117.)
2. Saturate a sporangium with a drop of 70% alcohol on the center of a sterile plain microscope slide.
3. As soon as the alcohol has evaporated, add a drop of distilled water and place another sterile slide over the wet sporangium.
4. Crush the sporangium with thumb pressure on the upper slide. Separate the two slides to expose the crushed sporangium.
5. Transfer a few loopfuls of crushed sporangial material to a drop of distilled water on a sterile cover glass.
6. Place the depression slide over the cover glass, make contact, and quickly invert to produce a completed hanging drop slide.
7. Examine under low and high power.

LABORATORY REPORT

Complete Laboratory Report 21.

LABORATORY REPORT

EXERCISE 21 Slime Mold Culture

A. Observations

1. What happened when the flow of protoplasm on a plasmodium was interrupted by cutting with a scalpel?

   _____

   _____

2. Describe your observations of the crushed spores on the hanging drop slide.

   _____

   _____

   _____

B. Questions

1. List two functions served by fructification (sporangia formation) in *Physarum*.

   a. _____

   b. _____

2. What is the principal function of the plasmodial stage of *Physarum*?

   _____

3. List two characteristics that the Myxobacterales and Gymnomycota have in common.

   a. _____

   b. _____

Slide Culture:
Molds

The isolation, culture, and microscopic examination of molds require the use of suitable selective media and special microscopic slide techniques. If simple wet mount slides of molds were attempted in Exercise 8, it became apparent that wet mount slides made from mold colonies usually don't reveal the arrangement of spores that is so necessary in identification. The process of merely transferring hyphae to a slide breaks up the hyphae and sporangiophores in such a way that identification becomes very difficult. In this exercise, a slide culture method will be used to prepare stained slides of molds. The method is superior to wet mounts in that the hyphae, sporangiophores, and spores remain more or less intact when stained.

When molds are collected from the environment, as in Exercise 7, Sabouraud's agar is most frequently used. It is a simple medium consisting of 1% peptone, 4% glucose, and 2% agar-agar. The pH of the medium is adjusted to 5.6 to inhibit bacterial growth.

Unfortunately, for some molds the pH of Sabouraud's agar is too low and the glucose content is too high. A better medium for these organisms is one suggested by C. W. Emmons that contains only 2% glucose, with 1% neopeptone, and an adjusted pH of 6.8–7.0. To inhibit bacterial growth, 40 mg of chloramphenicol is added to one liter of the medium.

In addition to the above two media, cornmeal agar, Czapek solution agar, and others are available for special applications in culturing molds.

Figure 22.2 illustrates the procedure that will be used to produce a mold culture on a slide that can be stained directly on the slide. Note that a sterile cube of Sabouraud's agar is inoculated on two sides with spores from a mold colony. Figure 22.1 illustrates how the cube is held with a scalpel blade as inoculation takes place. The cube is placed in the center of a microscope slide with one of the inoculated surfaces placed against the slide. On the other inoculated surface of the cube is placed a cover glass. The assembled slide is incubated at room temperature for 48 hours in a moist chamber (petri dish with a small amount of water). After incubation, the cube of medium is carefully separated from the slide and discarded.

During incubation the mold will grow over the glass surfaces of the slide and cover glass. By adding a little stain to the slide, a semipermanent slide can be made by placing a cover glass over it. The cover glass can also be used to make another slide by placing it on another clean slide with a drop of stain on it. Before the stain (lactophenol cotton blue) is used, it is desirable to add to the hyphae a drop of alcohol, which acts as a wetting agent.

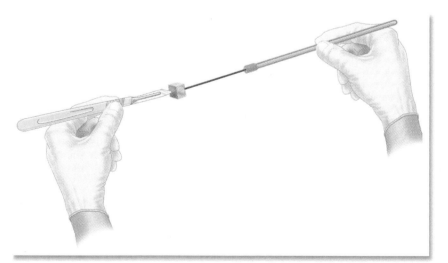

FIGURE 22.1 Inoculation technique

FIGURE 22.2 Procedure for making two stained slides from slide culture

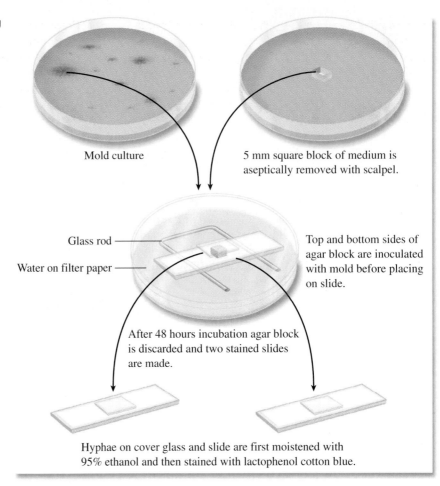

Mold culture

5 mm square block of medium is aseptically removed with scalpel.

Glass rod

Water on filter paper

Top and bottom sides of agar block are inoculated with mold before placing on slide.

After 48 hours incubation agar block is discarded and two stained slides are made.

Hyphae on cover glass and slide are first moistened with 95% ethanol and then stained with lactophenol cotton blue.

FIRST PERIOD

(Slide Culture Preparation)

Proceed as follows to make slide cultures of one or more mold colonies.

MATERIALS

- petri dishes, glass, sterile
- filter paper (9 cm dia, sterile)
- glass U-shaped rods
- mold culture plate (mixture)
- 1 petri plate of Sabouraud's agar or Emmons' medium per 4 students
- scalpels
- inoculating loop
- sterile water
- microscope slides and cover glasses (sterile)
- forceps

1. Aseptically, with a pair of forceps, place a sheet of sterile filter paper in a petri dish.

2. Place a sterile U-shaped glass rod on the filter paper. (Rod can be sterilized by flaming, if held by forceps.)
3. Pour enough sterile water (about 4 ml) on filter paper to completely moisten it.
4. With forceps, place a sterile slide on the U-shaped rod.
5. *Gently* flame a scalpel to sterilize, and cut a 5 mm square block of the medium from the plate of Sabouraud's agar or Emmons' medium.
6. Pick up the block of agar by inserting the scalpel into one side as illustrated in figure 22.1. Inoculate both top and bottom surfaces of the cube with spores from the mold colony. Be sure to flame and cool the loop prior to picking up spores.
7. Place the inoculated block of agar in the center of a microscope slide. Be sure to place one of the inoculated surfaces down.
8. Aseptically, place a sterile cover glass on the upper inoculated surface of the agar cube.

9. Place the cover on the petri dish and incubate at room temperature for 48 hours.

10. After 48 hours, examine the slide under low power. If growth has occurred, you should see hyphae and spores. If growth is inadequate and spores are not evident, allow the mold to grow another 24–48 hours before making the stained slides.

SECOND PERIOD

(Application of Stain)

As soon as there is evidence of spores on the slide, prepare two stained slides from the slide culture, using the following procedure:

MATERIALS

- microscope slides and cover glasses
- 95% ethanol
- lactophenol cotton blue stain
- forceps

1. Place a drop of lactophenol cotton blue stain on a clean microscope slide.

2. Remove the cover glass from the slide culture and discard the block of agar.

3. Add a drop of 95% ethanol to the hyphae on the cover glass. As soon as most of the alcohol has evaporated, place the cover glass, mold side down, on the drop of lactophenol cotton blue stain on the slide. This slide is ready for examination.

4. Remove the slide from the petri dish, add a drop of 95% ethanol to the hyphae, and follow this up with a drop of lactophenol cotton blue stain. Cover the entire preparation with a clean cover glass.

5. Compare both stained slides under the microscope; one slide may be better than the other one.

LABORATORY REPORT

There is no Laboratory Report for this exercise.

BACTERIAL VIRUSES

Viruses differ from bacteria in being much smaller and therefore below the resolution of the light microscope. The smallest virus is one million times smaller than a typical eukaryotic cell. Viruses are obligate intracellular parasites that require a host cell in order to replicate and reproduce, and hence they cannot be grown on laboratory media. Despite these obstacles, we can detect their presence by the effects that they have on their host cells.

Viruses infect all types of cells, eukaryotic and prokaryotic. They are composed of RNA or DNA but never both, and a protein coat, or capsid, that surrounds the nucleic acid. Their dependence on cells is due to their lack of metabolic machinery necessary for the synthesis of viral components. By invading a host cell, they can utilize the metabolic systems of the host cells to achieve their replication.

The study of viruses that parasitize plant and animal cells is time-consuming and requires special tissue culture techniques. Bacterial viruses are relatively simple to study, utilizing ordinary bacteriological techniques. It is for this reason that bacterial viruses will be studied here. However, the principles learned from studying the viruses that infect bacteria apply to viruses of eukaryotic cells.

Viruses that infect bacterial cells are called bacteriophages, or phages. They are diverse in their morphology and size. Some of the simplest ones have single-stranded DNA. Most phages are tadpole-shaped, with "heads" and "tails" as seen in figure 23.1. The capsid (head) may be round, oval, or polyhedral and is composed of individual protein subunits called capsomeres. It forms a protective covering around the viral genome. The tail structure or sheath is composed of a contractile protein that surrounds a hollow core, which is a conduit for the delivery of viral nucleic acid into the host cell. At the end of the tail is a base plate with tail fibers and spikes attached to it. The tail fibers bind to chemical groups on the surface of the bacterial cell and are responsible for recognition. Lysozyme associated with the tail portion of the virus erodes and weakens the cell wall of the host cell. This facilitates the injection of the viral nucleic acid by the sheath contracting and forcing the hollow core through the weakened area in the cell wall.

Infections by viruses can have two outcomes. The lytic cycle involves virulent phages that cause lysis and death of the host cell. The lysogenic cycle involves temperate phages, which can integrate their DNA into host cell DNA and alter the genetics of the host cell.

In the lytic cycle, the virus assumes control of cell metabolism and uses the cell's metabolic machinery to manufacture phage components (i.e., nucleic acid, capsid, sheath, tail fibers, spikes, and base plates). Mature phage particles are assembled and released from the cell where they can in turn invade new host cells. The result of a lytic infection for the host cell is almost always death.

In the lysogenic cycle, the viral DNA of the temperate virus is integrated into host DNA and no mature phages are made. Cells grow normally and are immune to further infections by the same phage. There is no visible evidence to indicate that a virus is even present in the cell. In some cases, the virus can carry genes that confer new genetic capabilities on the virally infected cell, or lysogen. For example, when *Corynebacterium diphtheriae* is infected with a certain lysogenic phage, because the phage carries a toxin gene in its genome, the host cells begin to produce a potent toxin responsible for many of the symptoms of diphtheria. This phenomenon is known as lysogenic conversion and is responsible for some of the toxins produced by various pathogens. Periodically, the lysogenic phage DNA can excise from the host DNA and initiate the lytic cycle and the production of mature phages. This results in the lysis of the host cell.

Visual evidence for lysis can be demonstrated by mixing phages with host cells and plating them onto media. The bacteria form a confluent lawn of growth, and where the phages cause lysis of the bacterial cells, there will be seen clear areas called plaques.

Some of the most studied bacteriophages are those that infect *Escherichia coli*, such as the T-even phages and lambda phage. They are known as the *coliphages*. Because *E. coli* is an intestinal bacterium, the coliphages can readily be isolated from raw sewage and coprophagous (dung-eating) insects such as flies. The exercises in this section will demonstrate some of the techniques for isolating, assaying, and determining the burst size of bacteriophages. It is recommended that you thoroughly understand the various stages in the lytic cycle before you begin the experiments in this section.

152

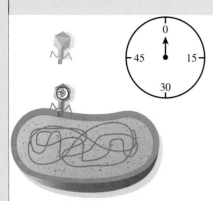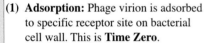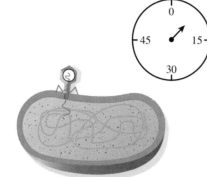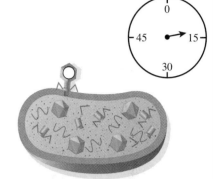

(1) **Adsorption:** Phage virion is adsorbed to specific receptor site on bacterial cell wall. This is **Time Zero**.

(2) Phage DNA enters cell to initiate **Eclipse Stage**. Bacterial DNA begins to disintegrate within minutes.

(3) Phage capsids, tails, and DNA begin to appear within 12 minutes as phage reorients cell metabolism to its own fabrication processes.

(4) Components of phage are assembled into mature infective virions. The eclipse period ends with first appearance of infective phage in cell.

(5) Cell wall opens up due to enzymatic action to release mature virions. **Burst size** is the number of units released by cell. Total time: 40 minutes.

FIGURE VI.1   The lytic cycle of a virulent bacteriophage

Determination of a Bacteriophage Titer

Bacteriophages are viruses that infect bacterial cells. They were first described by Twort and d'Herelle in 1915 when they both noted that bacterial cultures spontaneously cleared and the bacteria-free liquid that remained could cause new cultures of bacteria to also clear. Because it appeared that the cultures were being "eaten" by some unknown agent, d'Herelle coined the term *bateriophage,* which means "bacterial eater." Like all viruses, bacteriophages, or phages, for short, are **obligate intracellular parasites**, that is they must invade a host cell in order to replicate and reproduce. This is due to the fact that viruses are composed primarily of only a single kind of nucleic acid molecule encased in a protein coat, or **capsid,** that protects the nucleic acid. All viruses lack metabolic machinery, such as energy systems, and protein synthesis components necessary for independent replication. In order to replicate and reproduce, they must use the host cell's metabolic machinery to synthesize their various component parts.

Viruses also exhibit specificity for their hosts. For example, a certain bacteriophage may only infect a specific strain of a bacterium. Examples are the T-even bacteriophages that infect *Escherichia coli* B, whereas other phages infect *E. coli* K12, a different strain of the organism. A phage that infects *Staphylococcus aureus* does not infect *E. coli* and vice versa. This specificity can be used in phage typing of pathogens (Exercise 25).

The structure of a T_4 bacteriophage is shown in figure 23.1, A phage consists of a **nucleocapsid**, which is the nucleic acid and protein capsid. The nucleocapsid is attached to a protein **sheath** that is contractile and contains a hollow tube in its center. The sheath sits on a **base plate** to which **tail fibers** and **spikes** are attached. Most of the phage structure is necessary for delivery of the phage nucleic acid into its host. A single virus or phage particle is also called a **virion**.

The steps in a lytic phage infection of a bacterial cell are basically as follows:

Recognition   The bacteriophage recognizes its host by the tail fibers binding to chemical groups on the surface of the host cell. For example, these chem-

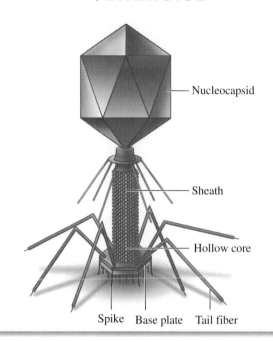

FIGURE 23.1   Bacteriophage

ical groups can be carbohydrate groups that are part of the lipopolysaccharide molecule in gram-negative cells. Part of the specificity in phage infections resides in the differences in cell surface structures of bacteria. If the groups that are recognized by tail fibers are not present on a bacterial cell, a phage cannot bind to the cell and cause an infection.

Penetration   The phage particle settles onto the surface of the host cell, and **lysozyme** that is associated with the phage tail begins to erode a localized area of the cell wall, thus weakening it. The sheath contracts, forcing a hollow core through the weakened area of the cell wall. As contraction occurs, the viral nucleic acid is injected through the core into the bacterial cytoplasm.

Replication   Only the phage nucleic acid enters the host cell, while the capsid and remainder of the phage structure remain on the outside of the bacterial cell. Some phages also inject a nuclease along with the viral nucleic acid. This nuclease specifically degrades host DNA. As a result, the host cell is incapacitated and its metabolic activities cannot interfere with viral replication. However, the virus leaves intact host cell machinery for making energy, synthesizing nucleic acids, and making proteins. The component parts of the virus are then manufactured using the host metabolic systems. The various components of the virus come together and are assembled into mature phage particles by the process of **self-assembly.** Genes on the viral genome also encode for the synthesis of lysozyme, which begins to degrade the cell wall and weaken it from inside the cell.

Release The combination of the weakened cell wall brought about by the action of lysozyme plus the pressure exerted by virus particles in the cell causes the cell to burst, releasing the phages into the environment where they can then infect other susceptible cells. One phage particle infecting one host cell can produce as many as 200 viral progeny. This number is called the **burst size.**

If a uniform layer of bacterial cells, called a **confluent lawn,** is immobilized in a soft agar overlay containing bacteriophages, the lysis and death of bacterial cells by the phages causes the formation in the agar of clear areas called **plaques.** Each plaque is formed by the progeny of a single phage that has replicated and lysed the bacterial cells in that area. Like colony-forming units, **plaque-forming units** can be counted to determine the number of viral particles in a suspension.

In the following exercise, you will work in pairs and determine the number of phage particles or plaque-forming units in a suspension of T_4 bacteriophage. You will use *E. coli* B as the host cell for this experiment.

MATERIALS

- 18–24 hour broth culture of *Escherichia coli* B
- 2 ml suspension of T_4 bacteriophages with a titer of at least 10,000 phages/ml
- 5 trypticase soy agar (TSA) plates. These should be warmed to 37°C before use.
- 5 tubes of soft agar (0.7% agar). Prior to use, melt and hold at 50°C in a water bath.
- 5 tubes of 9.0 ml trypticase soy (TS) broth
- 1 ml sterile pipettes
- pipette aids

1. Label the 5 TSA plates with your name and dilutions from 1:10 to 1:100,000.
2. Label 5 TS broth tubes with the dilutions 1:10 to 1:100,000 (figure 23.2)
3. Prepare serial 10-fold dilutions of the phage stock suspensions by transferring 1 ml of the phage suspension to the first dilution blank. Mix well and transfer 1 ml of the first dilution to the second dilution blank (1:100). Repeat this same procedure until the original phage stock has been diluted 1:100,000 (figure 23.2).
4. Aseptically transfer 2 drops of *E. coli* B broth culture to each of the 5 soft agar overlay tubes.
5. Transfer 1 ml of the first (1:10) phage dilution tube to a soft agar overlay and mix thoroughly but gently. After mixing, pour the contents of the soft agar tube onto the respective TSA plate. Make sure that the soft agar completely covers the surface of the TSA plate. This can be accomplished by gently swirling the plate several times after pouring and while the soft agar is still liquid.
6. Repeat this procedure for each dilution of the phage suspension.
7. Incubate the plates at 37°C for 24 hours. If the exercise cannot be completed at this time, refrigerate the plates until the next laboratory period.
8. Observe the plates. Plaques will appear as clear areas in the bacterial lawn. Count the plaques on the plates. Only include counts between 25 to 250 plaques. This can be facilitated with a bacterial colony counter. Multiply the number of plaques times the dilution factor to determine the number of phage particles in the original suspension of phages.
9. Record the phage titer in Laboratory Report 23.

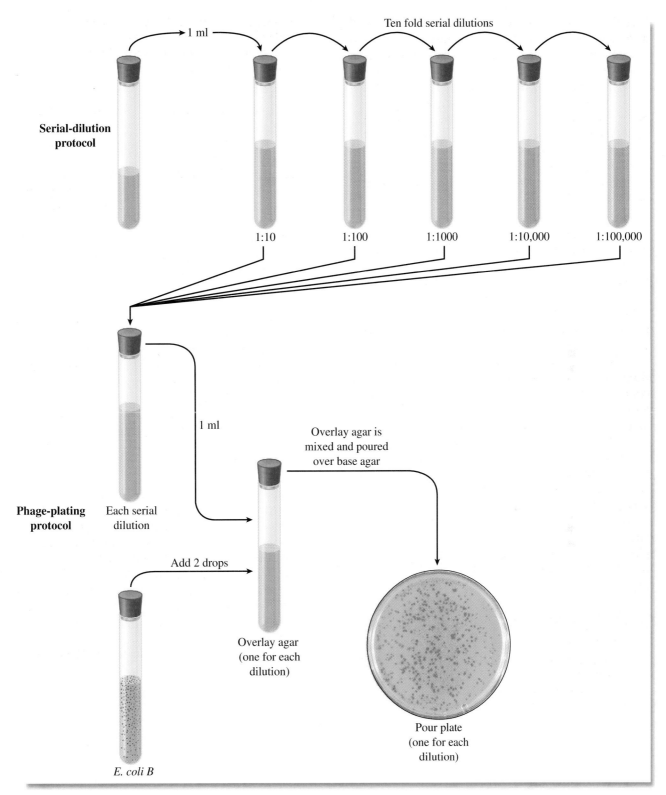

FIGURE 23.2   Procedure for determining the titer of bacteriophage

LABORATORY REPORT

Student: _____

Date: _____   Section: _____

EXERCISE 23   Determination of a Bacteriophage Titer

Dilution giving countable plaques between 25 and 250

Number of plaques counted on this plate

Number of phage particles in the original stock

Questions

1. On gram-positive cells, to what chemical groups might bacteriophages attach?

   _____

2. Besides a capsid, what other structures can be present on some viruses?

   _____

3. If you had a plate with plaques, how would you proceed to isolate and produce phage stocks from the plate?

   _____

Isolation of Phage from Flies

As stated earlier, coprophagous insects, as well as raw sewage, contain various kinds of bacterial viruses. Houseflies fall into the coprophagous category because they deposit their eggs in fecal material where the young larvae feed, grow, pupate, and emerge as adult flies. This type of environment is heavily populated by *E. coli* and its inseparable parasitic phages.

Figures 24.1 and 24.2 illustrate the procedure.

FLY COLLECTION

To increase the probability of success in isolating phage, it is desirable that one use 20 to 24 houseflies. A smaller number might be sufficient; the larger number, however, increases the probability of initial success. Houseflies should not be confused with the smaller blackfly or the larger blowfly. An ideal spot for collecting these insects is a barnyard or riding stable. One should not use a cyanide killing bottle or any other chemical means. Flies should be kept alive until just prior to crushing and placing them in the growth medium. There are many ways that one might use to capture them—use your ingenuity!

ENRICHMENT

Within the flies' digestive tracts are several different strains of *E. coli* and bacteriophage. Our first concern is to enhance the growth of both organisms to ensure an adequate supply of phage. To accomplish this the flies must be ground up with a mortar and pestle and then incubated in a special growth medium for a total of 48 hours. During the last 6 hours of incubation, a lysing agent, sodium cyanide, is included in the growth medium to augment the lysing properties of the phage.

MATERIALS

- bottle of phage growth medium* (50 ml)
- bottle of phage lysing medium* (50 ml)
- Erlenmeyer flask (125 ml capacity) with cotton plug
- mortar and pestle (glass)
- *see Appendix C for composition

1. Into a clean nonsterile mortar place 24 freshly killed houseflies. Pour half of the growth medium into the mortar and grind the flies to a fine pulp with the pestle.
2. Transfer this fly-broth mixture to an empty flask. Use the remainder of the growth medium to rinse out the mortar and pestle, pouring all the medium into the flask.
3. Wash the mortar and pestle with soap and hot water before returning them to the cabinet.
4. Incubate the fly-broth mixture for 42 hours at 37° C.
5. At the end of the 42-hour incubation period, add 50 ml of lysing medium to the fly-broth mixture. Incubate this mixture for another 6 hours.

CENTRIFUGATION

Before attempting filtration, you will find it necessary to separate the fly fragments and miscellaneous bacteria from the culture medium. If centrifugation is incomplete, the membrane filter will clog quickly and filtration will progress slowly. To minimize filter clogging, a triple centrifugation procedure will be used. To save time in the event filter clogging does occur, an extra filter assembly and an adequate supply of membrane filters should be available. These filters have a maximum pore size of 0.45 µm, which holds back all bacteria, allowing only the phage virions to pass through.

MATERIALS

- centrifuge
- 6–12 centrifuge tubes
- 2 sterile membrane filter assemblies (funnel, glass base, clamp, and vacuum flask)
- package of sterile membrane filters
- sterile Erlenmeyer flask with cotton plug (125 ml size)
- vacuum pump and rubber hose

1. Into 6 or 8 centrifuge tubes, dispense the enrichment mixture, filling each tube to within 1/2" of the top. Place the tubes in the centrifuge so that the load is balanced. Centrifuge the tubes at 2,500 rpm for 10 minutes.

(1) Twenty to twenty-four flies are ground up in phage growth medium with a mortar and pestle.

(2) Crushed flies are incubated in growth medium for 42 hours at 37° C. After adding lysing medium it is incubated for another 6 hours.

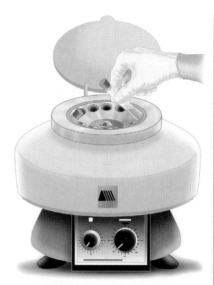

(3) Fly-broth culture is triple-centrifuged at 2,500 rpm.

(4) Membrane filter assembly is set up for filtration. This step must be done aseptically.

(5) Centrifuged supernatant is filtered to produce bacteria-free phage filtrate.

(6) Phage filtrate is dispensed to a sterile Erlenmeyer flask from which layered plates will be made. (Fig. 29.2)

FIGURE 24.1 **Procedure for preparation of bacteriophage filtrate from houseflies**

2. Without disturbing the material in the bottom of the tubes, decant all material from the tubes to within 1″ of the bottom into another set of tubes.
3. Centrifuge this second set of tubes at 2,500 rpm for another 10 minutes. While centrifugation is taking place, rinse out the first set of tubes.
4. When the second centrifugation is complete, pour off the top two-thirds of each tube into the clean set of tubes and centrifuge again in the same manner.

FILTRATION

While the third centrifugation is taking place, aseptically place a membrane filter on the glass base of a sterile filter assembly (illustration 4, figure 24.1). Use flamed forceps. Note that the filter is a thin sheet with grid lines on it. Place the glass funnel over the filter and fix the clamp in place. Hook up a rubber hose between the vacuum flask and pump.

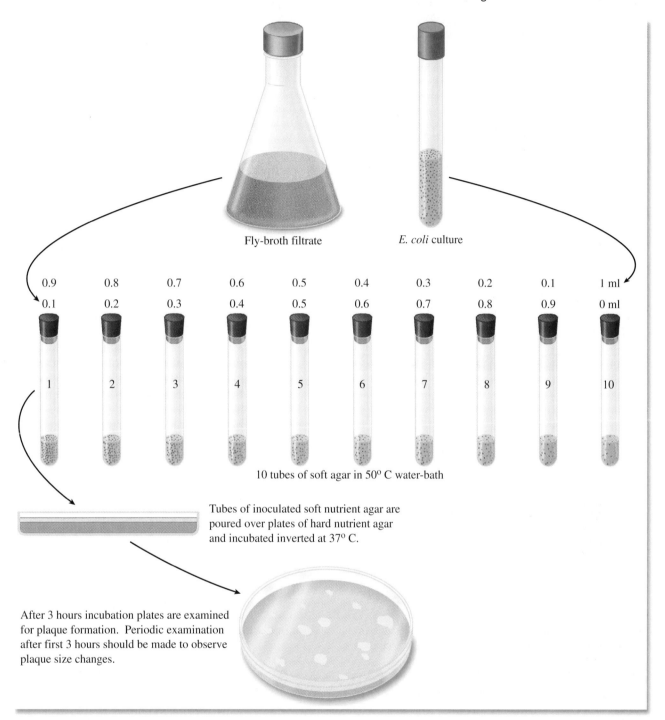

Fly-broth filtrate

E. coli culture

0.9	0.8	0.7	0.6	0.5	0.4	0.3	0.2	0.1	1 ml
0.1	0.2	0.3	0.4	0.5	0.6	0.7	0.8	0.9	0 ml
1	2	3	4	5	6	7	8	9	10

10 tubes of soft agar in 50° C water-bath

Tubes of inoculated soft nutrient agar are poured over plates of hard nutrient agar and incubated inverted at 37° C.

After 3 hours incubation plates are examined for plaque formation. Periodic examination after first 3 hours should be made to observe plaque size changes.

FIGURE 24.2 Inoculation of *Escherichia coli* with bacteriophage from fly-broth filtrate

Now, carefully decant the top three-fourths of each tube into the filter funnel. Take care not to disturb the material in the bottom of the tube. Turn on the vacuum pump. If centrifugation and decanting have been performed properly, filtration will occur almost instantly. If the filter clogs before you have enough filtrate, recentrifuge all material and pass it through the spare filter assembly.

Aseptically, transfer the final filtrate from the vacuum flask to a sterile 125 ml Erlenmeyer flask that has a sterile cotton plug. Putting the filtrate in a small flask is necessary to facilitate pipetting. Be sure to flame the necks of both flasks while pouring from one to the other.

INOCULATION AND INCUBATION

To demonstrate the presence of bacteriophage in the fly-broth filtrate, a strain of phage-susceptible *E. coli* will be used. To achieve an ideal proportion of phage to bacteria, a proportional dilution method will be used. The phage and bacteria will be added to tubes of soft nutrient agar that will be layered over plates of hard nutrient agar. Soft nutrient agar contains only half as much agar as ordinary nutrient agar. This medium and *E. coli* provide an ideal "lawn" for phage growth. Its jellylike consistency allows for better diffusion of phage particles; thus, more even development of plaques occurs.

Figure 24.2 illustrates the overall procedure. It is best to perform this inoculation procedure in the morning so that the plates can be examined in late afternoon. As plaques develop, one can watch them increase in size with the multiplication of phage and simultaneous destruction of *E. coli*.

MATERIALS

- nutrient broth cultures of *Escherchia coli* (ATCC #8677 phage host)
- flask of fly-broth filtrate
- 10 tubes of soft nutrient agar (5 ml per tube) with metal caps
- 10 plates of nutrient agar (15 ml per plate, and prewarmed at 37° C)
- 1 ml serological pipettes, sterile

1. Liquefy 10 tubes of soft nutrient agar and cool to 50° C. Keep tubes in water bath to prevent solidification.

2. With a marking pen, number the tubes of soft nutrient agar 1 through 10. Keep the tubes sequentially arranged in the test-tube rack.

3. Label 10 plates of prewarmed nutrient agar 1 through 10. Also, label plate 10 negative control. Prewarming these plates will allow the soft agar to solidify more evenly.

4. With a 1 ml serological pipette, deliver 0.1 ml of fly-broth filtrate to tube 1, 0.2 ml to tube 2, etc., until 0.9 ml has been delivered to tube 9. Refer to figure 24.2 for sequence. **Note that no fly-broth filtrate is added to tube 10.** This tube will be your negative control.

5. With a fresh 1 ml pipette, deliver 0.9 ml of *E. coli* to tube 1, 0.8 ml to tube 2, etc., as shown in figure 24.2. **Note that tube 10 receives 1.0 ml of *E. coli*.**

6. After flaming the necks of each of the tubes, pour them into similarly numbered plates.

7. When the agar has cooled completely, put the plates, inverted, into a 37° C incubator.

8. **After about 3 hours** incubation, examine the plates, looking for plaques. If some are visible, measure them and record their diameters on Laboratory Report 24.

9. If no plaques are visible, check the plates again in another **2 hours.**

10. Check the plaque size again at **12 hours,** if possible, recording your results. Incubate a total of 24 hours.

11. Complete Laboratory Report 24.

LABORATORY REPORT

Student: _____

Date: _____ Section: _____

EXERCISE 24 Isolation of Phage from Flies

A. Plaque Size Increase

With a china marking pencil, circle and label three plaques on one of the plates and record their sizes in millimeters at 1-hour intervals.

TIME	PLAQUE SIZE (millimeters)		
	Plaque No. 1	Plaque No. 2	Plaque No. 3
When first seen			
1 hour later			
2 hours later			
3 hours later			

1. Were any plaques seen on the negative control plate? _____

2. Do plates 1, 2, and 3 show a progressive increase in number of plaques with increased amount of fly-broth filtrate? _____

3. Did the phage completely "wipe out" all bacterial growth on any of the plates? _____
   If so, which plates? _____

B. Observations

Count all the plaques on each plate and record the counts in the following table. If the plaques are very numerous, use a Quebec colony counter and hand counting device. If this exercise was performed as a class project with individual students doing only one or two plates from a common fly-broth filtrate, record all counts on the chalkboard on a table similar to the one below.

Plate Number	1	2	3	4	5	6	7	8	9	10
E. coli (ml)	0.9	0.8	0.7	0.6	0.5	0.4	0.3	0.2	0.1	1.0
Filtrate (ml)	0.1	0.2	0.3	0.4	0.5	0.6	0.7	0.8	0.9	0
Number of plaques										

1. Were any plates completely "wiped out" by phage action? _____

2. If so, which ones? _____

C. Terminology

    1. Differentiate between the following:

       Lysis: _____

       Lysogeny: _____

    2. Differentiate between the following:

       Virulent phage: _____

       Temperate phage: _____

Phage Typing

The host specificity of bacteriophage is such that it is possible to delineate different strains of individual species of bacteria on the basis of their susceptibility to various kinds of bacteriophage. In epidemiological studies, where it is important to discover the source of a specific infection, determining the phage type of the causative organism can be an important tool in solving the riddle. For example, if it can be shown that the phage type of *S. typhi* in a patient with typhoid fever is the same as the phage type of an isolate from a suspected carrier, chances are excellent that the two cases are epidemiologically related. Since all bacteria are probably parasitized by bacteriophages, it is theoretically possible, through research, to classify each species into strains or groups according to their phage type susceptibility. This has been done for *Staphytococcus aureus, Salmonella typhi,* and several other pathogens. The following table illustrates the lytic groups of *S. aureus* as proposed by M. T. Parker.

Lytic Group	Phages in Group
I	29 52 52A 79 80
II	3A 3B 3C 55 71
III	6 7 42E 47 53 54 75 77 83A
IV	42D
not allotted	81 187

In bacteriophage typing, a suspension of the organism to be typed is swabbed over an agar surface. The bottom of the plate is marked off in squares and labeled to indicate which phage types are going to be used. To the organisms on the surface, a small drop of each phage type is added to their respective squares. After incubation, the plate is examined to see which phages were able to lyse the organisms. This is the procedure to be used in this exercise. See figure 25.1.

MATERIALS

- 1 petri plate of tryptone yeast extract agar
- bacteriophage cultures (available types) nutrient
- broth cultures of *S. aureus* with swabs

1. Mark the bottom of a plate of tryptone yeast extract agar with as many squares as there are phage

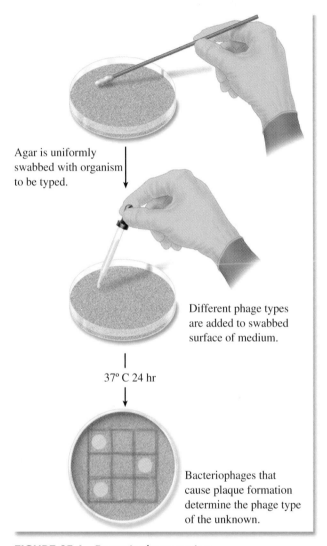

Agar is uniformly swabbed with organism to be typed.

Different phage types are added to swabbed surface of medium.

37° C 24 hr

Bacteriophages that cause plaque formation determine the phage type of the unknown.

FIGURE 25.1 Bacteriophage typing

types to be used. Label each square with the phage type numbers.
2. Swab the entire surface of the agar with the organisms.
3. Deposit 1 drop of each phage in its respective square.
4. Incubate the plate at 37° C for 24 hours and record the lytic group and phage type of the culture.
5. Record your results on Laboratory Report 25.

LABORATORY REPORT

Student: _____

Date: _____ Section: _____

EXERCISE 25   Phage Typing

1. To which phage types was this strain of *S. aureus* susceptible?

    _____

2. To what lytic group does this strain of staphylococcus belong?

    _____

3. In what way can bacteriophage alter the genetic structure of a bacterium?

    _____

    _____

    _____

    _____

    _____

ENVIRONMENTAL INFLUENCES AND CONTROL OF MICROBIAL GROWTH

The 10 exercises of this unit are concerned with two aspects of microbial growth: promotion and control. On the one hand, the microbiologist is concerned with providing optimum growth conditions to favor maximization of growth. The physician, nurse, and other members of the medical arts profession, on the other hand, are concerned with the limitation of microbial populations in disease prevention and treatment. An understanding of one of these facets of microbial existence enhances the other.

In part 4 we were primarily concerned with providing media for microbial growth that contain all the essential nutritional needs. Very little emphasis was placed on other limiting factors such as temperature, oxygen, or hydrogen ion concentration. An organism provided with all its nutritional needs may fail to grow if one or more of these essentials are not provided. The total environment must be sustained to achieve the desired growth of microorganisms.

Microbial control by chemical and physical means involves the use of antiseptics, disinfectants, antibiotics, ultraviolet light, and many other agents. The exercises of this unit that are related to these aspects are intended, primarily, to demonstrate methods of measurement; no attempt has been made to make in-depth evaluation.

Microorganisms grow over a broad temperature range that extends from below 0° C to greater than 100° C. Based on their temperature requirements, they can be divided into four groups that define their optimal growth:

psychrophiles: optimal growth between −5°C and 20° C; these bacteria can be found in the supercooled waters of the arctic and antarctic.

mesophiles: optimal growth between 20° C and 50° C; most bacteria fall into this class, for example most pathogens grow between 35° C and 40° C.

thermophiles: optimal growth between 50° C and 80° C; bacteria in this group occur in soils where the midday temperature can reach greater than 50° C or in compost piles where fermentation activity can cause temperatures to exceed 60–65° C.

hyperthermophiles: growth optimum above 80° C; many of the Archaea occupy environments that are heated by volcanic activity where water is superheated to above 100° C. These organisms have been isolated from thermal vents deep within the ocean floor and from volcanic heated hot springs.

It should be noted that one single organism is not capable of growth over the entire range but would be restricted to one of the temperature classes. However, some bacteria within the classes are capable of growth at temperatures lower or higher than their optima. For example, some mesophilic bacteria such as *Proteus, Pseudomonas, Campylobacter,* and *Leuconostoc* can grow at 4° C, refrigerator temperatures, and cause food spoilage. These bacteria are referred to as **psychrotrophs.**

Temperature can affect several metabolic factors in the cell. Enzymes are directly affected by temperature and any one enzyme will have a minimum, optimum, and maximum temperature for activity. Maximal enzyme activity will occur at the optimum temperature. At temperatures above the maximum, enzymes will begin to denature and loose activity. Below the minimum temperature, chemical activity slows down and some denaturation can also occur. In addition to the effects on

enzyme activity, temperature can also greatly affect cell membranes and transport. As temperature decreases, transport of nutrients into the cell also decreases due to fluidity changes in the membrane. If the temperature increases above the maximum of an organism, membrane lipids can be destroyed resulting in serious damage to the membrane and death of the organism. Last, ribosomes can be directly affected by temperature, and if extremes of temperature occur, they will cease to function adequately.

In this experiment, we will attempt to measure the effects of various temperatures on two physiological reactions: pigment production and growth rate. Nutrient broth and nutrient agar slants will be inoculated with three different organisms that have different optimum growth temperatures. One organism, *Serratia marcescens,* produces a red pigment called *prodigiosin,* which is produced only in a certain temperature range. It is our goal here to determine the optimum temperature for prodigiosin production and the approximate optimum growth temperatures for all three microorganisms. To determine optimum growth temperatures, we will be incubating cultures at five different temperatures. A spectrophotometer will be used to measure turbidity densities in the broth cultures after incubation.

FIRST PERIOD
(Inoculations)

To economize on time and media, it will be necessary for each student to work with only two organisms and seven tubes of media. Refer to table 26.1 to determine your assignment. Figure 26.1 illustrates the procedure.

MATERIALS

- nutrient broth cultures of *Serratia marcescens, Bacillus stearothermophilus,* and *Escherichia coli*

per student:
- 2 nutrient agar slants
- 5 tubes of nutrient broth

1. Label the tubes as follows:
   Slants: Label both of them *S. marcescens;* label one tube 25° C and the other tube 38° C.

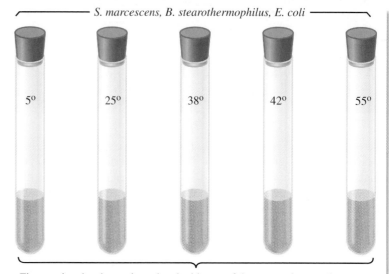

Two nutrient agar slants are streaked with *S. marcescens* and incubated at different temperatures for pigment production

Five nutrient broths are inoculated with one of three organisms and incubated at five different temperatures to determine optimum growth temperatures for each organism.

FIGURE 26.1 Inoculation procedure

Broths: Label each tube of nutrient broth with your other organism and one of the following five temperatures: 5° C, 25° C, 38° C, 42° C, or 55° C.
2. Inoculate each of the tubes with the appropriate organisms. Use a wire loop.
3. Place each tube in one of the five baskets that is labeled according to incubation temperature.
   Note: The instructor will see that the 5° C basket is placed in the refrigerator and the other four are placed in incubators that are set at the proper temperatures.

SECOND PERIOD
(Tabulation of Results)

MATERIALS

- slants and broth cultures that have been incubated at various temperatures
- spectrophotometer and cuvettes
- tube of sterile nutrient broth

1. Compare the nutrient agar slants of *S. marcescens*. Using colored pencils, draw the appearance of the growths on Laboratory Report 26.
2. Shake the broth cultures and compare them, noting the differences in turbidity. Those tubes that appear to have no growth should be compared with a tube of sterile nutrient broth.
3. If a spectrophotometer is available, determine the turbidity of each tube following the instructions on Laboratory Report 26.
4. If no spectrophotometer is available, record turbidity by visual observation. The Laboratory Report indicates how to do this.
5. Exchange results with other students to complete data collection for experiment.

LABORATORY REPORT

After recording all data, answer the questions on Laboratory Report 26.

TABLE 26.1 Inoculation Assignments

Student Number	S. marcescens	B. Stearothermophilus	E. coli
1, 4, 7, 10, 13, 16, 19, 22, 25	2 slants and 5 broths		
2, 5, 8, 11, 14, 17, 20, 23, 26	2 slants	5 broths	
3, 6, 9, 12, 15, 18, 21, 24, 27	2 slants		5 broths

LABORATORY REPORT

Student: _____

Date: _____   Section: _____

EXERCISE 26   Temperature: Effects on Growth

A. Pigment Formation and Temperature

1. Draw the appearance of the growth of *Serratia marcescens* on the nutrient agar slants using colored pencils.

2. Which temperature seems to be closest to the optimum temperature for pigment formation?

    _____

3. How is pigment production in *S. marcescens* controlled by temperature?

    _____

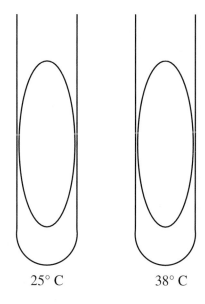

25° C   38° C

B. Growth Rate and Temperature

If a spectrophotometer is available, dispense the cultures into labeled cuvettes and determine the percent transmittance of each culture. Calculate the O.D. values

If no spectrophotometer is available, record only the visual readings as + , + + , + + + , and none.

Temp. °C	SERRATIA MARCESCENS		ESCHERICHIA COLI	
	Visual Reading	Spectrophotometer O.D.	Visual Reading	Spectrophotometer O.D.
5				
25				
38				
42				
55				

Growth curves of *Serratia marcescens* and *Escherichia coli* as related to temperature.

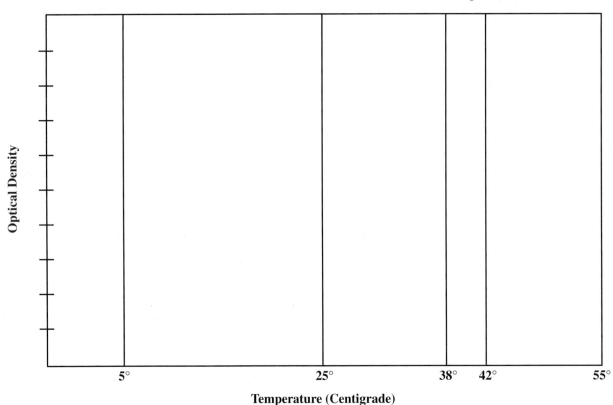

1. On the basis of the above graph, estimate the optimum growth temperature of the two organisms.

   Serratia marcescens: _____

   Escherichia coli: _____

2. To get more precise results for the above graph, what would you do?

   _____

   _____

3. Differentiate between the following:

   Thermophile: _____

   Mesophile: _____

   Psychrophile: _____

4. What is the optimum growth temperature range for most psychrophiles? _____

Temperature:
Lethal Effects

In attempting to compare the susceptibility of different organisms to elevated temperatures, it is necessary to use some yardstick of measure. Two methods of comparison are used: the thermal death point and the thermal death time. The thermal death point (TDP) is the temperature at which an organism is killed in 10 minutes. The thermal death time (TDT) is the time required to kill a suspension of cells or spores at a given temperature. Since various factors such as pH, moisture, composition of medium, and age of cells will greatly influence results, these variables must be clearly stated.

In this exercise, we will subject cultures of three different organisms to temperatures of 60°, 70°, 80°, 90°, and 100° C. At intervals of 10 minutes, organisms will be removed and plated out to test their viability. The spore-former *Bacillus megaterium* will be compared with the non-spore-formers *Staphylococ-*

cus aureus and *Escherichia coli.* The overall procedure is illustrated in figure 27.1.

Note in figure 27.1 that *before* the culture is heated, a **control plate** is inoculated with 0.1 ml of the organism. When the culture is placed in the water bath, a tube of nutrient broth with a thermometer inserted into it is placed in the bath at the same time. Timing of the experiment starts when the thermometer reaches the test temperature.

Due to the large number of plates that have to be inoculated to perform the entire experiment, it will be necessary for each member of the class to be assigned a specific temperature and organism to work with. Table 27.1 provides assignments by student number. After the plates have been incubated, each student's results will be tabulated on a Laboratory Report chart at the demonstration table. The instructor will have

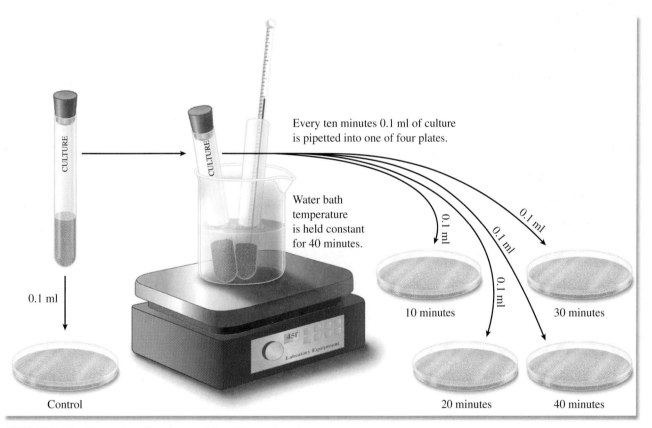

FIGURE 27.1   Procedure for determining thermal endurance

copies made of it to give each student so that everyone will have all the pertinent data needed to draw the essential conclusions.

Although this experiment is not difficult, it often fails to turn out the way it should because of student error. Common errors are (1) omission of the control plate inoculation, (2) putting the thermometer in the culture tube instead of in a tube of sterile broth, and (3) not using fresh sterile pipettes when instructed to do so.

MATERIALS

per student:
• 5 petri plates
• 5 pipettes (1 ml size)
• 1 tube of nutrient broth
• 1 bottle of nutrient agar (60 ml)
• 1 culture of organisms

class equipment:
• water baths set up at 60°, 70°, 80°, 90°, and 100° C

broth cultures:
• *Staphylococcus aureus, Escherichia coli,* and *Bacillus megaterium* (minimum of 5 cultures of each species per lab section)

1. Consult table 27.1 to determine what organism and temperature has been assigned to you. If several thermostatically controlled water baths have been provided in the lab, locate the one that you will use. If a bath is not available for your temperature, set up a bath on an electric hot plate or over a tripod and Bunsen burner.

    If your temperature is 100° C, a hot plate and beaker of water are the only way to go. When setting up a water bath use hot tap water to start with to save heating time.

2. Liquefy a bottle of 60 ml of nutrient agar and cool to 50° C. This can be done while the rest of the experiment is in progress.
3. Label five petri plates: **control, 10 min, 20 min, 30 min,** and **40 min.**
4. Shake the culture of organisms and transfer 0.1 ml of organisms with a 1 ml pipette to the control plate.
5. Place the culture and a tube of sterile nutrient broth into the water bath. Remove the cap from the tube of nutrient broth and insert a thermometer into the tube. *Don't make the mistake of inserting the thermometer into the culture of organisms!*
6. As soon as the temperature of the nutrient broth reaches the desired temperature, record the time here: _____.
   Watch the temperature carefully to make sure it does not vary appreciably.
7. After 10 minutes have elapsed, transfer 0.1 ml from the culture to the 10-minute plate with a fresh 1 ml pipette. Repeat this operation at 10-minute intervals until all the plates have been inoculated. *Use fresh pipettes each time and be sure to shake the culture before each delivery.*
8. Pour liquefied nutrient agar (50° C) into each plate, rotate, and cool.
9. Incubate at 37° C for 24 to 48 hours. After evaluating your plates, record your results on the chart in Laboratory Report 30 and on the chart on the demonstration table.

LABORATORY REPORT

Complete Laboratory Report 27 once you have a copy of the class results.

TABLE 27.1 Inoculation Assignments

Organism	Student Number				
	60°C	70°C	80°C	90°C	100°C
Staphylococcus aureus	1, 16	4, 19	7, 22	10, 25	13, 28
Escherichia coli	2, 17	5, 20	8, 23	11, 26	14, 29
Bacillus megaterium	3, 18	6, 21	9, 24	12, 27	15, 30f

LABORATORY REPORT

Student: _____

Date: _____ Section: _____

EXERCISE 27   Temperature: Lethal Effects

A. Tabulation of Results

Examine your five petri plates, looking for evidence of growth. Record on the chalkboard, using a chart similar to the one below, the presence or absence of growth as (+) or (−). When all members of the class have recorded their results, complete this chart.

ORGANISM	60° C					70° C					80° C					90° C					100° C				
	C*	10	20	30	40	C*	10	20	30	40	C*	10	20	30	40	C*	10	20	30	40	C*	10	20	30	40
S. aureus																									
E. coli																									
B. megaterium																									

1. If they can be determined from the above information, record the **thermal death point** for each of the organisms.

   S. aureus:_____   E. coli: _____   B. megaterium:_____

2. From the following table, determine the **thermal death time** for each organism at the tabulated temperatures.

ORGANISM	THERMAL DEATH TIME				
	60° C	70° C	80° C	90° C	100° C
S. aureus					
E. coli					
B. megaterium					

B. Questions

1. Give three reasons why endospores are much more resistant to heat than are vegetative cells.

   a. _____

   b. _____

   c. _____

2. Differentiate between the following:

   Thermoduric: _____

   _____

   Thermophilic: _____

   _____

3. List four diseases caused by spore-forming bacteria.

   a. _____

   b. _____

   c. _____

   d. _____

4. Since boiling water is unreliable in destroying endospores, how should one use heat in medical applications to ensure spore destruction? (three ways)

   a. _____

   b. _____

   c. _____

pH and Microbial Growth

Another factor that exerts a strong influence on growth is the hydrogen ion concentration designated by the term **pH** ($-\log 1/H^+$). The hydrogen ion concentration affects proteins and other charged molecules in the cell. Each organism will have an optimal pH at which it grows best. If pH values exceed the optimum for an organism, the solubility of charged molecules can be adversely affected and molecules can precipitate out of solution. For example, the pH can directly affect the charge on amino acids in proteins and result in denaturation and loss of enzyme activity. Most bacteria can grow over a range of 2–3 pH units. However, even when an organism is capable of growing over a range of 2–3 pH units, the cytoplasm is usually maintained at or near neutrality to prevent damage and destruction of charged species and macromolecules.

In this exercise, we will test the degree of inhibition of microorganisms that results from media containing different pH concentrations. Note in the materials list that tubes of six different hydrogen concentrations are listed. Your instructor will indicate which ones, if not all, will be tested.

FIRST PERIOD

MATERIALS

per student:
- 1 tube of nutrient broth of pH 3.0
- 1 tube of nutrient broth of pH 5.0
- 1 tube of nutrient broth of pH 7.0
- 1 tube of nutrient broth of pH 8.0
- 1 tube of nutrient broth of pH 9.0
- 1 tube of nutrient broth of pH 10.0

class materials:
- broth cultures of *Escherichia coli*
- broth cultures of *Staphylococcus aureus*
- broth cultures of *Alcaligenes faecalis**
- broth cultures of *Saccharomyces cerevisiae***

1. Inoculate a tube of each of these broths with one organism. Use the organism following your assigned number from the table below:

Student Number	Organism
1, 5, 9, 13, 17, 21, 25	*Escherichia coli*
2, 6, 10, 14, 18, 22, 26	*Staphylococcus aureus*
3, 7, 11, 15, 19, 23, 27	*Alcaligenes faecalis**
4, 8, 12, 16, 20, 24, 28	*Saccharomyces cerevisiae***

2. Incubate the tubes of *E. coli, S. aureus,* and *A. faecalis* at 37° C for 48 hours. Incubate the tubes of *S. ureae, C. glabrata,* and *S. cervisiae* at 20° C for 48 to 72 hours.

SECOND PERIOD

MATERIALS

- spectrophotometer
- 1 tube of sterile nutrient broth
- tubes of incubated cultures at various pHs

1. Use the tube of sterile broth to calibrate the spectrophotometer and measure the O.D. of each culture (page 134, Exercise 20). Record your results in the tables on Laboratory Report 28.
2. Plot the O.D. values in the graph on Laboratory Report 28 and answer all the questions.

---

*Sporosarcina ureae can be used as a substitute for *Alcaligenes faecalis.*
**Candida glabrata is a good substitute for *Saccharomyces cerevisiae.*

LABORATORY REPORT

Student: _____

Date: _____ Section: _____

EXERCISE 28   pH and Microbial Growth

A. Tabulation of Results

If a spectrophotometer is available, dispense the cultures into labeled cuvettes and determine the percent transmittance of each culture. Calculate the O.D. values. To complete the tables, get the results of the other three organisms from other members of the class, and delete the substitution organisms in the tables that were not used.

If no spectrophotometer is available, record only the visual reading as + , + + , + + + , and none.

pH	*Escherichia coli*		*Staphylococcus aureus*	
	Visual Reading	Spectrophotometer O.D.	Visual Reading	Spectrophotometer O.D.
3				
5				
7				
8				
9				
10				

pH	*Alcaligenes faecalis* or *Sporosarcina ureae*		*Saccharomyces cervisiae* or *Candida glabrata*	
	Visual Reading	Spectrophotometer O.D.	Visual Reading	Spectrophotometer O.D.
3				
5				
7				
8				
9				
10				

B. Growth Curves

Once you have computed all the O.D. values on the two tables, plot them on the following graph. Use different colored lines for each species.

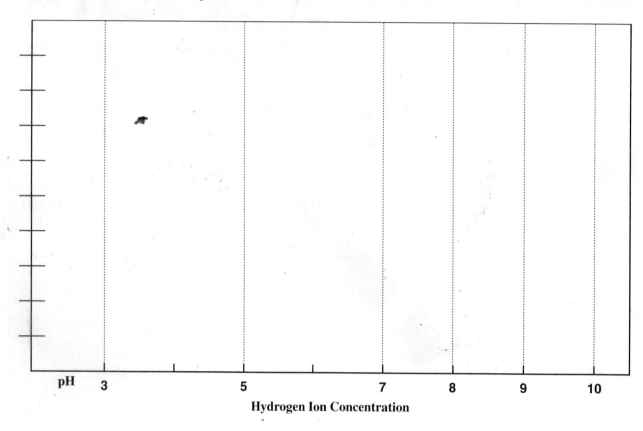

C. Questions

1. Which organism seems to grow best in acid media? _____

2. Which organism seems to grow best in alkaline media? _____

3. Which organism seems to tolerate the broadest pH range? _____

Water Activity and Osmotic Pressure

water can affect the growth of bacteria.

The growth of bacteria can be profoundly affected by the availability of water in an environment. The availability of water is defined by a physical parameter called the water activity, A_w. It is determined by measuring the ratio of the water vapor pressure of a solution to the water vapor pressure of pure water. The values for water activity vary between 0 and 1.0, and the closer the value is to 1.0, the more water is available to a cell for metabolic purposes. Water activity and hence its availability decrease with increases in the concentration of solutes such as salts. This results because water becomes involved in breaking ionic bonds and forming solvation shells around charged species to maintain them in solution.

In the process of **osmosis,** water diffuses from areas of low solute concentration where water is more plentiful to areas of high solute concentration where water is less available. Because there is normally a high concentration of nutrients in the cytoplasm relative to the outside of the cell, water will naturally diffuse into a cell. A medium where solute concentrations on the outside of the cell are lower than the cytoplasm is designated as **hypotonic** (figure 29.1). In general, bacteria are not harmed by hypotonic solutions because the rigid cell wall protects the membrane from being damaged by the osmotic pressure exerted against it. It also prevents the membrane from disrupting when water diffuses across the cell membrane into the cytoplasm.

Environments where the solute concentration is the same inside and outside the cell are termed **isotonic.** Animal cells require isotonic environments or else cells will undergo lysis because only the fragile cell membrane surrounds the cell. Tissue culture media for growing animal cells provides an isotonic environment to prevent cell lysis.

Hypertonic environments exist when the solute concentration is greater on the outside of the cell relative to the cytoplasm, and this causes water to diffuse out of the cytoplasm. When this develops, the cell undergoes **plasmolysis** resulting in a loss of water, dehydration of the cytoplasm, and shrinkage of the cell membrane away from the cell wall. In these situations, considerable and often irreversible damage can occur to the metabolic machinery of the cell. Low water activity and hypertonic environments have been used by humans for centuries to preserve food. Salted meat and fish, and fruit in jams and jellies with high sugar content resist contamination because very little water is available for cells to grow. Most bacteria that might contaminate these foods would undergo immediate plasmolysis.

Microorganisms can be grouped based on their ability to cope with extreme water activity and osmotic pressure. Most bacteria grow best when the water activity is around 0.9 to 1.0. In contrast, **halophiles** require high concentrations of sodium chloride to grow. Examples are the halophilic bacteria that require 15–30% sodium chloride to grow and maintain integrity of their cell walls. These bacteria, which belong to the Archaea, are found in salt lakes and brine solutions, and occasionally growing on salted fish. Some microorganisms are **halotolerant** and are capable of growth in moderate concentrations of salt. For example, *Staphylococcus aureus* can tolerate sodium chloride concentrations that approach 3 *M* or 11%.

Another group of organisms is the **osmophiles,** which are able to grow in environments where sugar concentrations are excessive. An example is *Xeromyces,* a yeast that can contaminate and spoil jams and jellies.

In this exercise, we will test the degree of inhibition of organisms that results with media containing different concentrations of sodium chloride. To accomplish this, you will streak three different organisms on four plates of media. The specific organisms used differ in their tolerance of salt concentrations. The salt concentrations will be 0.5, 5, 10, and 15%. After incubation for 48 hours and several more days,

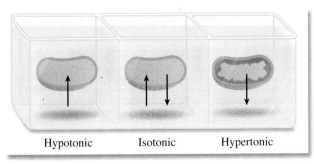

| Hypotonic | Isotonic | Hypertonic |

FIGURE 29.1   Osmotic variabilities

comparisons will be made of growth differences to determine their degrees of salt tolerances.

MATERIALS

per student:
- 1 petri plate of nutrient agar (0.5% NaCl)
- 1 petri plate of nutrient agar (5% NaCl)
- 1 petri plate of nutrient agar (10% NaCl)
- 1 petri plate of milk salt agar (15% NaCl)

cultures:
- *Escherichia coli* (nutrient broth)
- *Staphylococcus aureus* (nutrient broth)
- *Halobacterium salinarium* (slant culture)

1. Mark the bottoms of the four petri plates as indicated in figure 29.2.
2. Streak each organism in a straight line on the agar, using a wire loop.
3. Incubate all the plates for 48 hours at room temperature with exposure to light (the pigmentation of *H. salinarium* requires light to develop). Record your results on Laboratory Report 29.
4. Continue the incubation of the milk salt agar plate for several more days in the same manner, and record your results again on Laboratory Report 29.

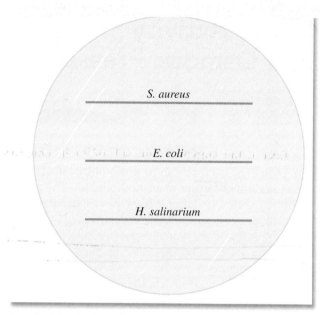

FIGURE 29.2 Streak pattern

LABORATORY REPORT

Student: _____

Date: _____   Section: _____

EXERCISE 29   Water Activity and Osmotic Pressure

A. Results

Record the amount of growth of each organism at the different salt concentrations, using, + , + + , + + + , and none to indicate degree of growth.

ORGANISM	SODIUM CHLORIDE CONCENTRATION							
	0.5%		5%		10%		15%	
	48 hr	96 hr	48 hr	96 hr	48 hr	96 hr	48 hr	96 hr
Escherichia coli								
Staphylococcus aureus								
Halobacterium salinarium								

B. Questions

1. Evaluate the salt tolerance of the above organisms. _____

   Tolerates very little salt: _____

   Tolerates a broad range of salt concentration: _____

   Grows only in the presence of high salt concentration: _____

2. How would you classify *Halobacterium salinarium* as to salt needs? Check one.

   Obligate halophile _____   Facultative halophile _____

3. Differentiate between the following:

   Halophile: Require High Concentrations of sodium Chloride to grow -
   Osmophile: grow in environments where sugar concentrations
   are excessive.

Ultraviolet Light:
Lethal Effects

Ultraviolet (UV) light is nonionizing short wavelength radiation that falls between 4 nm and 400 nm in the visible spectrum (figure 30.1). In general, for electromagnetic radiation, the shorter the wavelength, the more damaging it is to cells, thus UV light is much more germicidal than either visible light or infrared radiation. Most bacteria are killed by the effects of ultraviolet light and UV light is routinely used to sterilize surfaces, such as the work areas of transfer hoods used for the inoculation of cultures. The primary lethal effects of ultraviolet light are due to its mutagenic properties. UV radiation at **260 nm** is the most germicidal because this wavelength is the specific wavelength at which DNA maximally absorbs UV light. When DNA absorbs UV light, it causes the formation of **pyrimidine dimers.** These form when a covalent bond is formed between two adjacent thymine or cytosine molecules in a DNA strand (figure 30.2). Dimers essentially cause the DNA molecule to become deformed so that the DNA polymerase cannot replicate DNA strands past the site of dimer formation, nor can genes past this point be transcribed.

Cells have evolved various repair mechanisms to deal with mutational changes in DNA in order to insure that fidelity of replication occurs. One system is the **SOS system**, which enzymatically removes the dimers and inserts in their place new pyrimidine molecules. Unlike DNA polymerase, enzymes of the SOS system can move beyond the point where dimers occur in the molecule. However, if the exposure of UV light is sufficient to cause massive numbers of dimers

to form in the DNA of a cell, the SOS system is unable to effectively cope with this situation and it begins to make errors by inserting incorrect bases for the damaged bases, eventually resulting in cell death.

The killing properties of UV light depend on several factors. Time of exposure is important as well as the presence of materials that will block the radiation from reaching cells. For example, plastic can block UV radiation and plastic lenses are an effective means to protect the eyes from UV damage. Bacteria that form endospores are more resistant than vegetative cells to UV light because endospores are nondividing.

In Exercise 30, you will examine the germicidal effects for UV light on *Bacillus megaterium,* an endospore former, and *Staphylococcus aureus,* a non–endospore former. One-half of each plate will be shielded from the radiation to provide a control comparison. *Bacillus megaterium* and *Staphylococcus aureus* will be used to provide a comparison of the relative resistance of vegetative and spore types.

Exposure to ultraviolet light may be accomplished with a lamp as shown in figure 30.3 or with a UV box that has built-in ultraviolet lamps. The UV exposure effectiveness varies with the type of setup used. The exposure times given in table 30.1 work well for a specific type of mercury arc lamp. Note in the table that space is provided under the times for adding in different timing. Your instructor will inform you as to whether you should write in new times that will be more suited to the equipment in your lab. Proceed as follows to do this experiment.

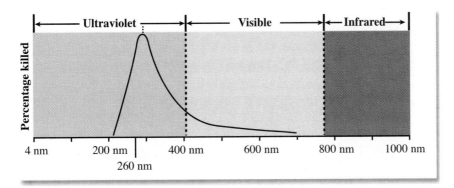

FIGURE 30.1 Lethal effectiveness of ultraviolet light

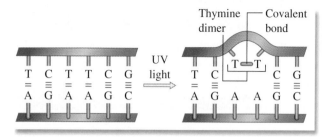

FIGURE 30.2 Formation of thymine dimers by UV light

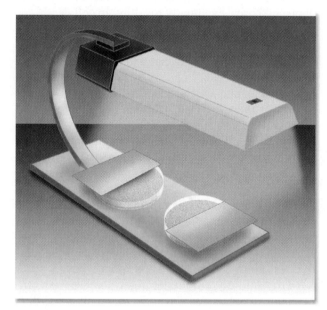

FIGURE 30.3 Plates are exposed to UV light with 50% coverage

MATERIALS

- petri plates of nutrient agar (one or more per student)
- ultraviolet lamp or UV exposure box
- timers (bell type)
- cards (3″ × 5″)
- nutrient broth cultures of *S. aureus* with swabs
- saline suspensions of *B. megaterium* with swabs

1. Refer to table 30.1 to determine which organism you will work with. You may be assigned more than one plate to inoculate. If different times are to be used, your instructor will inform you what times to write in. Since there are only 16 assignment numbers in the table, more student assignment numbers can be written in as designated by your instructor.
2. Label the bottoms of the plates with your assignment number and your initials.
3. Using a cotton-tipped swab that is in the culture tube, swab the entire surface of the agar in each plate. Before swabbing, express the excess culture from the swab against the inner wall of the tube.
4. Put on the protective goggles provided. Place the plates under the ultraviolet lamp *with the lids removed.* Cover one-half of each plate with a 3″ × 5″ card as shown in figure 30.3. Note that if your number is 8 or 16, you will not remove the lid from your plate. The purpose of this exposure is to determine to what extent plastic protects cells from the effects of UV light.

> **CAUTION** Before exposing the plates to UV light, put on protective goggles. Avoid looking directly into the UV light source. These rays can cause cataracts and eye injury.

5. After exposing the plates for the correct time durations, re-cover them with their lids, and incubate them inverted at 37° C for 48 hours.

LABORATORY REPORT

Record your observations in Laboratory Report 30 and answer all the questions.

TABLE 30.1 Student Inoculation Assignments

	Exposure Times (Student Assignments)							
S. aureus	1	2	3	4	5	6	7	8
	10 sec	20 sec	40 sec	80 sec	2.5 min	5 min	10 min	20 min*
B. megaterium	9	10	11	12	13	14	15	16
	1 min	2 min	4 min	8 min	15 min	30 min	60 min	60 min*

*These petri plates will be covered with dish covers during exposure.

LABORATORY REPORT

Student: _____

Date: _____ Section: _____

EXERCISE 30 Ultraviolet Light: Lethal Effects

A. Tabulation of Results

Your instructor will construct a table similar to the one below on the chalkboard for you to record your results. If substantial growth is present in the exposed area, record your results as + + + . If three or fewer colonies survived, record + . Moderate survival should be indicated as + + . No growth should be recorded as − . Record all information in the table.

ORGANISMS	EXPOSURE TIMES							
S. aureus	10 sec	20 sec	40 sec	80 sec	2.5 min	5 min	10 min	• 20 min
Survival								
B. megaterium	1 min	2 min	4 min	8 min	15 min	30 min	60 min	• 6 min
Survival								

* plates covered during exposure

B. Questions

1. What length of time is required for the destruction of non-spore-forming bacteria such as *Staphylococcus aureus?*

   _____

2. Can you express, quantitatively, how much more resistant *B. megaterium* spores are to ultraviolet light than *S. aureus* vegetative cells (i.e., how many *times* more resistant are they)?

   _____

   _____

   _____

3. Why is it desirable to remove the cover from the petri dish when making exposures?

   _____

4. In what specific way does ultraviolet light kill microorganisms?

   _____

   _____

5. What adverse effect can result from overexposure of human tissues to ultraviolet light?

   _____

   _____

6. How do organisms deal with the effects of UV light?

   _____

   _____

7. What wavelength of ultraviolet is most germicidal and why? _____

8. List several practical applications of ultraviolet light to microbial control.

_____

_____

_____

_____

_____

_____

_____

_____

Oligodynamic Action

The ability of small amounts of heavy metals to exert a lethal effect on bacteria is designated as oligodynamic action (Greek: *oligos,* small: *dynamis,* power). The effectiveness of these small amounts of metal is probably due to the high affinity of cellular proteins for the metallic ions. Although the concentration of ions in solution may be miniscule (a few parts per million), cells die due to the cumulative effects of ions within the cell.

The success of silver amalgam fillings to prevent secondary dental decay in teeth over long periods of time is due to the small amounts of silver and mercury ions that diffuse into adjacent tooth dentin. Its success in this respect has led to much debated concern that its toxicity may cause long-term injury to patients. In addition to its value (or harm) as a dental restoration material, oligodynamic action of certain other heavy metals has been applied to water purification, ointment manufacture, and the treatment of bandages and fabrics.

In this exercise, we will compare the oligodynamic action of three metals (copper, silver, and aluminum) to note the differences.

MATERIALS

- 1 petri plate
- 1 nutrient agar pour
- forceps and Bunsen burner
- acid-alcohol
- broth culture of *E. coli* and *S. aureus*
- 3 metallic disks (copper, silver, aluminum)
- water bath at student station (beaker of water and electric hot plate)

1. Liquefy a tube of nutrient agar, cool to 50° C, and inoculate with either *E. coli* or *S. aureus* (odd: *E. coli;* even: *S. aureus*).
2. Pour half of the medium from each tube into a sterile petri plate and leave the other half in a water bath (50° C). Allow agar to solidify in the plate.
3. Clean three metallic disks, one at a time, and place them on the agar, evenly spaced, as soon as they are cleaned. Use this routine:
   - Wash first with soap and water; then rinse with water.
   - With flamed forceps dip in acid-alcohol and rinse with distilled water.
4. Pour the remaining seeded agar from the tube over the metal disks. Incubate for 48 hours at 37° C.

LABORATORY REPORT

After incubation, compare the zones of inhibition and record your results on the last portion of Laboratory Report 31.

LABORATORY REPORT

Student: _____

Date: _____   Section: _____

EXERCISE 31   Oligodynamic Action

A. Tabulation of Results

Measure the zone of inhibition from the edge of each piece of metal with a millimeter ruler. Record the measurements in the table. Spaces are provided for write-in of additional metals.

METAL	MILLIMETERS OF INHIBITION	
	E. coli	*S. aureus*
Copper		
Silver		
Aluminum		

B. Questions

1. Which metal seems to exhibit the greatest amount of oligodynamic action?

   _____

2. Which metal or metals seem to be ineffective? _____

3. Do these two organisms seem to differ in their susceptibility to oligodynamic action?

   _____

   Explain: _____

   _____

   _____

4. What specific chemical substances in bacterial cells are inactivated by heavy metals, affecting growth?

   _____

   _____

As a skin disinfectant, 70% alcohol is a widely used agent. The ubiquitous prepackaged alcohol swabs used by nurses and technicians are evidence that these items are indispensible. The question that often arises is: How really effective is alcohol in routine use? When the skin is swabbed prior to penetration, are all, or mostly all, of the surface bacteria killed? To determine alcohol effectiveness, as it might be used in routine skin disinfection, we are going to perform a very simple experiment that utilizes four thumbprints and a plate of enriched agar. Class results will be pooled to arrive at a statistical analysis.

Figure 32.1 illustrates the various steps in this test. Note that the petri plate is divided into four parts. On the left side of the plate an unwashed left thumb is first pressed down on the agar in the lower quadrant of the plate. Next, the left thumb is pressed down on the upper left quadrant. With the left thumb, we are trying to establish the percentage of bacteria that are removed by simple contact with the agar.

On the right side of the plate, an unwashed right thumb is pressed down on the lower right quadrant of the plate. The next step is to either dip the right thumb into alcohol or to scrub it with an alcohol swab and dry it. Half of the class will use the dipping method and the other half will use alcohol swabs. Your instructor will indicate what your assignment will be.

The last step is to press the dried right thumb on the upper right quadrant of the plate.

After the plate is inoculated, it is incubated at 37° C for 24–48 hours. Colony counts will establish the effectiveness of the alcohol.

MATERIALS
- 1 petri plate of veal infusion agar
- small beaker
- 70% ethanol
- alcohol swab

1. Perform this experiment with unwashed hands.
2. With a marking pencil, mark the bottom of the petri plate with two perpendicular lines that divide it into four quadrants. Label the left quadrants **A** and **B** and the right quadrants **C** and **D** as shown in figure 32.1. (*Keep in mind that when you turn the plates over to label them, the A and B quadrants will be on the right and C and D will be on the left.*)
3. Press the pad of your left thumb against the agar surface in the A quadrant.
4. Without touching any other surface, press the left thumb into the B quadrant.
5. Press the pad of your right thumb against the agar surface of the C quadrant.

FIGURE 32.1 Procedure for testing the effectiveness of alcohol on the skin

Without touching any other surface the left thumb is pressed against the agar in quadrant B.

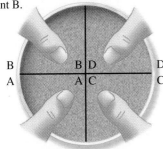

The pad of the treated right thumb is pressed against the agar in the D quadrant.

5
The alcohol-treated right thumb is allowed to completely air-dry.

The pad of the right thumb is immersed in 70% alcohol or scrubbed with an alcohol swab for 10 seconds.

1
The pad of the unwashed left thumb is momentarily pressed against the agar in quadrant A.

3
The pad of the unwashed right thumb is momentarily pressed against the agar in quadrant C.

6. Disinfect the right thumb by one of the two following methods:
   - dip the thumb into a beaker of 70% ethanol for 5 seconds, or
   - scrub the entire pad surface of the right thumb with an alcohol swab.
7. Allow the alcohol to completely evaporate from the skin.

8. Press the right thumb against the agar in the D quadrant.
9. Incubate the plate at 37° C for 24–48 hours.
10. Follow the instructions in Laboratory Report 32 for evaluating the plate and answer all of the questions.

LABORATORY REPORT

Student: _____

Date: _____   Section: _____

EXERCISE 32   Evaluation of Alcohol: Its Effectiveness as an Antiseptic

A. Tabulation of Results

Count the number of colonies that appear on each of the thumbprints and record them in the following table. If the number of colonies has increased in the second press, record a 0 in percent reduction. Calculate the percentages of reduction and record these data in the appropriate column. Use this formula:

$$\text{Percent reduction} = \frac{(\text{Colony count 1st press}) - (\text{Colony count 2nd press})}{(\text{Colony count 1st press})} \times 100$$

LEFT THUMB (Control)			RIGHT THUMB (Dipped)			RIGHT THUMB (Swabbed)		
Colony Count 1st Press	Colony Count 2nd Press	Percent Reduction	Colony Count 1st Press	Colony Count 2nd Press	Percent Reduction	Colony Count 1st Press	Colony Count 2nd Press	Percent Reduction
Av. % Reduction, Left (C)			Av. % Reduction, Right (D)			Av. % Reduction, Right (S)		

B. Questions

1. In general, what effect does alcohol have on the level of skin contaminants? _____

   _____

2. Is there any difference between the effects of dipping versus swabbing? _____

   Which method appears to be more effective? _____

3. There is definitely survival of some microorganisms even after alcohol treatment. Without staining or microscopic scrutiny, predict what types of microbes are growing on the medium where you made the right

   thumb impression after treatment. _____

   _____

   _____

   _____

   _____

4. Why is it impossible to completely sterilize the human skin? _____

   _____

Once the causative organism of a specific disease in a patient has been isolated, it is up to the attending physician to administer a chemotherapeutic agent that will inhibit or kill the pathogen without causing serious harm to the individual. The method must be relatively simple to use, be very reliable, and yield results in as short a time as possible. The Kirby-Bauer method of sensitivity testing is such a method. It is used for testing both antibiotics and drugs. **Antibiotics** are chemotherapeutic agents of low molecular weight produced by microorganisms that inhibit or kill other microorganisms. **Drugs,** on the other hand, are antimicrobic agents that are man-made. Both types of agents will be tested in this laboratory session according to the procedure shown in figure 33.1.

The effectiveness of an antimicrobic in sensitivity testing is based on the size of the zone of inhibition that surrounds a disk that has been impregnated with a specific concentration of the agent. The zone of inhibition, however, varies with the diffusibility of the agent, the size of the inoculum, the type of medium, and many other factors. Only by taking all these variables into consideration can a reliable method be worked out.

The **Kirby-Bauer method** is a standardized system that takes all variables into consideration. It is sanctioned by the U.S. FDA and the Subcommittee on Antimicrobial Susceptibility Testing of the National Committee for Clinical Laboratory Standards. Although time is insufficient here to consider all facets of this test, its basic procedure will be followed.

The recommended medium in this test is Mueller-Hinton II agar. Its pH should be between 7.2 and 7.4, and it should be poured to a uniform thickness of 4 mm in the petri plate. This requires 60 ml in a 150 mm plate and 25 ml in a 100 mm plate. For certain fastidious microorganisms, 5% defibrinated sheep blood is added to the medium.

Inoculation of the surface of the medium is made with a cotton swab from a broth culture. In clinical applications, the broth turbidity has to match a defined standard. Care must also be taken to express excess broth from the swab prior to inoculation.

High-potency disks are used that may be placed on the agar with a mechanical dispenser or sterile forceps. To secure the disks to the medium, it is necessary to press them down onto the agar.

After 16 to 18 hours incubation, the plates are examined and the diameters of the zones are measured to the nearest millimeter. To determine the significance of the zone diameters, one must consult a table (Appendix A).

In this exercise, we will work with four microorganisms: *Staphylococcus aureus, Escherichia coli, Proteus vulgaris,* and *Pseudomonas aeruginosa.* Each student will inoculate one plate with one of the four organisms and place the disks on the medium by whichever method is available. Since each student will be doing only a portion of the total experiment, student assignments will be made. Proceed as follows:

FIRST PERIOD
(Plate Preparation)

MATERIALS

- 1 petri plate of Mueller-Hinton II agar nutrient broth cultures (with swabs) of *S. aureus, E. coli, P. vulgaris,* and *P. aeruginosa*
- disk dispenser (BBL or Difco)
- cartridges of disks (BBL or Difco)
- forceps and Bunsen burner
- zone interpretation charts (Difco or BBL)

1. Select the organisms you are going to work with from the following table.

Organism	Student Number
S. aureus	1, 5, 9, 13, 17, 21, 25
E. coli	2, 6, 10, 14, 18, 22, 26
P. vulgaris	3, 7, 11, 15, 19, 23, 27
P. aeruginosa	4, 8, 12, 16, 20, 24, 28

2. Label your plate with the name of your organism.
3. Inoculate the surface of the medium with the swab after expressing excess fluid from the swab by pressing and rotating the swab against the inside walls of the tube above the fluid level. Cover the surface of the agar evenly by swabbing in three directions. A final sweep should be made of the agar rim with the swab.

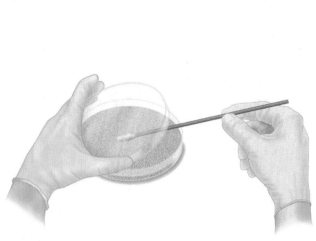

(1) The entire surface of a plate of nutrient medium is swabbed with organism to be tested.

(2) Handle of dispenser is pushed down to place 12 disks on the medium. In addition to dispensing disks, this dispenser also tamps disks onto medium.

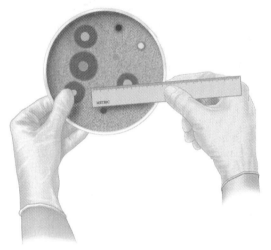

(3) Cartridges (Difco) can be used to dispense individual disks. Only 4 or 5 disks should be placed on small (100 mm) plates.

(4) After 18 hours incubation, the zones of inhibition (diameters) are measured in millimeters. Significance of zones is determined from Kirby-Bauer chart.

FIGURE 33.1 Antimicrobic sensitivity testing

4. Allow **3 to 5 minutes** for the agar surface to dry before applying disks.
5. Dispense disks as follows:
   a. If an automatic dispenser is used, remove the lid from the plate, place the dispenser over the plate, and push down firmly on the plunger. With the sterile tip of forceps, tap each disk lightly to secure it to medium.
   b. If forceps are used, sterilize them first by flaming before picking up the disks. Keep each disk at least 15 mm from the edge of the plate. Place no more than 13 on a 150 mm plate, no more than 5 on a 100 mm plate. Apply light pressure to each disk on the agar with the tip of a sterile forceps or inoculating loop to secure it to medium.
6. Invert and incubate the plate for 16 to 18 hours at 37°C.

SECOND PERIOD
(Interpretation)

After incubation, measure the zone diameters with a metric ruler to the nearest whole millimeter. The zone of complete inhibition is determined without magnification. Ignore faint growth or tiny colonies that can be detected by very close scrutiny. Large colonies growing within the clear zone might represent resistant variants or a mixed inoculum and may require reidentification and retesting in clinical situations. Ignore the "swarming" characteristics of *Proteus,* measuring only to the margin of heavy growth.

TABLE 33.1   Antibiotic Susceptibility Test Discs (courtesy Becton Dickenson)

Antibiotic Agent	Concentration	Individual / 10 Pack
Amikacin	30 µg	231596 / 231597
Amoxicillin / Clavulanic Acig	30 µg	231628 / 23629
Ampicillin	10 µg	230705 / 231264
Ampicillin / Subactam	10/10 µg	231659 / 231660
Azlocillin	75 µg	231624 / 231625
Bacitracin	2 units	230717 / 231267
Carbenicillin	100 µg	231235 / 231555
Cefaclor	30 µg	231652 / 231653
Cefazolin	30 µg	231592 / 231593
Cefixime	5 µg	231663 / NA
Cefoperazone	75 µg	231612 / 231613
Cefotaxime	30 µg	231606 / 231607
Cefotetan	30 µg	231655 / 231656
Cefoxitin	30 µg	231590 / 231591
Ceftazidime	30 µg	231632 / 231633
Ceftriaxone	30 µg	231634 / 231635
Cefuroxime	30 µg	231620 / 231621
Cephalothin	30 µg	230725 / 231271
Chloramphenical	30 µg	230733 / 231274
Clindamycin	2 µg	231213 / 231275
Doxycycline	30 µg	230777 / 231286
Erythromycin	15 µg	230793 / 231290
Gentimicin	10 µg	231227 / 231299
Imipenem	10 µg	231644 / 231645
Kanamycin	30 µg	230825 / 230829
Mezlocillin	75 µg	231614 / 231615
Minocycline	30 µg	231250 / 231251

(continued)

TABLE 33.1 (Antibiotic Susceptibility Test Discs continued)

Antibiotic Agent	Concentration	Individual / 10 Pack
Moxalactam	30 μg	231610 / 231611
Nafcillin	1 μg	230866 / 231309
Nalidixic Acid	30 μg	230870 / 230874
Netilimicin	30 μg	231602 / 231603
Nitrofurantoin	100 μg	230801 / 231292
Penicillin	2 units	230914 / 231320
Piperacillin	100 μg	231608 / 231609
Rifampin	5 μg	231541 / 231544
Streptomycin	10 μg	230942 / 231328
Sulfisoxazole	0.25 mg	230813 / 231296
Tetracycline	5 μg	230994 / 231343
Ticarcillin	75 μg	231618 / 231619
Tobramycin	10 μg	231568 / 213569
Trimethoprim	5 μg	231600 / 231601
Vancomycin	30 μg	231034 / 231353

Record the zone measurements on the table of Laboratory Report 33 and on the chart on the demonstration table, which has been provided by the instructor.

Use table 33.1 for identifying the various disks.

To determine which antibiotics your organism is sensitive to (S), or resistant to (R), or intermediate (I), consult table VII in Appendix A. It is important to note that the significance of a zone of inhibition varies with the type of organism. If you cannot find your antibiotic on the chart, consult a chart that is supplied by BBL or Difco that is on the demonstration table or bulletin board. Table VII is incomplete.

LABORATORY REPORT

Student: _____

Date: _____ Section: _____

EXERCISE 33   Antimicrobic Sensitivity Testing: The Kirby-Bauer Method

A. Tabulation

List the antimicrobics that were used for each organism. Consult table 33.1 to identify the various disks. After measuring and recording the zone diameters, consult table VII in Appendix A for interpretation. Record the degrees of sensitivity (R, I, or S) in the sensitivity column. Exchange data with other class members to complete the entire chart.

	ANTIMICROBIC	ZONE DIA.	RATING (R, I, S)	ANTIMICROBIC	ZONE DIA.	RATING (R, I, S)
S. aureus						
P. aeruginosa						
Proteus vulgaris						
E. coli						

B. Questions

1. Which antimicrobics would be suitable for the control of the following organisms?

   S. aureus: _____

   E. coli: _____

   P. vulgaris: _____

   P. aeruginosa: _____

2. Differentiate between the following:

   Narrow spectrum antibiotic: _____

   _____

   Broad spectrum antibiotic: _____

   _____

3. Which antimicrobics used in this experiment would qualify as being excellent broad-spectrum antimicrobics?

   _____

   _____

4. Differentiate between the following:

   Antibiotic: _____

   _____

   Antimicrobic: _____

   _____

5. How can drug resistance in microorganisms be circumvented?

   _____

   _____

   _____

6. Why might there be a differential effect of an antibiotic on a gram-positive verses a gram-negative bacterium?

   _____

Evaluation of Antiseptics:
The Filter Paper Disk Method

The term *antiseptic* has, unfortunately, been somewhat ill-defined. Originally, the term was applied to any agent that prevents sepsis, or putrefaction. Since sepsis is caused by growing microorganisms, it follows that an antiseptic inhibits microbial multiplication without necessarily killing them. By this definition, we can assume that antiseptics are essentially bacteriostatic agents. Part of the confusion that has resulted in its definition is that the United States Food and Drug Administration rates antiseptics essentially the same as disinfectants. Only when an agent is to be used in contact with the body for a long period of time do they rate its bacteriostatic properties instead of its bactericidal properties.

If we are to compare antiseptics on the basis of their bacteriostatic properties, the filter paper disk method (figure 34.1) is a simple, satisfactory method to use. In this method, a disk of filter paper (1/2″ diameter) is impregnated with the chemical agent and placed on a seeded nutrient agar plate. The plate is incubated for 48 hours. If the substance is inhibitory, a clear zone of inhibition will surround the disk. The size of this zone is an expression of the agent's effectiveness and can be compared quantitatively against other substances.

In this exercise, we will measure the relative effectiveness of three agents (phenol, formaldehyde, and iodine) against two organisms: *Staphylococcus aureus* (gram-positive) and *Pseudomonas aeruginosa* (gram-negative). Table 34.1 will be used to assign each student one chemical agent to be tested against one organism. Note that space has been provided in the table for different agents to be written in as substitutes for the three agents listed. Your instructor may wish to make substitutions. Proceed as follows:

FIRST PERIOD
(Disk Application)

MATERIALS

per student:
- 1 nutrient agar pour and 1 petri plate
- broth culture of *S. aureus* or *P. aeruginosa* on demonstration table
- petri dish containing sterile disks of filter paper (1/2″ dia)
- forceps and Bunsen burner
- chemical agents in small beakers (5% phenol, 5% formaldehyde, 5% aqueous iodine)

1. Consult table 34.1 to determine your assignment.
2. Liquefy a nutrient agar pour in a water bath and cool to 50° C.
3. Label the bottom of a petri plate with the names of the organism and chemical agent.

TABLE 34.1 Student Assignments

Chemical Agent		Student Number	
	Substitution	*S. aureus*	*P. aeruginosa*
5% Phenol		1, 7, 13, 19, 25	2, 8, 14, 20, 26
5% Formaldehyde		3, 9, 15, 21, 27	4, 10, 16, 22, 28
5% Iodine		5, 11, 17, 23, 29	6, 12, 18, 24, 30

4. Inoculate the agar pour with one loopful of the organism and pour into the plate.
5. After the medium has solidified in the plate, pick up a sterile disk with *lightly flamed* forceps, dip the disk *halfway* into a beaker of the chemical agent, and place the disk in the center of the medium.

    To secure the disk to the medium, press lightly on it with the forceps.
6. Incubate the plate at 37° C for 48 hours.

SECOND PERIOD

(Evaluation)

1. Measure the zone of inhibition from the edge of the disk to the edge of the growth (see illustration 5, figure 34.1).
2. Exchange plates with other members of the class so that you will have an opportunity to complete the table in Laboratory Report 34.

(1) Liquefied nutrient agar is inoculated with one loopful of organisms.

(2) Seeded nutrient agar is poured into plate and allowed to solidify.

(3) Sterile disk is dipped halfway into agent. If completely submerged it will be too wet.

(4) Impregnated disk is placed in center of nutrient agar and pressed down lightly to secure it.

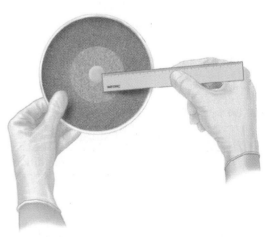

(5) After 24–48 hours incubation the zone of inhibition is measured on bottom of plate. Note that measurement is between disk edge and growth.

FIGURE 34.1 Filter paper disk method of evaluating an antiseptic

LABORATORY REPORT

Student: _____

Date: _____ Section: _____

EXERCISE 34   Evaluation of Antiseptics: The Filter Paper Disk Method

A. Tabulation of Results

With a millimeter scale, measure the zones of inhibition between the edge of the filter paper disk and the organisms. Record this information. Exchange your plates with other students' plates to complete the measurements for all chemical agents.

DISINFECTANT	MILLIMETERS OF INHIBITION	
	Staphylococcus aureus	*Pseudomonas aeruginosa*
5% phenol		
5% formaldehyde		
5% iodine		

B. Questions

1. What conclusions can be derived from these results? _____

    _____

    _____

    _____

    _____

    _____

    _____

    _____

2. What factors influence the size of the zone of inhibition? _____

    _____

    _____

    _____

    _____

Effectiveness of Hand Scrubbing

The importance of hand washing in preventing the spread of disease is accredited to the observations of Ignaz Semmelweis at the Lying-In Hospital in Vienna in 1846 and 1847. Semmelweis was the head of obstetrics and he noted that the number of cases of childbirth fever (puerperal sepsis) was primarily the result of the lack of sanitary practices. He observed that medical students and physicians would go directly from dissection and autopsy rooms to a patient's bedside and assist in deliveries without washing their hands. In these wards, the death rate from childbirth fever was very high, approaching 20% in some cases. In contrast, some women were attended to by midwives and nurses who were not allowed in autopsy rooms and who were more sanitary in handling patients. When women were assisted in deliveries by midwives and nurses, the death rate was considerably less. As a result of his observations, Semmelweis instituted a policy whereby physicians and medical students had to disinfect their hands with a solution of chloride of lime (bleach) prior to examining obstetric patients or assisting in deliveries. His efforts resulted in a significant decrease of puerperal sepsis, down to about 1% for women treated by physicians and medical students. But this success was short lived, as complaints by doctors and medical students to the director of the hospital forced Semmelweis to abandon this practice, and death rates for childbirth fever once again increased. Despondent, Semmelweis resigned his post and returned to his native Hungary.

Today, it is routine practice to wash one's hands prior to the examination of any patient and to do a complete surgical scrub prior to surgery. Scrubbing is the most important way to prevent infections in a hospital and physicians' offices, and is an effective way to remove some opportunistic pathogens that occur on the skin, such as *Staphylococcus aureus.*

The human skin is inhabited by a diverse group of microorganisms. They protect us from invasion by pathogens and hence contribute to our overall health. The normal flora that dwells on the skin can be placed into three main groups:

Diphtheroids: these are gram-positive bacteria that are similar to *Corynebacterium diphtheriae* owing to their variable morphology. However, they are nonpathogenic.

An example is *Propionibacterium acnes,* an anaerobic diphtheroid that lives in hair follicles and breaks down sebum, the oily secretion of the follicle that prevents the skin from drying out.

Staphylococci: a second major group of bacteria on the skin is the staphylococci. An example is *Staphylococcus epidermiditis,* a nonpathogenic, coagulase-negative staphylococcus. These organisms inhibit pathogens from establishing a presence on the skin because they effectively compete for nutrients on the skin and produce inhibitory substances. Some individuals do harbor *Staphylococcus aureus* on the skin, and this bacterium can be an opportunistic pathogen.

Yeasts and **Fungi:** various yeasts and fungi are also found on the skin. They primarily degrade lipid secretions of the secretory glands of the skin, and they are nonpathogenic.

In addition to the normal flora, there are transient bacteria that can occur on the skin. Organisms such as endospore formers may be cultured by swabbing the skin because their endospores are present. However, they are not part of the persistent population that inhabits the skin. Other bacteria may be present because the skin has become temporarily contaminated with them. Most are easily removed by washing because they are contaminants on the skin and antiseptic soaps are effective in killing them. The normal flora is much more difficult to remove by washing because these organisms reside in hair follicles and are entrenched in the skin, making them very difficult to remove or kill.

In this exercise, the class will evaluate the effectiveness of length of time in removal of organisms from the hands using a surgical scrub technique. One member of the class will be selected to perform the scrub. Another student will assist by supplying the soap, brushes, and basins, as needed. During the scrub, at 2-minute intervals, the hands will be scrubbed into a basin of sterile water. Bacterial counts will be made of these basins to determine the effectiveness of the previous 2-minute scrub in reducing the bacterial flora of the hands. Members of the class not involved in the scrub procedure will make the inoculations from the basins for the plate counts.

SCRUB PROCEDURE

The two members of the class who are chosen to perform the surgical scrub will set up their materials near a sink for convenience. As one student performs the scrub, the other will assist in reading the instructions and providing materials as needed. The basic steps, which are illustrated in figure 35.1, are also described in detail below. Before beginning the scrub, both students should read all the steps carefully.

MATERIALS

- 5 sterile surgical scrub brushes, individually wrapped
- 5 basins (or 2000 ml beakers), containing 1000 ml each of sterile water. These basins should be covered to prevent contamination.
- 1 dispenser of green soap
- 1 tube of hand lotion

(1) Sixty-second hand scrub into Basin A. No soap.

(2) Two-minute soap scrub with running water.

(3) Sixty-second hand scrub into Basin B. No soap.

(4) Same as 2.

(5) Sixty-second hand scrub into Basin C. No soap.

(6) Same as 2.

(7) Sixty-second hand scrub into Basin D. No soap.

(8) Same as 2.

(9) Sixty-second hand scrub into Basin E. No soap.

FIGURE 35.1 Hand scrubbing routine

Step 1 To get some idea of the number of transient organisms on the hands, the scrubber will scrub all surfaces of each hand with a sterile surgical scrub brush for 30 seconds into basin A. No green soap will be used for this step. The successful performance of this step will depend on

- spending the same amount of time on each hand (30 seconds),
- maintaining the same amount of activity on each hand, and
- scrubbing under the fingernails, as well as working over their surfaces.

After completion of this 60-second scrub, notify Group A that their basin is ready for the inoculations.

Step 2 Using the *same* brush as above, begin scrubbing with green soap for 2 minutes, using cool tap water to moisten and rinse the hands. One minute is devoted to each hand.

The assistant will make one application of green soap to each hand as it is being scrubbed.

Rinse both hands for 5 seconds under tap water at the completion of the scrub.

Discard the brush.

Note: This same procedure will be followed exactly in steps 4, 6, and 8 of figure 35.1.

Step 3 With a *fresh* sterile brush, scrub the hands into basin B in a manner that is identical to step 1. Don't use soap. Notify Group B when this basin is ready.

Note: Exactly the same procedure is used in steps 5, 7, and 9 of figure 35.1, using Basins C, D, and E.

Remember: It is important to use a fresh sterile brush for the preparation of each of these basins.

After Scrubbing After all scrubbing has been completed, the scrubber should dry his or her hands and apply hand lotion.

MAKING THE POUR PLATES

While the scrub is being performed, the rest of the class will be divided into five groups (A, B, C, D, and E) by the instructor. Each group will make six plate inoculations from one of the five basins (A, B, C, D, or E). It is the function of these groups to determine the bacterial count per milliliter in each basin. In this way, we hope to determine, in a relative way, the effectiveness of scrubbing in bringing down the total bacterial count of the skin.

MATERIALS

- 30 veal infusion agar pours—6 per group
- 1 ml pipettes
- 30 sterile petri plates—6 per group
- 70% alcohol
- L-shaped glass stirring rod (optional)

1. Liquefy six pours of veal infusion agar and cool to 50° C. While the medium is being liquefied, label two plates each: 0.1 ml, 0.2 ml, and 0.4 ml. Also, indicate your group designation on the plate.
2. As soon as the scrubber has prepared your basin, take it to your table and make your inoculations as follows:
   a. Stir the water in the basin with a pipette or an L-shaped stirring rod for 15 seconds. If the stirring rod is used (figure 35.2), sterilize it before using by immersing it in 70% alcohol and flaming. *For consistency of results all groups should use the same method of stirring.*

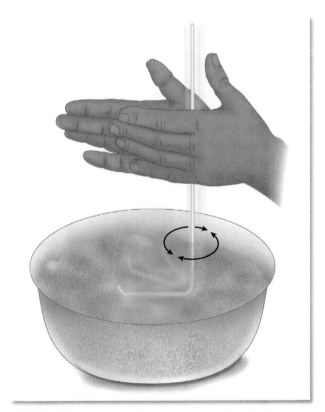

FIGURE 35.2 An alternative method of stirring utilizes an L-shaped glass stirring rod

b. Deliver the proper amounts of water from the basin to the six petri plates with a sterile serological pipette. Refer to figure 35.3. If a pipette was used for stirring, it may be used for the deliveries.

c. Pour a tube of veal infusion agar, cooled to 50° C, into each plate, rotate to get good distribution of organisms, and allow to cool.

d. Incubate the plates at 37° C for 24 hours.

3. After the plates have been incubated, select the pair that has the best colony distribution with no fewer than 30 or more than 300 colonies. Count the colonies on the two plates and record your counts on the chart on the chalkboard.

4. After all data are on the chalkboard, complete the table and graph on Laboratory Report 35.

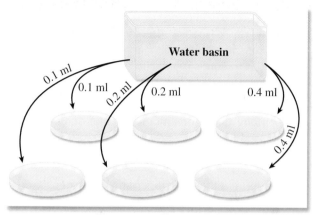

FIGURE 35.3 Scrub water for count is distributed to six petri plates in amounts as shown

LABORATORY REPORT

Student: _____

Date: _____ Section: _____

EXERCISE 35 Effectiveness of Hand Scrubbing

A. Tabulation of Results

The instructor will draw a table on the chalkboard similar to the one below. Examine the six plates that your group inoculated from the basin of water. Select the two plates of a specific dilution that have approximately 30 to 300 colonies and count all of the colonies of each plate with a Quebec colony counter. Record the counts for each plate and their averages on the chalkboard. Once all the groups have recorded their counts, record the dilution factors for each group in the proper column. To calculate the organisms per milliliter, multiply the average count by the dilution factor.

GROUP	0.1 ml COUNT		0.2 ml COUNT		0.4 ml COUNT		DILUTION FACTOR*	ORGANISMS PER MILLILITER
	Per Plate	Average	Per Plate	Average	Per Plate	Average		
A								
B								
C								
D								
E								

*Dilution factors: 0.1 ml = 10; 0.2 ml = 5; 0.4 ml = 2.5

Effectiveness of Hand Scrubbing (continued)

B. Graph

After you have completed this tabulation, plot the number of organisms per milliliter that was present in each basin.

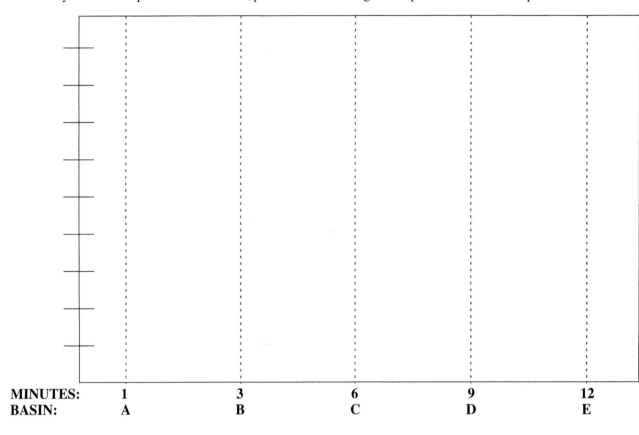

MINUTES:	1	3	6	9	12
BASIN:	A	B	C	D	E

C. Questions

1. What conclusions can be derived from this exercise?

2. What might be an explanation of a higher count in basin D than in B, ruling out contamination or faulty techniques?

3. Why is it so important that surgeons scrub their hands prior to surgery even though they wear rubber gloves?

IDENTIFICATION OF UNKNOWN BACTERIA

One of the most interesting experiences in introductory microbiology is to attempt to identify an unknown microorganism that has been assigned to you as a laboratory problem. The next six exercises pertain to this phase of microbiological work. You will be given one or more cultures of bacteria to identify. The only information that might be given to you about your unknowns will pertain to their sources and habitats. All the information needed for identification will have to be acquired by you through independent study.

Although you will be engrossed in trying to identify an unknown organism, there is a more fundamental underlying objective of this series of exercises that goes far beyond simply identifying an unknown. That objective is to gain an understanding of the cultural and physiological characteristics of bacteria. Physiological characteristics will be determined with a series of biochemical tests that you will perform on the organisms. Although correctly identifying the unknowns that are given to you is very important, it is just as important that you thoroughly understand the chemistry of the tests that you perform on the organisms.

The first step in the identification procedure is to accumulate information that pertains to the organisms' morphological, cultural, and physiological (biochemical) characteristics. This involves making different kinds of slides for cellular studies and the inoculation of various types of media to note the growth characteristics and types of enzymes produced. As this information is accumulated, it is recorded in an orderly manner on descriptive charts, which are located in the back of the manual.

After sufficient information has been recorded, the next step is to consult a taxonomic key, which enables one to identify the organism. For this final step, *Bergey's Manual of Systematic Bacteriology* will be used. Copies of volumes 1 and 2 of this book will be available in the laboratory, library, or both.

Success in this endeavor will require meticulous techniques, intelligent interpretation, and careful recordkeeping. Your mastery of aseptic methods in the handling of cultures and the performance of inoculations will show up clearly in your results. Contamination of your cultures with unwanted organisms will yield false results, making identification hazardous speculation. If you have reason to doubt the validity of the results of a specific test, repeat it; *don't rely on chance!* As soon as you have made an observation or completed a test, record the information on the descriptive chart. Do not trust your memory—record data immediately.

217

Morphological Study of Unknown Bacterium

Malam 3 → Uet 2 uic 1

The first step in the study of your unknown bacterium is to set up stock cultures that will be used in the subsequent exercises. A reserve stock culture will not be used for making slides or inoculating tests. It will be stored in the refrigerator in case your working stock becomes contaminated and you need to make a fresh working stock. A working stock will be used to inoculate the various tests that you will perform to identify your unknown bacterium. It is crucial that you practice good aseptic technique when inoculating from your working stock in order to avoid contaminating the culture. If it becomes contaminated or loses viability, you can prepare a fresh culture from the reserve stock culture that you have maintained in the refrigerator.

Identifying your unknown will be a kind of "microbiological adventure" that will test the skills and knowledge that you have acquired thus far. You will gather a great deal of information regarding your unknown by performing staining reactions and numerous metabolic tests. The Gram stain will play a very critical role in the process because it will eliminate thousands of possible organisms. The results of these tests will be compared to flow charts provided in this manual and to information in *Bergey's Manual*. From your "detective" work, you will be able to ascertain the identity of the unknown that you were given. To set up the stock cultures, proceed in the following way (see figure 36.1).

STOCK CULTURES

You will receive a broth culture or an agar slant of your unknown bacterium. From this culture you will prepare your working stock and your reserve stock cultures. From the working stock, you will be able to determine such things as cell morphology, the Gram reaction of the unknown, and, in some cases, whether the culture forms any pigment. You can also determine other morphological characteristics such as the presence of a glycocalyx, endospores, or cytoplasmic granules.

MATERIALS

FIRST PERIOD

- nutrient agar or tryptone agar slants

1. Label the agar slants with the code number of the unknown, your name, lab section, and date.
2. Inoculate both slants with your unknown organism. Begin your streak at the bottom of the slant and move the inoculating loop toward the top of the slant in a straight motion. Remember to practice good aseptic technique.
3. Place the respective tubes in the appropriate baskets labeled with the two incubation temperatures, 20°C and 37°C (figure 36.2). Incubate the slants for 18–24 hours.

SECOND PERIOD

1. Examine the slants. Look for growth. Some organisms produce sparse growth and you must examine the cultures closely to determine if growth is present. Is either culture producing a pigment and, if so, is the pigment associated with the cells or has it diffused into the agar? Remember, however, that pigment production could require longer incubation times.
2. Determine which incubation temperature produced the best growth. If no growth occurred on either slant, your original culture could be nonviable or more time is needed for growth of the culture to occur. A third possibility is that neither temperature supported growth. Think through the possibilities and decide what course of action you need to take.
3. If growth occurred on the slant, pick the tube with the best growth and designate it as your **reserve stock culture.** Store the reserve stock in the refrigerator. Cultures stored in this manner are viable for several weeks. Do not use the reserve stock to make inoculations of the various media you will employ or to make stains. **Do not store the culture in your desk.**

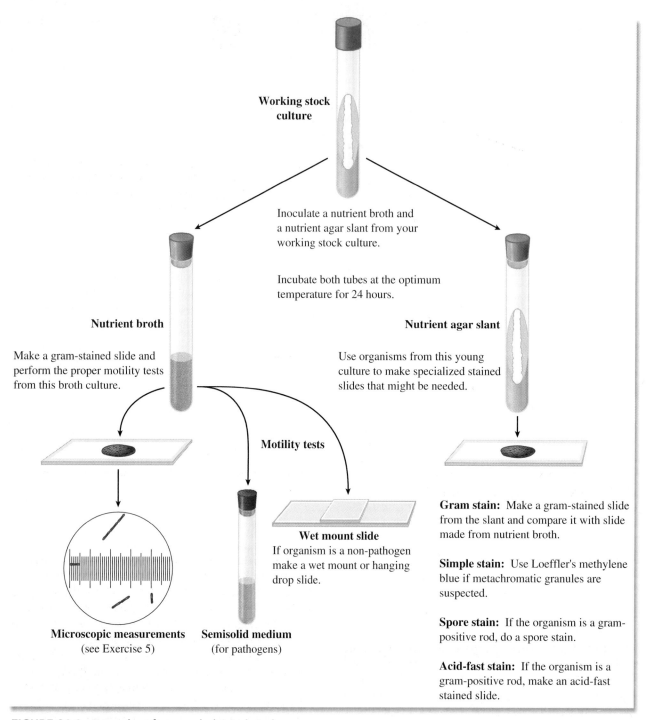

Working stock culture

Inoculate a nutrient broth and a nutrient agar slant from your working stock culture.

Incubate both tubes at the optimum temperature for 24 hours.

Nutrient broth

Make a gram-stained slide and perform the proper motility tests from this broth culture.

Nutrient agar slant

Use organisms from this young culture to make specialized stained slides that might be needed.

Motility tests

Microscopic measurements
(see Exercise 5)

Semisolid medium
(for pathogens)

Wet mount slide
If organism is a non-pathogen make a wet mount or hanging drop slide.

Gram stain: Make a gram-stained slide from the slant and compare it with slide made from nutrient broth.

Simple stain: Use Loeffler's methylene blue if metachromatic granules are suspected.

Spore stain: If the organism is a gram-positive rod, do a spore stain.

Acid-fast stain: If the organism is a gram-positive rod, make an acid-fast stained slide.

FIGURE 36.1 Procedure for morphological study

4. Designate the second culture as your **working stock culture.** This culture will be the source of the inoculum for the various tests and stains that you will perform. If the culture is 18–24 hours old, it can be used to perform the Gram stain. If not, you will have to prepare a fresh slant from the working stock to do the Gram stain.

As soon as morphological information is acquired, be sure to record your observations on the descriptive chart at the back of the manual. Proceed as follows:

MATERIALS

- Gram-staining kit
- spore-staining kit
- acid-fast staining kit
- Loeffler's methylene blue stain

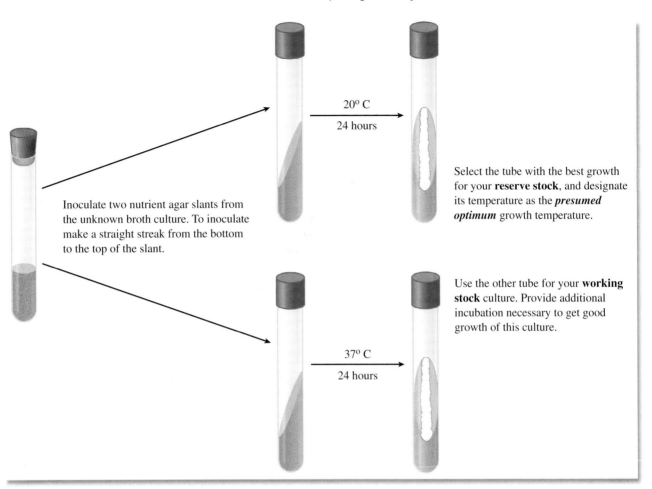

Inoculate two nutrient agar slants from the unknown broth culture. To inoculate make a straight streak from the bottom to the top of the slant.

20° C
24 hours

Select the tube with the best growth for your **reserve stock**, and designate its temperature as the *presumed optimum* growth temperature.

37° C
24 hours

Use the other tube for your **working stock** culture. Provide additional incubation necessary to get good growth of this culture.

FIGURE 36.2 Stock culture procedure

- nigrosine or india ink
- tubes of nutrient broth and nutrient agar
- gummed labels for test tubes

New Inoculations

For all of these staining techniques, you will need 24–48 hour cultures of your unknown. If your working stock slant is a fresh culture, use it. If you don't have a fresh broth culture of your unknown, inoculate a tube of nutrient broth and incubate it at its estimated optimum temperature for 24 hours.

Gram's Stain

Since a good Gram-stained slide will provide you with more valuable information than any other slide, this is the place to start. Make Gram-stained slides from both the broth and agar slants, and compare them under oil immersion.

Two questions must be answered at this time: (1) Is the organism gram-positive, or is it gram-

negative? and (2) Is the organism rod- or coccus-shaped? If your staining technique is correct, you should have no problem with the Gram reaction. If the organism is a long rod, the morphology question is easily settled; however, if your organism is a very short rod, you may incorrectly decide it is coccus-shaped.

Keep in mind that short rods with round ends (coccobacilli) look like cocci. If you have what seems to be a coccobacillus, examine many cells before you make a final decision. Also, keep in mind that *while rod-shaped organisms frequently appear as cocci under certain growth conditions, cocci rarely appear as rods.* (*Streptococcus mutans* is unique in forming rods under certain conditions.) Thus, it is generally safe to assume that if you have a slide on which you see both coccuslike cells and short rods, the organism is probably rod-shaped. This assumption is valid, however, only if you are not working with a contaminated culture!

Record the shape of the organism and its reaction to the stain on the descriptive chart on page 223.

Cell Size

Once you have a good Gram-stained slide, determine the size of the organism with an ocular micrometer. Refer to Exercise 4. If the size is variable, determine the size range. Record this information on the descriptive chart.

Motility and Cellular Arrangement

If your organism is a nonpathogen, make a wet mount or hanging drop slide from the broth culture. Refer to Exercise 17. This will enable you to determine whether the organism is motile, and it will allow you to confirm the cellular arrangement. By making this slide from broth instead of the agar slant, the cells will be well dispersed in natural clumps. Note whether the cells occur singly, in pairs, masses, or chains. *Remember to place the slide preparation in a beaker of disinfectant when finished with it.*

If your organism happens to be a pathogen, do not make a slide preparation of the organisms; instead, stab the organism into a tube of semisolid or SIM medium to determine motility (Exercise 17). Incubate for 48 hours.

Be sure to record your observations on the descriptive chart.

Endospores

If your unknown is a gram-positive rod, check for endospores. *Only rarely is a coccus or gram-negative rod a spore-former.* Examination of your gram-stained slide made from the agar slant should provide a clue, since endospores show up as transparent holes in gram-stained spore-formers. Endospores can also be seen on unstained organisms if studied with phase-contrast optics.

If there seems to be evidence that the organism is a spore-former, make a slide using one of the spore-staining techniques you used in Exercise 15. *Since some spore-formers require at least a week's time of incubation before forming spores, it is prudent to double-check for spores in older cultures.*

Record on the descriptive chart whether the spore is terminal, subterminal, or in the middle of the rod.

Acid-Fast Staining

If your organism is a gram-positive, non-spore-forming rod, you should determine whether or not it is acid-fast. Although some bacteria require 4 or 5 days growth to exhibit acid-fastness, most species become acid-fast within 2 days. For best results, therefore, do not use cultures that are too old.

Another point to keep in mind is that most acid-fast bacteria do not produce cells that are 100% acid-fast. An organism is considered acid-fast if only portions of the cells exhibit this characteristic. Refer to Exercise 16 for this staining technique.

A final bit of advice: If you feel insecure about your adeptness at Gram staining and think that you might *possibly* have a gram-positive organism, even though your organism seems to be gram-negative, make an acid-fast stained slide. Many students find (much to their chagrin later) that they didn't do acid-fast staining because their organism seemed to be gram-negative. An improperly Gram-stained slide can be very misleading when it comes to unknown identification.

Other Structures

If the protoplast in Gram-stained slides stains unevenly, you might wish to do a simple stain with Loeffler's methylene blue (Exercise 11) for evidence of metachromatic granules.

Although a capsule stain (Exercise 13) may be performed at this time, it might be better to wait until a later date when you have the organism growing on blood agar. Capsules usually are more apparent when the organisms are grown on this medium.

LABORATORY REPORT

There is no Laboratory Report to fill out for this exercise. All information is recorded on the descriptive chart.

Descriptive Chart

STUDENT: _____

LAB SECTION: _____

Habitat: _____ Culture No.: _____

Source: _____

Organism: _____

MORPHOLOGICAL CHARACTERISTICS	PHYSIOLOGICAL CHARACTERISTICS	

MORPHOLOGICAL CHARACTERISTICS

Cell Shape:

Arrangement:

Size:

Spores:

Gram's Stain:

Motility:

Capsules:

Special Stains:

CULTURAL CHARACTERISTICS

Colonies:

   Nutrient Agar:

   Blood Agar:

Agar Slant:

Nutrient Broth:

Gelatin Stab:

Oxygen Requirements:

Optimum Temp.:

PHYSIOLOGICAL CHARACTERISTICS

	TESTS	RESULTS
Fermentation	Glucose	
	Lactose	
	Sucrose	
	Mannitol	
Hydrolysis	Gelatin Liquefaction	
	Starch	
	Casein	
	Fat	
IMViC	Indole	
	Methyl Red	
	V-P (acetylmethylcarbinol)	
	Citrate Utilization	
	Nitrate Reduction	
	H_2S Production	
	Urease	
	Catalase	
	Oxidase	
	DNase	
	Phenylalanase	

	REACTION	TIME
Litmus Milk	Acid	_____
	Alkaline	_____
	Coagulation	_____
	Reduction	_____
	Peptonization	_____
	No Change	_____

Cultural Characteristics

The cultural characteristics of an organism pertain to its macroscopic appearance on different kinds of media. Descriptive terms, which are familiar to all bacteriologists and are used in *Bergey's Manual,* must be used in recording cultural characteristics. The most frequently used media for a cultural study are nutrient agar, nutrient broth, and nutrient gelatin. For certain types of unknowns, it is also desirable to inoculate a blood agar plate; if necessary, this plate can be inoculated later. In addition to these media, you will be inoculating a fluid thioglycollate medium to determine the oxygen requirements of your unknown.

FIRST PERIOD

(Inoculations)

During this period, one nutrient agar plate, one nutrient gelatin deep, two nutrient broths, and one tube of fluid thioglycollate medium will be inoculated. Inoculations will be made with the original broth culture of your unknown. The reason for inoculating two tubes of nutrient broth here is to recheck the optimum growth temperature of your unknown. In Exercise 36, you incubated your nutrient agar slants at 20° C and 37° C. It may well be that the optimum growth temperature is closer to 30° C. It is to check out this intermediate temperature that an extra nutrient broth is being inoculated. Proceed as follows:

MATERIALS

for each unknown:
- 1 nutrient agar pour
- 1 nutrient gelatin deep
- 2 nutrient broths
- 1 fluid thioglycollate medium (FTM)
- 1 petri plate

1. Pour a petri plate of nutrient agar for each unknown and streak it with a method that will give good isolation of colonies. Use the original broth culture for streaking.
2. Inoculate the tubes of nutrient broth with a loop.
3. Make a stab inoculation into the gelatin deep by stabbing the inoculating needle (straight wire) directly down into the medium to the bottom of the

tube and pulling it straight out. The medium must not be disturbed laterally.
4. Inoculate the tube of FTM with a loopful of your unknown. Mix the organisms throughout the tube by rolling the tube between your palms.
5. Place all tubes except one nutrient broth into a basket and incubate for 24 hours at the temperature that seemed best in Exercise 36. Incubate the remaining tube of nutrient broth separately at 30° C. Incubate the agar plate, inverted, at the presumed best temperature.

SECOND PERIOD

(Evaluation)

After the cultures have been properly incubated, *carry them to your desk in a careful manner* to avoid disturbing the growth pattern in the nutrient broths and FTM. Before studying any of the tubes or plates, place the tube of nutrient gelatin in an ice water bath. It will be studied later. Proceed as follows to study each type of medium and record the proper descriptive terminology on the descriptive chart on page 365.

MATERIALS

- reserve stock agar slant of unknown
- spectrophotometer and cuvettes
- hand lens
- ice water bath near sink

Nutrient Agar Slant (Reserve Stock)

Examine your reserve stock agar slant of your unknown that has been stored in the refrigerator since the last laboratory period. Evaluate it in terms of the following criteria:

Amount of Growth   The abundance of growth may be described as *none, slight, moderate,* and *abundant.*

Color   Pigments can be associated with the colony or diffuse into the growth medium. Most bacteria do not produce pigments, and therefore their colonies are white or buff colored. However, some are chromogenic, producing various shades of different colors.

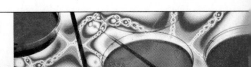

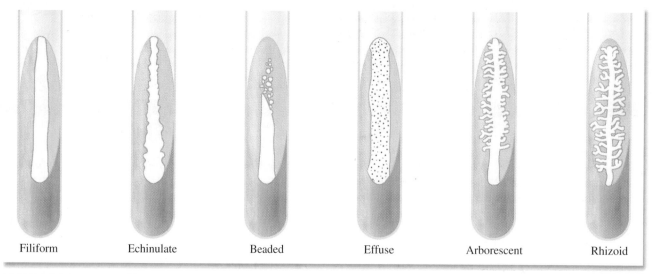

FIGURE 37.1 Types of bacterial growth on nutrient agar slants

When soluble pigments are produced, they diffuse into the medium. To examine for the production of soluble pigments, hold the slant up to a strong light and look for diffuse pigmentation in the medium.

Opacity Organisms that grow prolifically on the surface of a medium will appear more opaque than those that exhibit a small amount of growth. Degrees of opacity may be expressed in terms of *opaque*, *transparent*, and *translucent* (partially transparent).

Form The gross appearance of different types of growth are illustrated in figure 37.1. The following descriptions of each type will help in differentiation:

> *Filiform:* characterized by uniform growth along the line of inoculation
> *Echinulate:* margins of growth exhibit toothed appearance
> *Beaded:* separate or semiconfluent colonies along the line of inoculation
> *Effuse:* growth is thin, veil-like, unusually spreading
> *Arborescent:* branched, treelike growth
> *Rhizoid:* rootlike appearance

Nutrient Broth

The nature of growth on the surface, subsurface, and bottom of the tube is significant in nutrient broth cultures. Describe your cultures as thoroughly as possible on the descriptive chart with respect to these characteristics:

Surface Figure 37.2 illustrates different types of surface growth. A *pellicle* type of surface differs from the *membranous* type in that the latter is much thinner. A *flocculent* surface is made up of floating adherent masses of bacteria.

Subsurface Below the surface, the broth may be described as *turbid* if it is cloudy, *granular* if specific small particles can be seen, *flocculent* if small masses are floating around, and *flaky* if large particles are in suspension.

Sediment The amount of sediment in the bottom of the tube may vary from none to a great deal. To describe the type of sediment, agitate the tube, putting the material in suspension. The type of sediment can be described as *granular, flocculent, flaky*, and *viscid*. Test for viscosity by probing the bottom of the tube with a sterile inoculating loop.

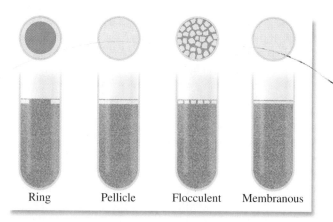

FIGURE 37.2 Types of surface growth in nutrient broth

Amount of Growth To determine the amount of growth, it is necessary to shake the tube to disperse the organisms. Terms such as *slight* (scanty), *moderate*, and *abundant adequately* describe the amount.

Optimum Temperature To determine which temperature produced the best growth, pour the contents from each tube of nutrient broth into separate cuvettes and measure their percent transmittances on the spectrophotometer. If the percent transmittance is less at 30° C than at the other presumed optimum temperature, revise the optimum temperature on your descriptive chart.

Fluid Thioglycollate Medium

Since the primary purpose of inoculating a tube of fluid thioglycollate medium is to determine oxygen requirements of your unknown, examine the tube to note the position of growth in the tube. Compare your tube with figure 19.5 on page 128 to make your analysis. Designate your organism as being *aerobic, microaerophilic, facultative,* or *anaerobic* on the descriptive chart.

Gelatin Stab Culture

Remove your tube of nutrient gelatin from the ice water bath and examine it. Check first to see if liquefaction has occurred. Organisms that are able to liquefy gelatin produce the enzyme *protease.*

Liquefaction Tilt the tube from side to side to see if a portion of the medium is still liquid. If liquefaction has occurred, check the configuration with figure 37.3 to see if any of the illustrations match your tube. A description of each type follows:

> *Crateriform:* saucer-shaped liquefaction
> *Napiform:* turniplike
> *Infundibular:* funnel-like or inverted cone
> *Saccate:* elongate sac, tubular, cylindrical
> *Stratiform:* liquefied to the walls of the tube in the upper region

Note: The configuration of liquefaction is not as significant as the mere fact that liquefaction takes place. If your organism liquefies gelatin, but you are unable to determine the exact configuration, don't worry about it. However, be sure to record on the descriptive chart the *presence* or *absence* of protease production.

Another important point: Some organisms produce protease at a very slow rate. Tubes that are neg-

Growth without liquefaction

Filiform Beaded Papilate Vilous Aborescent

Liquefaction configurations

Crateriform Napiform Infundibular Saccate Stratiform

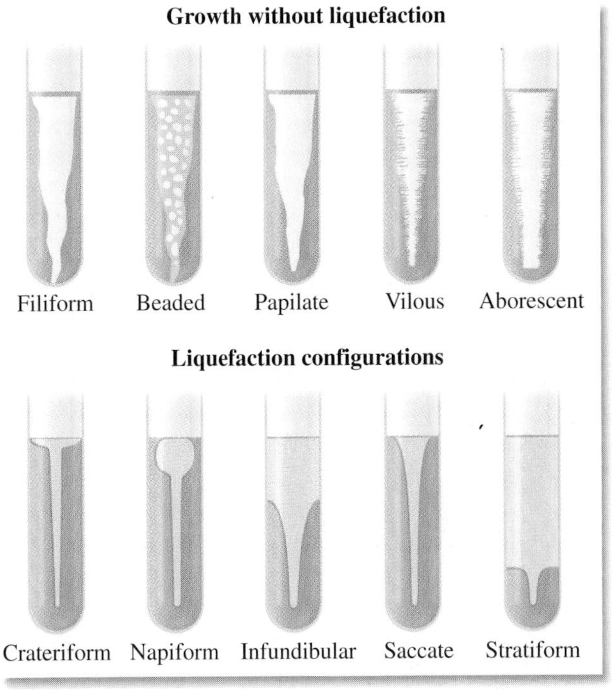

FIGURE 37.3 Growth in gelatin stabs

ative should be incubated for another 4 or 5 days to see if protease is produced slowly.

Type of Growth (No Liquefaction) If no liquefaction has occurred, check the tube to see if the organism grows in nutrient gelatin (some do, some don't). If growth has occurred, compare the growth with the bottom of the illustration in figure 37.3. It should be pointed out, however, that, from a categorization standpoint, the nature of growth in gelatin is not very important.

Nutrient Agar Plate

Colonies grown on plates of nutrient agar should be studied with respect to size, color, opacity, form, elevation, and margin. With a dissecting microscope or hand lens study individual colonies carefully. Refer to figure 37.4 for descriptive terminology. Record your observations on the descriptive chart.

LABORATORY REPORT

There is no Laboratory Report for this exercise. Record all information on the descriptive chart on page 365.

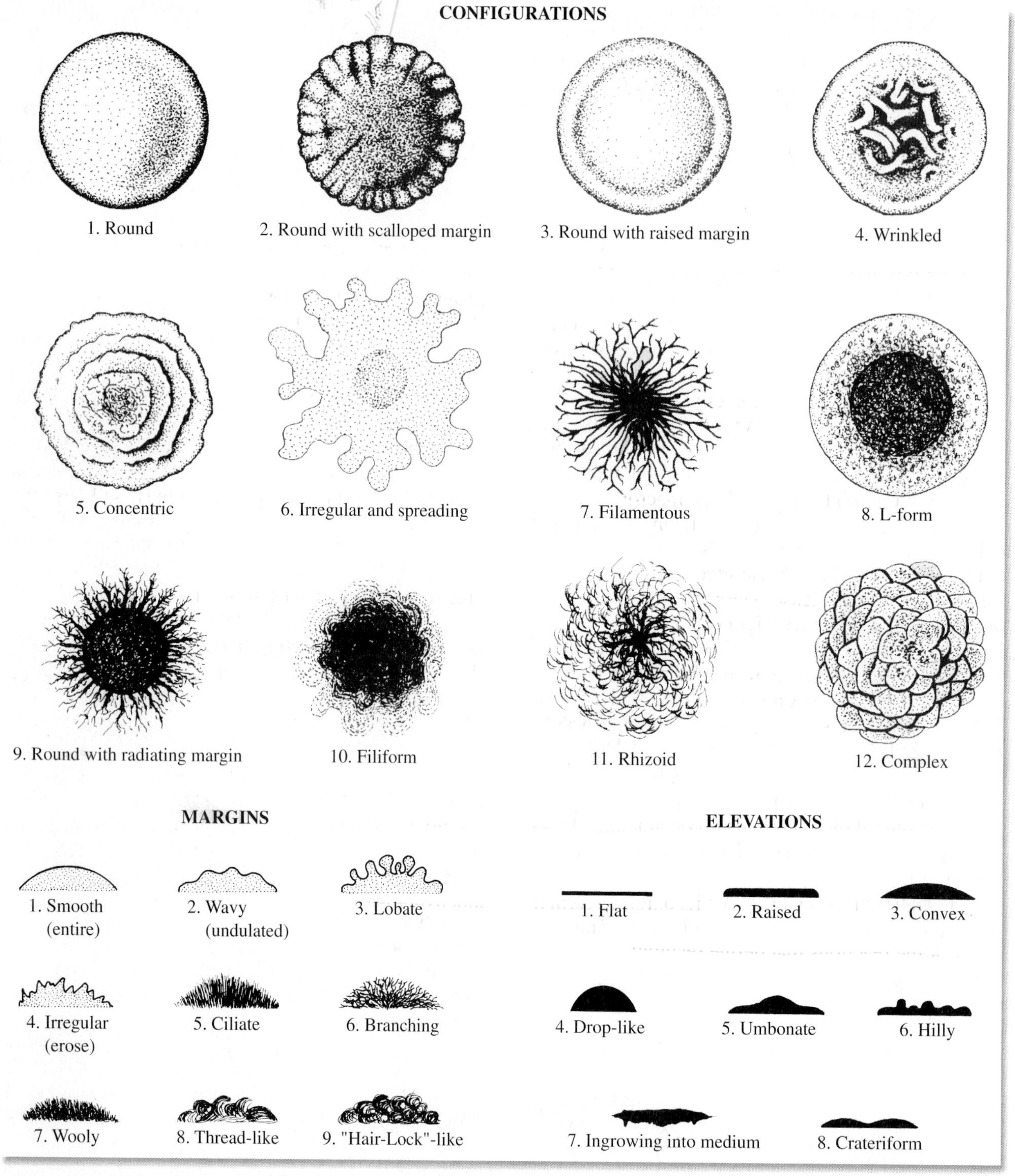

CONFIGURATIONS

1. Round

2. Round with scalloped margin

3. Round with raised margin

4. Wrinkled

5. Concentric

6. Irregular and spreading

7. Filamentous

8. L-form

9. Round with radiating margin

10. Filiform

11. Rhizoid

12. Complex

MARGINS

1. Smooth (entire)

2. Wavy (undulated)

3. Lobate

4. Irregular (erose)

5. Ciliate

6. Branching

7. Wooly

8. Thread-like

9. "Hair-Lock"-like

ELEVATIONS

1. Flat

2. Raised

3. Convex

4. Drop-like

5. Umbonate

6. Hilly

7. Ingrowing into medium

8. Crateriform

FIGURE 37.4 Growth in gelatin stabs

Physiological Characteristics:
Oxidation and Fermentation Tests

The sum total of the chemical reactions that occur in a cell are referred to as metabolism, and the individual chemical reactions that make up the metabolic pathways in a cell are catalyzed by protein molecules called **enzymes**. Most enzymes function inside the cell where metabolic pathways carry out the breakdown (**catabolism**) of food materials and the biosynthesis of cell constituents (**anabolism**). Because bacteria cannot carry out phagocytosis owing to their rigid cell walls, they excrete **exoenzymes** that function outside the cell to degrade large macromolecules. For example, exoenzymes break down proteins and polysaccharides into amino acids and monosaccharides, which are then transported into the cell for metabolic needs. Protease, DNAse, and amylase are examples of exoenzymes (figure 38.1).

Some enzymes are assisted in catalytic reactions by **coenzymes.** The latter transfer small molecules from one substrate to another, for example, NAD, FAD, and coenzyme A transfer protons and acetate groups, respectively. Most coenzymes are derivatives of vitamins. As examples, NAD is synthesized from niacin, and FAD comes from folic acid. Coenzymes are only required by a cell in catalytic amounts, however, and when an enzymatic reaction catalyzes an oxidation step that converts NAD to NADH, the coenzyme must be converted back into its oxidized form if the metabolic pathway is to continue to function. Many of the reactions that define respiration and fermentation are concerned with regenerating coenzymes such as NAD and FAD.

The primary goal of catabolism is the production of energy, which is needed for biosynthesis and growth. Bacteria can obtain their energy needs by two different metabolic means, respiration or fermentation. In respiration, organic molecules are completely degraded to carbon dioxide and water. ATP is generated by the energy created from a proton gradient that is established across the cell membrane when protons are transported from the cytoplasm to the outside of the cell. The shuttling of electrons down an electron transport chain involving cytochromes facilitates the movement of the protons to the outside of the cell. This process is called **oxidative phosphorylation** and, in the process, reduced coenzyme NADH generated in metabolic reactions is converted back to NAD because oxygen acts as the terminal electron acceptor and is converted to water. In contrast, fermentation is the partial breakdown of organic molecules to alcohols, aldehydes, acids, and gases such as carbon dioxide and hydrogen. In this process, organic molecules in metabolic pathways serve as terminal electron acceptors and become the end products in a fermentation pathway. Reactions that carry out oxidation steps and utilize NAD in metabolic pathways are coupled to reactions that use NADH to reduce the metabolic intermediates. An example is the oxidation of glyceraldehyde-3-phosphate in glycolysis being coupled with the formation of lactate from pyruvate when *Streptococcus lactis* ferments glucose.

In fermentation, ATP is synthesized by **substrate level phosphorylation** in which metabolic

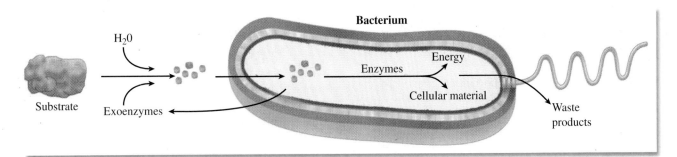

FIGURE 38.1 Note that the hydrolytic exoenzymes split larger molecules into smaller ones, utilizing water in the process. The smaller molecules are then assimilated by the cell to be acted upon by endoenzymes to produce energy and cellular material

intermediates in pathways directly transfer high-energy phosphates to ADP to synthesize ATP. In the glycolytic fermentation of glucose, ATP is formed when phosphoenol pyruvate transfers a high-energy phosphate to ADP and pyruvate is formed. In general, fermentation is much less efficient in producing energy relative to respiration because the use of metabolic intermediates as electron acceptors leaves most of the available energy in molecules that form the end products. Some bacteria are capable of growing both by respiration and fermentation. *Escherichia coli* is a facultative anaerobe that will grow by respiratory means if oxygen is present but will switch metabolic gears in anaerobic conditions and grow by fermentation.

Sugars, particularly glucose, are compounds most widely used by fermenting organisms. However, other compounds such as organic acids, amino acids, and fats are fermented by bacteria. Butter becomes rancid because bacteria ferment butter fat producing volatile and odoriferous organic acids. The end products of a particular fermentation are like a "fingerprint" for an organism and can be used in its identification. For example, *Escherichia coli* can be differentiated from *Enterobacter aerogenes* because the primary fermentation end products for *E. coli* are mixed organic acids, whereas *E. aerogenes* produces acetylmethylcarbinol, a neutral end product.

TESTS TO BE PERFORMED

Six types of reactions will be studied in this exercise: (1) Durham-tube sugar fermentations, (2) mixed acid fermentation, (3) butanediol fermentation, (4) catalase production, (5) oxidase production, and (6) nitrate reduction. The performance of all these tests on your unknown will involve a considerable number of inoculations because a set of positive test controls will also be needed. Although photographs of positive test results are provided in this exercise, seeing the actual test results in test tubes will make it more meaningful.

As you perform these various tests, attempt to keep in mind what groups of bacteria relate to each test. Although some tests are not very specific in pointing the way to unknown identification, others are very narrow in application.

One last comment of importance: *It is not routine practice to perform all these tests in identifying every unknown.* Remember that although it might appear that our prime concern here is to identify an organism, our most important goal is to learn about the various types of tests for enzymes that are available. The use of unknown bacteria to learn about them simply makes it more of a challenge. In actual practice, phys-iological tests are used very selectively. The "shotgun approach" employed here is used to expose you to the multitude of tests that are available.

FIRST PERIOD

(Inoculations)

The following two sets of inoculations (unknown and test controls) may be done separately or combined into one operation. The media for each set of inoculations are listed separately under each heading.

Unknown Inoculations

Figure 38.2 illustrates the procedure for inoculating seven test tubes and one petri plate with your unknown. Since your instructor may want you to inoculate some different sugar broths, blanks have been provided in the materials list for write-ins. *If different media are distinguished from each other with different-colored tube caps, write down the colors after each medium below.*

MATERIALS: (for each unknown)

- Durham tubes with phenol red indicator
- 1 glucose broth
- 1 lactose broth
- 1 mannitol broth

_____ _____

_____ _____

- 2 MR-VP medium
- 1 nitrate broth
- 1 nutrient agar slant
- 1 petri plate of trypticase soy agar (TSA)

1. Label each tube with the number of your unknown and an identifying letter as designated in figure 38.2.
2. Label one half of the petri plate UNKNOWN and the other half *P. AERUGINOSA*.
3. Inoculate all broths and the slant with a loop. Inoculate one half of the TSA plate with your unknown, using an isolation technique.

Test Control Inoculations

Figure 38.4 on page 233 illustrates the procedure that will be used for inoculating five test tubes to be used for positive test controls. The petri plate shown on the right side is the same one that is shown in figure 38.2; thus, it will not be listed in the materials below.

MATERIALS

- 1 glucose broth (Durham tube)
- 2 MR-VP medium

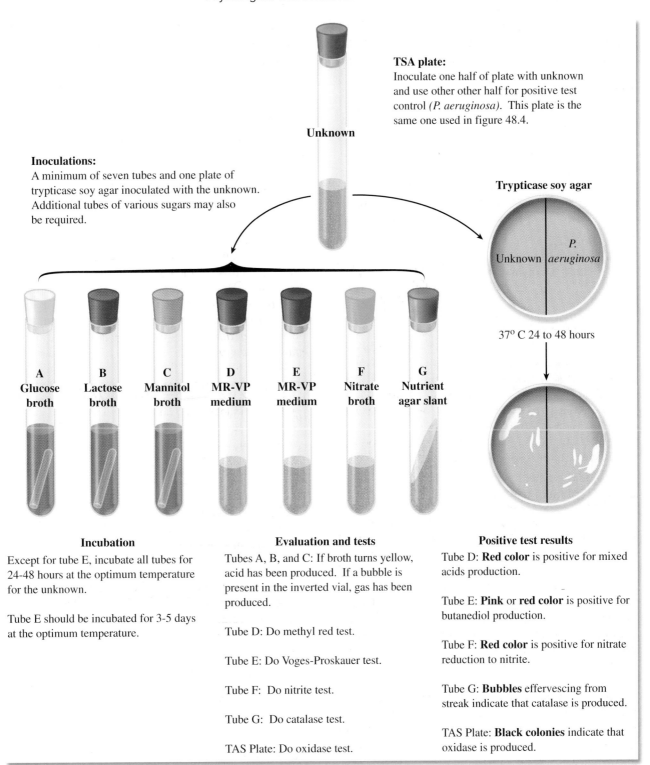

TSA plate:
Inoculate one half of plate with unknown and use other other half for positive test control *(P. aeruginosa)*. This plate is the same one used in figure 48.4.

Inoculations:
A minimum of seven tubes and one plate of trypticase soy agar inoculated with the unknown. Additional tubes of various sugars may also be required.

Unknown

Trypticase soy agar

Unknown | *P. aeruginosa*

37° C 24 to 48 hours

| A Glucose broth | B Lactose broth | C Mannitol broth | D MR-VP medium | E MR-VP medium | F Nitrate broth | G Nutrient agar slant |

Incubation

Except for tube E, incubate all tubes for 24-48 hours at the optimum temperature for the unknown.

Tube E should be incubated for 3-5 days at the optimum temperature.

Evaluation and tests

Tubes A, B, and C: If broth turns yellow, acid has been produced. If a bubble is present in the inverted vial, gas has been produced.

Tube D: Do methyl red test.

Tube E: Do Voges-Proskauer test.

Tube F: Do nitrite test.

Tube G: Do catalase test.

TAS Plate: Do oxidase test.

Positive test results

Tube D: **Red color** is positive for mixed acids production.

Tube E: **Pink** or **red color** is positive for butanediol production.

Tube F: **Red color** is positive for nitrate reduction to nitrite.

Tube G: **Bubbles** effervescing from streak indicate that catalase is produced.

TAS Plate: **Black colonies** indicate that oxidase is produced.

FIGURE 38.2 **Procedure for performing oxidation and fermentation tests**

- 1 nitrate broth
- 1 nutrient agar slant
- nutrient broth cultures of *Escherichia coli,* *Enterobacter aerogenes, Staphylococcus aureus,* and *Pseudomonas aeruginosa*

1. Label each tube with the code letter assigned to it as listed:

glucose broth   A[1]
MR-VP medium   D[1]

Inverted vial
(with gas)

FIGURE 38.3 Durham fermentation tube

MR-VP medium	E[1]
nitrate broth	F[1]
nutrient agar slant	G[1]

2. Inoculate each of these tubes with a loopful of the appropriate test organism according to figure 38.4.
3. Inoculate the other half of the TSA plate with *P. aeruginosa.*

Incubation

Except for tube E (MR-VP), all the unknown inoculations should be incubated for 24–48 hours at the unknown's optimum temperature. Tube E should be incubated for 3–5 days at the optimum temperature.

Except for Tube E[1] of the test controls, incubate all the test-control tubes and the TSA plate at 37°C for 24–48 hours. Tube E[1] should be incubated at 37°C for 3–5 days.

SECOND PERIOD

(Test Evaluations)

After 24 to 48 hours incubation, arrange all your tubes (except tubes E and E[1]) in a test-tube rack in alphabetical order, with the unknown tubes in one row and the test controls in another row. As you interpret the results, record the information on the descriptive chart on page 367 immediately. Don't trust your memory. Any result that is not properly recorded will have to be repeated.

Durham Tube Sugar Fermentations

When we use a bank of tubes containing various sugars, we are able to determine what sugars an organism is able to ferment. If an organism is able to ferment a particular sugar, acid will be produced and gas *may* be produced. The presence of acid is detectable with the color change of a pH indicator in the medium. Gas production is revealed by the formation of a void in the inverted vial of the Durham tube (figure 38.3).

Media The sugar broths used here contain 0.5% of the specific carbohydrate plus sufficient amounts of beef extract and peptone to satisfy the nitrogen and mineral needs of most bacteria. The pH indicator phenol red is included for acid detection. This indicator is red when the pH is above 7 and yellow below this point.

Although there are many sugars that one might use, glucose, lactose, and mannitol are logical ones to begin with. Your instructor may have had you include one or more additional kinds, and it is very likely that you may wish to use some others later.

Interpretation Examine the glucose test control tube (tube A[1]), which you inoculated with *E. coli.* Note that the phenol red has turned yellow, indicating acid production. Also, note that the inverted vial has a gas bubble in it. These observations tell us that *E. coli* ferments glucose to produce acid and gas. The illustration in figure 38.4, page 233 illustrates how this positive tube compares with a negative tube and an uninoculated one.

Now examine the three sugar broths (tubes A, B, and C) that were inoculated with your unknown and record your observations on the descriptive chart. If there is no color change, record NONE after the specific sugar. If the tube is yellow with no gas, record ACID. If the inverted vial contains gas and the tube is yellow, record ACID AND GAS.

An important point to keep in mind at this time is that *a negative result on an unknown is as important as a positive result.* Don't feel that you have failed in your technique if many of your tubes are negative!

Mixed Acid Fermentation

(Methyl-Red Test)

A considerable number of gram-negative intestinal bacteria can be differentiated on the basis of the end products produced when they ferment the glucose in MR-VP medium. Genera of bacteria such as *Escherichia, Salmonella, Proteus,* and *Aeromonas* ferment glucose to produce large amounts of lactic, acetic, succinic, and formic acids, plus CO_2, H_2, and ethanol. The accumulation of these acids lowers the pH of the medium to 5.0 and less.

If methyl red is added to such a culture, the indicator turns red, an indication that the organism is a *mixed acid fermenter.* These organisms are generally great gas producers, too, because they produce the enzyme *formic hydrogenlyase,* which splits formic acid into equal parts of CO_2 and H_2.

$$HCOOH \xrightarrow{\text{formic hydrogenlyase}} CO_2 + H_2$$

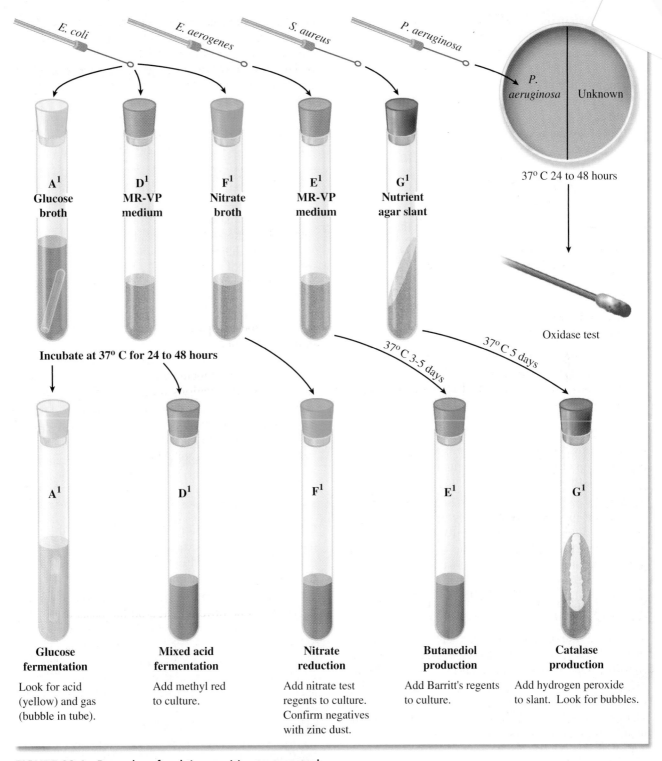

FIGURE 38.4 Procedure for doing positive test controls

Medium MR-VP medium is essentially a glucose broth with some buffered peptone and dipotassium phosphate.

Test Procedure Perform the methyl-red test first on your test-control tube (D¹) and then on your unknown (tube D). Proceed as follows:

MATERIALS

- dropping bottle of methyl-red indicator

1. Add 3 or 4 drops of methyl red to test-control tube D¹, which was inoculated with *E. coli*. The tube should become red immediately. A **reddish color,**

as shown in the left-hand tube of the middle illustration of figure 38.5 is a positive methyl-red test.

2. Repeat the same procedure with your unknown culture (tube D) of MR-VP medium. If your unknown culture becomes yellow like the tube in figure 38.5, your unknown is negative for this test.

3. Record your results on the descriptive chart on page 366.

Butanediol Fermentation

(Voges-Proskauer Test)

A negative methyl red test may indicate that the organism produced the neutral compound 2,3-butanediol rather than the organic acids that produce a positive methyl red test. All species of *Enterobacter* and *Serratia* as well as some species of *Erwinia, Bacillus,* and *Aeromonas* produce butanediol and are therefore negative for the methyl red test.

The 2,3-butanediol cannot be detected directly but the precursor of this compound, acetoin (acetylmethylcarbinol), is easily detected with Barritt's reagent. This reagent consists of alpha naphthol and KOH. The reagent is added to a 3- to 5-day-old culture grown in MR-VP medium, and the culture is vigorously shaken, which results in the oxidation of butanediol back to acetoin. The tube is allowed to stand at room temperature for a maximum of 30 minutes in which time the medium will turn pink to red if acetoin is present. This indirect method for testing for butanediol is called the *Voges-Proskauer* test.

Test Procedure Perform the Voges-Proskauer test on your unknown and test-control tubes of MR-VP medium (tubes E and E¹). Note that the test-control tube was inoculated with *E. aerogenes.* Follow this procedure:

MATERIALS

- Barritt's reagents
- 2 pipettes (1 ml size)
- 2 empty test tubes

1. Label one empty test tube E (for unknown) and the other E¹ (for control).
2. Pipette 1 ml from culture tube E to the empty tube E and 1 ml from culture tube E¹ to the empty tube E¹. Use separate pipettes for each tube.
3. Add 18 drops (about 0.5 ml) of Barritt's solution A (alpha naphthol) to each of the tubes that contain 1 ml of culture.
4. Add an equal amount of Barritt's solution B (KOH) to the same tubes.
5. Shake the tubes vigorously and allow them to stand for 30 minutes. In this time, the control tube E¹ should turn pink to red. Compare the control to the unknown. *Vigorous shaking is necessary to oxidize the butanediol to acetoin.*

   A positive Voges-Proskauer reaction is **pink** or **red.** The left-hand tube in the right-hand illustration of figure 38.5 shows what a positive result looks like.
6. Record your results on the descriptive chart on page 366.

Catalase

When aerobic bacteria grow by respiration, they produce hydrogen peroxide as a by-product of reducing oxygen to water. Hydrogen peroxide is highly reactive and will damage enzymes, nucleic acids, and

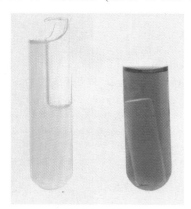

Durham Tubes

Tube on left is positive;
tube on right is negative.

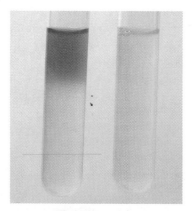

Methyl-Red Test

Tube on left is positive (*E. coli*);
tube on right is negative.

Voges-Proskauer Test

Tube on left is positive (*E. aerogenes*);
tube on right is negative.

FIGURE 38.5 Durham tubes, mixed acid, and butanediol fermentation tests
© The McGraw-Hill Companies/Auburn University Photographic Service

small molecules in the cell. To avoid this damage, aerobes produce the enzyme catalase, which converts hydrogen peroxide to harmless oxygen and water.

$$2H_2O \xrightarrow{\text{catalase}} 2H_2O + O_2$$

Strict anaerobes and aerotolerant bacteria such as *Streptococcus* lack this enzyme and hence they are unable to deal with the hydrogen peroxide produced in aerobic environments. The presence of catalase is one way to differentiate these bacteria from aerobes or facultative anaerobes. For example, catalase production can be used to differentiate the aerobic staphylococci, which possess catalase from the aerotolerant streptococci, which lack catalase.

Test Procedure To determine if catalase is produced, a small amount of growth is transferred from a plate or slant to a microscope slide using a wooden stick. A couple of drops of 3% hydrogen peroxide are added to the cells on the slide. If catalase is present, there will be vigorous bubbling due to the release of oxygen.

Note: do not use a wire inoculating loop to transfer the cells, as iron can cause the hydrogen peroxide to break down releasing oxygen. Also, do not perform the catalase test on cells growing on blood agar, as blood contains catalase.

MATERIALS

- 3% hydrogen peroxide
- test-control tube G^1 with *Staphylococcus aureus* and unknown tube G

1. Using a wooden stick, transfer a small amount of G^1 cells from the culture to the surface of a clean microscope slide. Spread the cells in a circular motion as if you are making a smear.
2. Add 2 drops of 3% hydrogen peroxide and observe for vigorous bubbling.
3. Repeat the same procedure for your unknown organism.

Oxidase Test

The oxidase test assays for the presence of cytochrome oxidase. This enzyme catalyzes the transfer of electrons from reduced cytochrome c to molecular oxygen producing oxidized cytochrome c and water. Oxidase occurs in bacteria that carry out respiration where oxygen is the terminal electron acceptor, and the test differentiates those bacteria from bacteria that do not use oxygen as a terminal electron acceptor. The

enzyme is detected by using an artificial electron acceptor, N,N,N,N,-tetramethyl-*p*-phenylenediamine, which changes from yellow to purple when electrons are transferred from reduced cytochrome c to the chemical acceptor.

The oxidase test will differentiate most species of oxidase-positive *Pseudomonas* from the Enterobacteriaceae, which are oxidase negative. The method you will employ involves the use of a commercial reagent in ampules that are broken just prior to use.

MATERIALS

- TSA slants or plates streaked with *Pseudomonas aeruginosa*
- ampules of 1% oxidase reagent, N,N,N,N,-tetramethyl-*p*-phenylenediaminedihydrochloride (Difco)
- Whatman no. 2 filter paper

1. Grasp an oxidase reagent ampule between your thumb and forefinger. Hold the ampule so that the tip is pointed away from you. Squeeze the ampule until the glass breaks and the reagent is released into the plastic dropper. Tap the ampule on the tabletop several times and invert the ampule.
2. Touch a sterile swab to colonies on a plate or growth on a slant. Remove the cap from an oxidase ampule and deliver several drops of reagent to the growth on the swab.
3. A positive culture will turn from yellow to purple in 10–30 seconds. Changes after 30 seconds are considered negative reactions (figure 38.6).
4. Repeat the test for your unknown organism.

Note: use a wooden stick to transfer the culture as iron loops can interfere with the test.

Nitrate Reduction

Some facultative anaerobes, such as *Escherichia coli,* can use nitrate as a terminal electron acceptor under anaerobic conditions to carry out nitrate respiration. The enzyme nitrate reductase catalyzes the transfer of electrons from cytochrome b to nitrate reducing it to nitrite. This is an inducible enzyme that is repressed when oxygen is present. The chemical reaction for the enzymatic reaction is as follows:

$$NO_3^- + 2e^- + 2H^+ \xrightarrow{\text{nitrate reductase}} NO_2^- + H_2O$$

Test Procedure The ability of bacteria to reduce nitrate can be determined by assaying for the presence of nitrite. Cultures are grown in a beef extract,

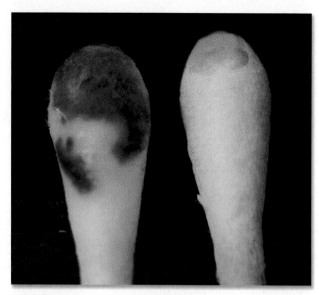

FIGURE 38.6 Left hand swab shows a purple reaction due to oxidase production. Right hand swab shows a culture that is oxidase negative © The McGraw-Hill Companies/Auburn University Photographic Service

peptone broth containing potassium nitrate. Reagent A, sulfanilic acid, is added to a culture, followed by reagent B, dimethyl-alpha-naphthylamine. If nitrite is present, the medium will turn red due to the formation of a chemical complex. A negative test could mean, however, that nitrate was not reduced (negative test) or that nitrate was further reduced by the organism to nitrogen gas. To differentiate between the two possibilities, zinc dust is added to a negative culture. Zinc will chemically reduce the nitrate causing the formation of the red color, confirming that nitrate was not reduced. However, if further reduction of nitrite has occurred, no color will develop in the culture tube.

MATERIALS

- nitrate broth cultures of the unknown organism (tube F) and the test control *E. coli* (tube F¹)
- nitrate test reagents: reagent A—sulfanilic acid; reagent B—dimethyl-alpha-naphthylamine
- zinc dust

1. Add 2–3 drops of reagent A and an equal amount of reagent B to the nitrate broth culture of *E. coli*.

A red color will appear immediately (figure 38.7) indicating that nitrate was reduced.

> **CAUTION** Avoid skin contact with solution B. Dimethyl-alpha-naphthylamine is carcinogenic.

2. Repeat this same procedure for the unknown organism. If a red color fails to appear, the organism is most probably negative for nitrate reduction. To confirm this, perform the zinc test.

    Zinc test: To the negative culture, add a pinch of zinc dust and shake the tube vigorously. If the tube becomes red, the zinc has reduced the nitrate and the test is confirmed as negative.

3. Record your results in Laboratory Report 38–40.

LABORATORY REPORT

Answer the questions in Laboratory Report 38–40 on page 245 that pertain to this exercise.

FIGURE 38.7 Nitrate test: Left hand tube shows red color due to nitrate reduction. Middle tube shows reduction of nitrate to nitrogen gas which is trapped in the Durham tube. Right tube is an uninoculated control © The McGraw-Hill Companies/Auburn University Photographic Service

Physiological Characteristics:
Hydrolytic Reactions

Bacteria are unable to carry out phagocytosis for acquiring food materials, and therefore they excrete exoenzymes that split larger molecules into smaller units. For example, amylases and cellulases degrade polysaccharides such as starch and cellulose into simple sugars, which are then transported into the cell for metabolic purposes. A variety of proteases degrade protein molecules, such as casein and gelatin, into amino acids, and fats or triglycerides are split into fatty acids and glycerol by various lipases. Bacteria also hydrolyze small molecules, producing signature compounds that can be used in identifying them. Tryptophane is split into pyruvate and indole, which can be detected by a biochemical test specific for the indole ring. Likewise, some bacteria degrade urea producing carbon dioxide and ammonia, which causes a color change in pH indicators owing to the alkaline conditions produced by ammonia. In this exercise, you will perform some of these tests. In each case, you will compare your unknown to known organisms.

Figure 39.1 illustrates the general procedure to be used. Three agar plates and four test tubes will be inoculated. After incubation, some of the plates and tubes will have test reagents added to them; others will reveal the presence of hydrolysis by changes that have occurred during incubation. Proceed as follows:

FIRST PERIOD

(Inoculations)

If each student is working with only one unknown, students can work in pairs to share petri plates. Note in figure 39.1 how each plate can serve for two unknowns with the test-control organism streaked down the middle. If each student is working with two unknowns, the plates will not be shared. Whether or not the two tubes for test controls will be shared depends on the availability of materials.

MATERIALS

per pair of students with one unknown each, or for one student with two unknowns:
- 1 starch agar plate
- 1 skim milk agar plate
- 1 spirit blue agar plate
- 3 urea slants

- 3 tryptone broths
- nutrient broth cultures of *B. subtilis, E. coli, S. aureus,* and *P. vulgaris.*

1. Label and streak the three different agar plates in the manner shown in figure 39.1. Note that straight line streaks are made on each plate. Indicate, also, the type of medium in each plate.
2. Label a tube of urea broth *P. vulgaris* and a tube of tryptone broth *E. coli*. These will be your test controls for urea and tryptophan hydrolysis. Inoculate each tube accordingly.
3. For each unknown, label one tube of urea broth and one tube of tryptone broth with the code number of your unknown. Inoculate each tube with the appropriate unknown.
4. Incubate the plates and two test-control tubes at 37° C. Incubate the unknown tubes of urea broth and tryptone broth at the optimum temperatures for the unknowns.

SECOND PERIOD

(Evaluation of Tests)

After 24 to 48 hours incubation of unknowns and test controls, compare your unknowns with the test controls, recording all data on the descriptive chart on page 367.

Starch Hydrolysis

Since many bacteria are capable of hydrolyzing starch, this test has fairly wide application. The starch molecule is a large one consisting of two constituents: amylose, a straight chain polymer of 200 to 300 glucose units, and amylopectin, a larger branched polymer with phosphate groups. Bacteria that hydrolyze starch produce *amylases* that yield molecules of maltose, glucose, and dextrins.

MATERIALS

- Gram's iodine
- starch agar culture plate

Iodine solution (Gram's) is an indicator of starch. When iodine comes in contact with a medium containing starch, it turns blue. If starch is hydrolyzed and

Inoculations:
Three agar plates, a urea broth, and a tryptone broth are inoculated with the unknown. Note that straight-line streaks are used on the plates and that test control organisms are also used on the plates.

Test control tube inoculations:
A tryptone broth is inoculated with *E. coli*, and a urea slant is inoculated with *P. vulgaris*. Incubate at 37° C for 24-48 hours.

Incubation:
The three plates and the two tubes should be incubated at the optimum temperature of the unknown for 24-48 hours.

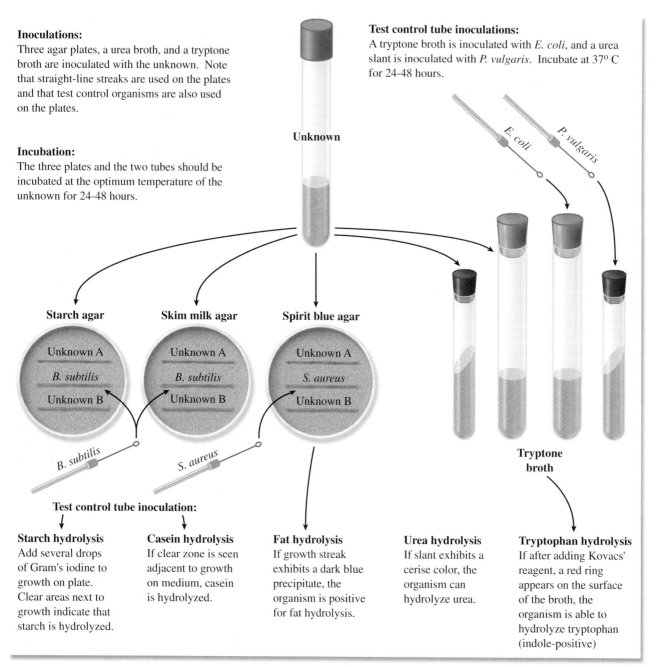

FIGURE 39.1   Procedure for doing hydrolysis tests on unknowns

starch is no longer present, the medium will have a **clear zone** next to the growth.

By pouring Gram's iodine over the growth on the medium, one can see clearly where starch has been hydrolyzed. If the area immediately adjacent to the growth is clear, amylase is produced.

Pour enough iodine over each streak to completely wet the entire surface of the plate. Rotate and tilt the plate gently to spread the iodine. Compare your unknowns with the positive result seen along the growth of *B. subtilis*. The left-hand illustration of figure 39.2 illustrates what it looks like.

Casein Hydrolysis

Casein is the predominant protein in milk and its presence causes milk to have its characteristic white color. Many bacteria produce *proteases* that degrade proteins such as casein into peptides and individual amino acids. This is referred to as *proteolysis.*

Examine the streaks on the skim milk agar plates. Note that a **clear zone** exists adjacent to the growth of *B. subtilis*. This is evidence of casein hydrolysis. The middle illustration in figure 39.2 shows what it looks like. Compare your unknown

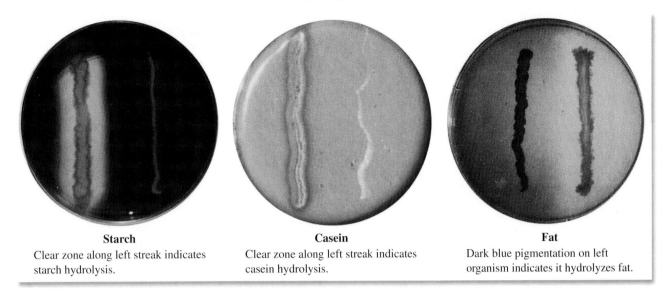

Starch
Clear zone along left streak indicates starch hydrolysis.

Casein
Clear zone along left streak indicates casein hydrolysis.

Fat
Dark blue pigmentation on left organism indicates it hydrolyzes fat.

FIGURE 39.2   Hydrolysis test plates: Starch, casein, fat

with this positive result and record the results on the descriptive chart on page 369.

Fat Hydrolysis

Fats or triglycerides are composed of three fatty acid molecules covalently bonded to glycerol. In animals, triglycerides are storage products. Some bacteria produce enzymes called *lipases* that cleave the fatty acids from the glycerol. The fatty acids and glycerol can then be used for various metabolic purposes such as synthesizing phospholipids for membrane construction or for catabolism to attain energy. The decomposition of fats plays a role in food spoilage such as butter or margarine becoming rancid.

$$
\begin{array}{c}
CH_2-O-\overset{\displaystyle O}{\overset{\|}{C}}-R \\[4pt]
CH-O-\overset{\displaystyle O}{\overset{\|}{C}}-R' \xrightarrow[\text{lipase}]{+\,3\,H_2O} \\[4pt]
CH_2-O-\overset{\displaystyle O}{\overset{\|}{C}}-R''
\end{array}
\qquad
\begin{array}{cc}
CH_2OH & RCOOH \\[4pt]
CHOH\; + & R'COOH \\[4pt]
CH_2OH & R''COOH
\end{array}
$$

Triglyceride   Glycerol   Fatty acids

Spirit blue agar contains a vegetable oil that, when hydrolyzed by most organisms, results in the lowering of the pH sufficiently to produce a **dark blue precipitate.** Unfortunately, the hydrolytic action of some organisms on this medium does not produce a blue precipitate because the pH is not lowered sufficiently.

Examine the *S. aureus* growth carefully. You should be able to see this dark blue reaction. The right-hand illustration in figure 39.2 exhibits what it should look like.

Compare the positive reaction of *S. aureus* with the reaction of your unknown. *If your unknown appears to be negative, hold the plate up toward the light and look for a region near growth where oil droplets are depleted.* If you see depletion of oil drops, consider your organism to be positive for this test. Record the results on the descriptive chart.

Tryptophan Hydrolysis

Certain bacteria, such as *E. coli*, have the ability to split the amino acid tryptophan into indole and pyruvic acid. The enzyme that causes this hydrolysis is *tryptophanase*. Indole can be easily detected with Kovacs' reagent. This test is particularly useful in differentiating *E. coli* from some closely related enteric bacteria.

$$
\text{Tryptophan} + H_2O \xrightarrow{\text{tryptophanase}} \text{Indole} + \underset{\text{COOH}}{CH_3-C=O} + NH_3
$$

Tryptophan   Indole   Pyruvic acid

Tryptone broth (1%) is used for this test because it contains a great deal of tryptophan. Tryptone is a peptone derived from casein by pancreatic digestion.

FIGURE 39.3 Tryptophane hydrolysis. The left tube shows the presence of indole (red band) at the top of the tube. The right hand tube is an unoculated control © The McGraw-Hill Companies/Auburn University Photographic Service

MATERIALS

- Kovacs' reagent
- tryptone broth cultures of unknown and *E. coli*

To test for indole, add 10 or 12 drops of Kovacs' reagent to the *E. coli* culture in tryptone broth. A **red layer** should form at the top of the culture, as shown in figure 39.3. Repeat the test on your unknown and record the results on the descriptive chart.

Urea Hydrolysis

Urea is waste product excreted in urine by animals. Some gram-negative enteric bacteria produce the enzyme *urease,* which splits the urea molecule into carbon dioxide and ammonia. The urease test is useful in identifying the genera *Proteus, Providentia,* and *Morganella,* which elaborate this enzyme. Refer to the separation outline in figure 41.1.

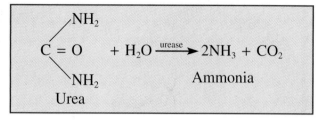

$$\begin{array}{c} \text{NH}_2 \\ | \\ \text{C} = \text{O} \\ | \\ \text{NH}_2 \end{array} + \text{H}_2\text{O} \xrightarrow{\text{urease}} 2\text{NH}_3 + \text{CO}_2$$

Urea → Ammonia

Urea slants contain yeast extract, urea, and a buffer. They also contain phenol red as the pH indicator. Urea is unstable and is broken down at 15 psi of pressure. It cannot be added to the medium for autoclaving and is therefore filter sterilized and added to the medium after autoclaving.

When urease is produced by an organism, ammonia is released causing the pH of the urea slant to become alkaline. As the pH increases, phenol red changes from yellow (pH 6.8) to a bright pink or cerise (pH 8.1. or higher). See figure 39.4.

Examine the urea slants inoculated with *Proteus vulgaris* and compare them to your unknown. *If your unknown is negative, continue to incubate the slant for 7 days to check for slow urease production.* Record your result in the descriptive chart on page 369.

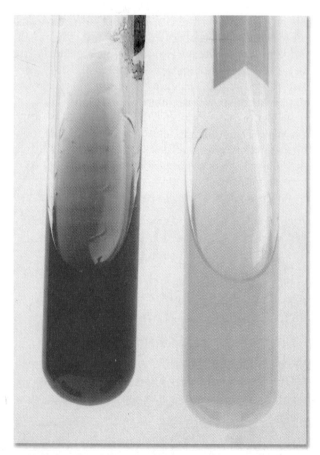

FIGURE 39.4 Urease test. Tube on the left is positive (Proteus); tube on the right is negative. © The McGraw-Hill Companies/Auburn University Photographic Service

Physiological Characteristics:
Biochemical Tests

There are several additional physiological tests used in unknown identification. They include tests for hydrogen sulfide production, citrate utilization, phenylalanine deaminization, and litmus milk reactions. During the first period, inoculations of four kinds of media will be made for these tests. An explanation of the value of the IMViC tests will also be included.

FIRST PERIOD

(Inoculations)

Since test controls are included in this exercise, two sets of inoculations will be made. For economy of materials, one set of test controls will be made by students working in pairs.

MATERIALS

for test controls, per pair of students:
- 1 Kligler's iron agar deep or SIM medium
- 1 Simmons citrate agar slant
- 1 phenylalanine agar slant
- nutrient broth cultures of *Proteus vulgaris*, *Staphylococcus aureus*, and *Enterobacter aerogenes*

per unknown, per student:
- 1 Kligler's iron agar deep or SIM medium
- 1 Simmons citrate agar slant
- 1 phenylalanine agar slant
- 1 litmus milk

1. Label one tube of Kligler's iron agar (or SIM medium) *P. vulgaris* and additional tubes with your unknown numbers. Inoculate each tube by stabbing with a straight wire.
2. Label one tube of Simmons citrate agar *E. aerogenes* and additional tubes with your unknown numbers. Use a straight wire to streak-stab each slant; i.e., streak the slant first, and then stab into the middle of the slant.
3. Label one tube of phenylalanine agar slant *P. vulgaris* and the other with your unknown code number. Streak each slant with the appropriate organisms.
4. With a loop, inoculate one tube of litmus milk with your unknown. (**Note:** A test control for this medium will not be made. Figure 40.2 will take its place.)
5. Incubate the unknowns at their optimum temperatures. Incubate the test controls at 37° C for 24–48 hours. 773-40ŀ 563e

SECOND PERIOD

(Evaluation of Tests)

After 24 to 48 hours incubation, examine the tubes to evaluate according to the following discussion. Record all results on the descriptive chart on page 368.

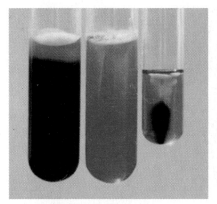

Hydrogen sulfide test

a) Positive tubes have black precipitate. Large tubes: Kligler; Small tube is SIM.

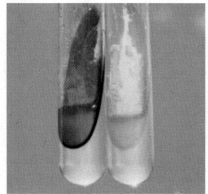

PPA test

b) Left-hand tube exhibits a positive reaction (green). Other tube is negative.

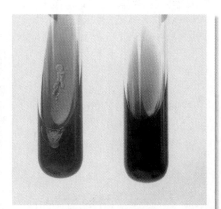

Citrate utilization

c) Left hand tube exhibits citrate utilization (Prussian blue color). Right hand tube is uninoculated or negative (green).

FIGURE 40.1 Hydrogen sulfide, PPA, and citrate utilization tests; (c) © The McGraw-Hill Companies/Auburn University Photographic Service

Hydrogen Sulfide Production

Certain bacteria, such as *Proteus vulgaris,* produce hydrogen sulfide from the amino acid cysteine. These organisms produce the enzyme *cysteine desulfurase,* which works in conjunction with the coenzyme pyridoxyl phosphate. The production of H_2S is the initial step in the deamination of cysteine as indicated below:

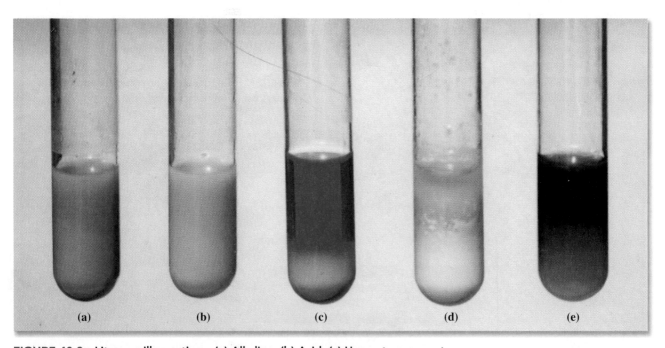

CH₂SH		CH₂		CH₃		CH₃

$$
\begin{array}{c}
\text{CH}_2\text{SH} \\
\overset{\text{H}}{\underset{\text{NH}_2}{\text{C}}} \\
\text{COOH}
\end{array}
\longrightarrow
\text{H}_2\text{S} +
\begin{array}{c}
\text{CH}_2 \\
\overset{\parallel}{\text{C}} - \text{NH}_3 \\
\text{COOH}
\end{array}
\longrightarrow
\begin{array}{c}
\text{CH}_3 \\
\text{C} = \text{NH} \\
\text{COOH}
\end{array}
\overset{\text{H}_2\text{O}}{\longrightarrow}
\begin{array}{c}
\text{CH}_3 \\
\text{C} = \text{O} + \text{NH}_3 \\
\text{COOH}
\end{array}
$$

Cysteine | α Amino acrylic acid | Imino acid | Pyruvic acid

Kligler's iron agar or SIM medium is used here to detect hydrogen sulfide production. Both of these media contain iron salts that react with H_2S to form a **dark precipitate** of iron sulfide.

Kligler's iron agar also contains glucose, lactose, and phenol red. When this medium is used in slants it is an excellent medium for detecting glucose and lactose fermentation. SIM medium, on the other hand, can also be used for determining motility and testing for indole production.

Examine the tube of one of these media that was inoculated with *P. vulgaris.* If it is Kligler's iron agar it will look like the left-hand tube in figure 40.1. A positive reaction in SIM medium will look like the small tube on the right.

Compare your unknown with this control tube and record your results on the descriptive chart.

Citrate Utilization

The ability of some organisms, such as *E. aerogenes* and *Salmonella typhimurium,* to utilize citrate as a sole source of carbon can be a very useful differentiation characteristic in working with intestinal bacteria. Koser's citrate medium and Simmons citrate agar are two media that are used to detect this ability in bacteria. In both of these synthetic media sodium citrate is the sole carbon source; nitrogen is supplied by ammonium salts instead of amino acids.

Examine the test control slant of this medium that was inoculated with *E. aerogenes.* Note the distinct **Prussian blue color change** that has occurred. Refer to the right-hand illustration in figure 40.1. Record your results on the descriptive chart.

Phenylalanine Deamination

A few bacteria, such as *Proteus, Morganella,* and *Providencia,* produce the *phenylalanine* deaminase, which deaminizes the amino acid phenylalanine to produce phenylpyruvic acid (PPA). This characteristic is used to help differentiate these three genera from other genera of the Enterobacteriaceae. The reaction is as follows:

FIGURE 40.2 Litmus milk reactions: (a) Alkaline. (b) Acid. (c) Upper transparent portion is peptonization; solid white portion in bottom is coagulation and litmus reduction; overall redness is interpreted as acid. (d) Coagulation and litmus reduction in lower half; some peptonization (transparency) and acid in top portion. (e) Litmus indicator is masked by production of soluble pigment (Pseudomonas); some peptonization is present but difficult to see in photo.

Proceed as follows to test for the production of phenylpyruvic acid, which is evidence that the enzyme *phenylalanine* deaminase has been produced:

MATERIALS

- dropping bottle of 10% ferric chloride

Allow 5–10 drops of 10% ferric chloride to flow down over the slants of the test control (*P. vulgaris*) and your unknowns. To hasten the reaction, use a loop to emulsify the organisms into solution. A deep **green color** should appear on the test control slant in 1–5 minutes. Refer to the middle illustration in figure 40.1. Compare your unknown with the control and record your results in Laboratory Report 38–40.

The IMViC Tests

In the differentiation of *E. aerogenes* and *E. coli,* as well as some other related species, four physiological tests have been grouped together into what are called the IMViC tests. The *I* stands for indole; the *M* and *V* stand for methyl red and Voges-Proskauer tests; *i* simply facilitates pronunciation; and the *C* signifies citrate utilization. In the differentiation of the two coliforms *E. coli* and *E. aerogenes,* the test results appear as charted below, revealing completely opposite reactions for the two organisms on all tests.

	I	M	V	C
E. coli	+	+	−	−
E. aerogenes	−	−	+	+

The significance of these tests is that when testing drinking water for the presence of the sewage indicator *E. coli,* one must be able to rule out *E. aerogenes,* which has many of the morphological and physiological characteristics of *E. coli.* Since *E. aerogenes* is not always associated with sewage, its presence in water would not necessarily indicate sewage contamination.

If you are attempting to identify a gram-negative, facultative, rod-shaped bacterial organism, group these series of tests together in this manner to see how your unknown fits this combination of tests.

Litmus Milk Reactions

Litmus milk contains 10% powdered skim milk and a small amount of litmus as a pH indicator. When the medium is made up, its pH is adjusted to 6.8. It is an excellent growth medium for many organisms and can be very helpful in unknown characterization. In addition to revealing the presence or absence of fermentation, it can detect certain proteolytic characteristics in bacteria. A number of facultative bacteria with strong reducing powers are able to utilize litmus as an alternative electron acceptor to render it colorless. Figure 40.2 reveals the color changes that cover the spectrum of litmus milk changes. Since some of the reactions take 4 to 5 days to occur, the cultures should be incubated for at least this period of time; they should be examined every 24 hours, however. Look for the following reactions:

Acid Reaction   Litmus becomes pink. Typical of fermentative bacteria.

Alkaline Reaction   Litmus turns blue or purple. Many proteolytic bacteria cause this reaction in the first 24 hours.

Litmus Reduction   Culture becomes white; actively reproducing bacteria reduce the O/R potential of medium.

Coagulation   Curd formation. Solidification is due to protein coagulation. Tilting tube at 45° will indicate whether or not this has occurred.

Peptonization   Medium becomes translucent. It often turns brown at this stage. Caused by proteolytic bacteria.

Ropiness   Thick, slimy residue in bottom of tube. Ropiness can be demonstrated with sterile loop.

Record the litmus milk reactions of your unknown on the descriptive chart on page 368.

LABORATORY REPORT

Complete Laboratory Report 38–40, which reviews all physiological tests performed in the last three exercises.

LABORATORY REPORT

Student: _____

Date: _____ Section: _____

EXERCISES 38–40   Physiological Characteristics of Bacteria

A. Media

List the media that are used for the following tests:

1. Butanediol production
2. Hydrogen sulfide production
3. Indole production
4. Starch hydrolysis
5. Urease production
6. Citrate utilization
7. Fat hydrolysis
8. Casein hydrolysis
9. Catalase production
10. Mixed acid fermentation
11. Glucose fermentation
12. Nitrate reduction

B. Reagents

Select the reagents that are used for the following tests:

1. Indole test   Barritt's reagent—1
2. Voges-Proskauer test Gram's iodine—2
3. Catalase test   Hydrogen peroxide—3
4. Starch hydrolysis   Kovacs' reagent—4
                        None of these—5

ANSWERS

Media

1. _____
2. _____
3. _____
4. _____
5. _____
6. _____
7. _____
8. _____
9. _____
10. _____
11. _____
12. _____

Reagents

1. _____
2. _____
3. _____
4. _____

C. Ingredients

Select the ingredients of the reagents for the following tests. Consult Appendix B. More than one ingredient may be present in a particular reagent.

1. Oxidase test
2. Voges-Proskauer test
3. Indole test
4. Nitrite test

α-naphthol—1

Dimethyl-α-naphthylamine—2

Dimethyl-p-phenylenediamine hydrochloride—3

p-dimethylamine benzaldehyde—4

Potassium hydroxide—5

Sulfanilic acid—6

D. Enzymes

What enzymes are involved in the following reactions?

1. Urea hydrolysis
2. Hydrogen gas production from formic acid
3. Casein hydrolysis
4. Indole production
5. Nitrate reduction
6. Starch hydrolysis
7. Fat hydrolysis
8. Gelatin hydrolysis (Ex. 39)
9. Hydrogen sulfide production

ANSWERS

Ingredients

1. _____
2. _____
3. _____
4. _____

Enzymes

1. _____
2. _____
3. _____
4. _____
5. _____
6. _____
7. _____
8. _____
9. _____

Physiological Characteristics of Bacteria (continued)

E. **Test Results**

Indicate the appearance of the following positive test results.

1. Glucose fermentation, no gas
2. Citrate utilization
3. Urease production
4. Indole production
5. Acetoin production
6. Hydrogen sulfide production
7. Coagulation of milk
8. Peptonization in milk
9. Litmus reduction in milk
10. Nitrate reduction
11. Catalase production
12. Casein hydrolysis
13. Fat hydrolysis

ANSWERS

1. _____
2. _____
3. _____
4. _____
5. _____
6. _____
7. _____
8. _____
9. _____
10. _____
11. _____
12. _____
13. _____

F. **General Questions**

    1. Differentiate between the following:

        Respiration: _____

        Fermentation: _____

        Oxidation: _____

        Reduction: _____

        Catalase: _____

        Peroxidase: _____

    2. List two or three difficulties one encounters in trying to differentiate bacteria on the basis of physiological

        characteristics. _____

        _____

    3. Now that you have determined the morphological, cultural, and physiological characteristics of your un-
        known, what other kinds of tests might you perform on the organism to assist in identification?

        _____

        _____

Use of Bergey's Manual

Once you have recorded all the data on your descriptive chart pertaining to morphological, cultural, and physiological characteristics of your unknown, you are ready to determine its genus and species. Determination of the genus should be relatively easy; species differentiation, however, is considerably more difficult.

The most important single source of information we have for the identification of bacteria is *Bergey's Manual of Systematic Bacteriology.* This monumental achievement, which consists of four volumes, replaced a single-volume eighth edition of *Bergey's Manual of Determinative Bacteriology.* Although the more recent publication consists of four volumes, only volumes 1 and 2 will be used for the identification of the unknowns in this course.

Bergey's Manual is a worldwide collaborative effort that has an editorial board of 13 trustees. Over 200 specialists from 19 countries are listed as contributors to the first two volumes. One of the purposes of this exercise is to help you glean the information from these two volumes that is needed to identify your unknown. Before we get into the mechanics of using *Bergey's Manual,* a few comments are in order pertaining to the problems of bacterial classification.

CLASSIFICATION PROBLEMS

Compared with the classification of bacteria, the classification of plants and animals has been relatively easy. In these higher forms, a hierarchy of orders, families, and genera is based, primarily, on evolutionary evidence revealed by fossils laid down in sedimentary layers of earth's crust. Some of the earlier editions of *Bergey's Manual* attempted to use the same hierarchial system, but the attempt had to be abandoned when the eighth edition was published; without paleontological information to support the system, it literally fell apart.

The present system of classification in *Bergey's Manual* uses a list of "sections" that separate the various groups. Each section is described in common terms so that it is easily understood (even for beginners). For example, section 1 is entitled **The Spirochaetes**. Section 4 pertains to **Gram-Negative Aerobic Rods and Cocci**. If one scans the table of contents in each volume after having completed all tests, it is possible, usually, to find a section that contains the unknown being studied.

A perusal of these sections will reveal that some sections have a semblance of hierarchy in the form of orders, families, and genera. Other sections list only genera.

Thus, we see that the classification system of bacteria, as developed in *Bergey's Manual,* is not the tidy system we see in higher forms of life. The important thing is that it works.

Our dependency over the years on *Bergey's Manual* has led many to think of its classification system as the "official classification." Staley and Krieg in their overview in volume 1 emphasize that no official classification of bacteria exists; in other words, the system offered in *Bergey's Manual* is simply a workable system, but in no sense of the word should it be designated as the official classification system.

PRESUMPTIVE IDENTIFICATION

The place to start in identifying your unknown is to determine what genus it fits into. If *Bergey's Manual* is available, scan the tables of contents in volumes 1 and 2 to find the section that seems to describe your unknown. If these books are not immediately available, you can determine the genus by referring to the separation outlines in figures 41.1 and 41.2. Note that seven groups of gram-positive bacteria are winnowed out in figure 41.1 and four groups of gram-negative bacteria in figure 41.2.

To determine which genus in the group best fits the description of your unknown, compare the genera descriptions provided below. Note that each group has a section designation to identify its position in *Bergey's Manual.*

Group I (section 13, vol. 2) Although there are only three genera listed in this group, section 13 in *Bergey's Manual* lists three additional genera, one of which is *Sporosarcina,* a coccus-shaped organism

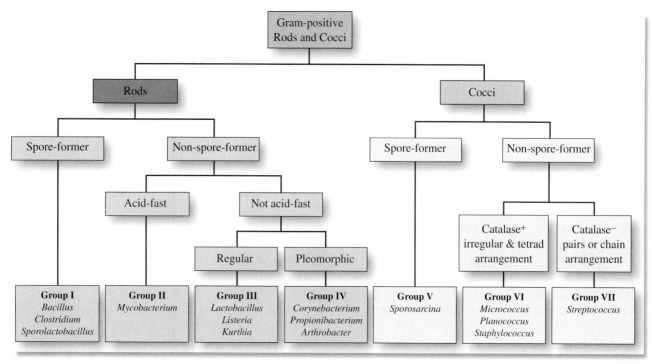

FIGURE 41.1 Separation outline for gram-positive rods and cocci

(see Group V). Most members of Group I are motile and differentiation is based primarily on oxygen needs.

Bacillus Although most of these organisms are aerobic, some are facultative anaerobes. Catalase is usually produced. For comparative characteristics of the 34 species in this genus refer to Table 13.4.

Clostridium While most of members of this genus are strict anaerobes, some may grow in the presence of oxygen. Catalase is not usually produced. An excellent key for presumptive species identification is provided on pages 1143–1148. Species characterization tables are also provided.

Sporolactobacillus Microaerophilic and catalase-negative. Nitrates are not reduced and indole is not formed. Spore formation occurs very infrequently (1% of cells).

Since there is only one species in this genus, one needs only to be certain that the unknown is definitely of this genus. Table 13.11 can be used to compare other genera that are similar to this one.

Group II (section 16, vol. 2) This group consists of Family Mycobacteriaceae, with only one genus:

Mycobacterium. Fifty-four species are listed in section 16. Differentiation of species within this group depends to some extent on

whether the organism is classified as a slow or a fast grower. Tables in this section can be used for comparing the characteristics of the various species.

Group III (section 14, vol. 2) Of the seven diverse genera listed in section 14, only three have been included here in this group.

Lactobacillus Non-spore-forming rods, varying from long and slender to coryneform (club-shaped) coccobacilli. Chain formation is common. Only rarely motile. Facultative anaerobic or microaerophilic. Catalase-negative. Nitrate usually not reduced. Gelatin not liquefied. Indole and H_2S not produced.

Listeria Regular, short rods with rounded ends; occur singly and in short chains. Aerobic and facultative anaerobic. Motile when grown at 20–25° C. Catalase-positive and oxidase-negative. Methyl red positive. Voges-Proskauer positive. Negative for citrate utilization, indole production, urea hydrolysis, gelatinase production, and casein hydrolysis. Table 14.12 provides information pertaining to species differentiation in this genus.

Kurthia Regular rods, 2–4 micrometers long with rounded ends; in chains in young cultures; coccoidal in older cultures. Strictly aerobic. Catalase-positive, oxidase-negative.

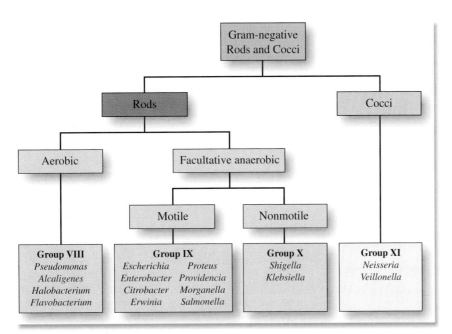

FIGURE 41.2 Separation outline
for gram-negative rods and cocci

Also negative for gelatinase production
and nitrate reduction. Only two species
in this genus.

Group IV (section 15, vol. 2) Although there are 21
genera listed in this section of *Bergey's Manual*, only
three genera are described here.

> *Corynebacterium* Straight to slightly curved rods
> with tapered ends. Sometimes club-shaped.
> Palisade arrangements common due to snap-
> ping division of cells. Metachromatic gran-
> ules formed. Facultative anaerobic. Cata-
> lase-positive. Most species produce acid
> from glucose and some other sugars.
> Often produce pellicle in broth. Table 15.3
> provides information for species
> characterization.
>
> *Proprionibacterium* Pleomorphic rods, often
> diphtheroid or club-shaped with one end
> rounded and the other tapered or pointed.
> Cells may be coccoid, bifid (forked, di-
> vided), or even branched. Nonmotile. Some
> produce clumps of cells with "Chinese char-
> acter" arrangements. Anaerobic to aerotoler-
> ant. Generally catalase-positive. Produce
> large amounts of proprionic and acetic acids.
> All produce acid from glucose.
>
> *Arthrobacter* Gram-positive rod and coccoid
> forms. Pleomorphic. Growth often starts out
> as rods, followed by shortening as growth
> continues, and finally becoming coccoidal.
> Some V and angular forms; branching by

some. Rods usually nonmotile; some motile.
Oxidative, never fermentative. Catalase-
positive. Little or no gas produced from
glucose or other sugars. Type species is
Arthrobacter globiformis.

Group V (section 13, vol. 2) This group, which
has only one genus in it, is closely related to genus
Bacillus.

> *Sporosarcina* Cells are spherical or oval when
> single. Cells may adhere to each other when
> dividing to produce tetrads or packets of
> eight or more. Endospores formed (see
> photomicrographs). Strictly aerobic. Gener-
> ally motile. Only two species: *S. ureae* and
> *S. halophila.*

Group VI (section 12, vol. 2) This section contains
two families and 15 genera. Our concern here is with
only three genera in this group. Oxygen requirements
and cellular arrangement are the principal factors in
differentiating the genera. Most of these genera are
not closely related.

> *Micrococcus* Spheres, occurring as singles,
> pairs, irregular clusters, tetrads, or cubical
> packets. Usually nonmotile. Strict aerobes
> (one species is facultative anaerobic). Cata-
> lase- and oxidase-positive. Most species pro-
> duce carotenoid pigments. All species will
> grow in media containing 5% NaCl. For
> species differentiation, see table 12.4.

Planococcus Spheres, occurring singly, in pairs, in groups of three cells, occasionally in tetrads. Although cells are generally gram-positive, they may be gram-variable. Motility is present. Catalase- and gelatinase-positive. Carbohydrates not attacked. Do not hydrolyze starch or reduce nitrate. Refer to table 12.9 for species differentiation.

Staphylococcus Spheres, occurring as singles, pairs, and irregular clusters. Nonmotile. Facultative anaerobes. Usually catalase-positive. Most strains grow in media with 10% NaCl. Susceptible to lysis by lysostaphin. Glucose fermentation: acid, no gas. Coagulase production by some. Refer to Exercise 70 for species differentiation, or to table 12.10.

Group VII
(section 12, vol. 2) Note that the single genus of this group is included in the same section of *Bergey's Manual* as the three genera in group VI. Members of the genus *Streptococcus* have spherical to ovoid cells that occur in pairs or chains when grown in liquid media. Some species, notably, *S. mutans,* will develop short rods when grown under certain circumstances. Facultative anaerobes. Catalase-negative. Carbohydrates are fermented to produce lactic acid without gas production. Many species are commensals or parasites of humans or animals. Refer to Exercise 53 for species differentiation of pathogens. Several tables in *Bergey's Manual* provide differentiation characteristics of all the streptococci.

Group VIII
(section 4, vol. 1) Although there are many genera of gram-negative aerobic rod-shaped bacteria, only four genera are likely to be encountered here.

Pseudomonas Generally motile. Strict aerobes. Catalase-positive. Some species produce soluble fluorescent pigments that diffuse into the agar of a slant. Many tables are available in *Bergey's Manual* for species differentiation.

Alcaligenes Rods, coccal rods, or cocci. Motile. Obligate aerobes with some strains capable of anaerobic respiration in presence of nitrate or nitrite.

Halobacterium Cells may be rod- or disk-shaped. Cells divide by constriction. Most are strict aerobes; a few are facultative anaerobes. Catalase- and oxidase-positive. Colonies are pink, red, or red to orange. Gelatinase not produced. Most species re-quire high NaCl concentrations in media. Cell lysis occurs in hypotonic solutions.

Flavobacterium Gram-negative rods with parallel sides and rounded ends. Nonmotile. Oxidative. Catalase-, oxidase-, and phosphatase-positive. Growth on solid media is typically pigmented yellow or orange. Nonpigmented strains do exist. See tables differentiation.

Groups IX and X
(section 5, vol. 1) Section 5 in *Bergey's Manual* lists three families and 34 genera; of these 34, only 10 genera of Family *Enterobacteriaceae* have been included in these two groups. If your unknown appears to fall into one of these groups, use the separation outline in figure 41.3 to determine the genus. Another useful separation outline is provided in figure 54.1. *Keep in mind, when using these separation outlines, that there are some minor exceptions in the applications of these tests.* The diversity of species within a particular genus often presents some problematical exceptions to the rule. Your final decision can be made only after checking the species characteristics tables for each genus in *Bergey's Manual.*

Group XI
These genera are morphologically quite similar, yet physiologically quite different.

Neisseria (section 4, vol. 1) Cocci, occurring singly, but more often in pairs (diplococci); adjacent sides are flattened. One species (*N. elongata*) consists of short rods. Nonmotile. Except for *N. elongata*, all species are oxidase- and catalase-positive. Aerobic.

Veillonella (section 8, vol. 1) Cocci, appearing as diplococci, masses, and short chains. Diplococci have flattening at adjacent surfaces. Nonmotile. All are oxidase- and catalase-negative. Nitrate is reduced to nitrite. Anaerobic.

PROBLEM ANALYSIS

If you have identified your unknown by following the above procedures, congratulations! Not everyone succeeds at first attempt. If you are having difficulty, consider the following possibilities:

- You may have been given the wrong unknown! Although this is a remote possibility, it does happen at times. Occasionally, clerical errors are made when unknowns are put together.
- Your organism may be giving you a "false-negative" result on a test. This may be due to an in-

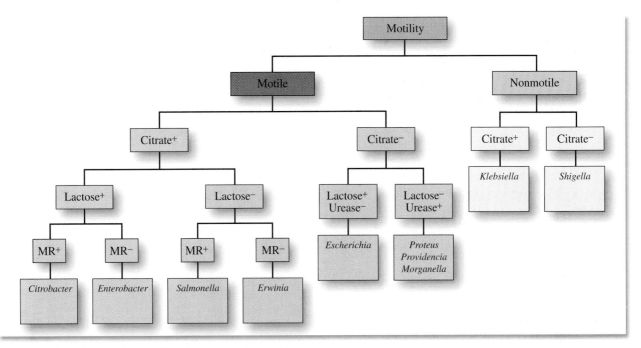

FIGURE 41.3 Separation outline for groups IX and X

correctly prepared medium, faulty test reagents, or improper testing technique.

- Your unknown organisms may not match the description *exactly* as stated in *Bergey's Manual.* By now you are aware that the words *generally, usually,* and *sometimes* are frequently used in the book. It is entirely possible for one of these words to be inadvertently left out in Bergey's assignment of certain test results to a species. *In other words, test results, as stated in the manual, may not always apply!*
- Your culture may be contaminated. If you are not working with a pure culture, all tests are unreliable.
- You may not have performed enough tests. Check the various tables in *Bergey's Manual* to see if

there is some other test that will be helpful. In addition, double-check the tables to make sure that you have read them correctly.

CONFIRMATION OF RESULTS

There are several ways to confirm your presumptive identification. One method is to apply serological techniques, if your organism is one for which typing serum is available. Another alternative is to use one of the miniature multitest systems that are described in the next section of this manual. Your instructor will indicate which of these alternatives, if any, will be available.

LABORATORY REPORT

There is no Laboratory Report for this exercise.

MINIATURIZED MULTITEST SYSTEMS

9

Having run a multitude of tests in Exercises 36 through 41 in an attempt to identify an unknown, you undoubtedly have become aware of the tremendous amount of media, glassware, and preparation time that is involved just to set up the tests. And then, after performing all of the tests and meticulously following all the instructions, you discover that finding the specific organism in "Encyclopedia Bergey" is not exactly the simplest task you have accomplished in this course. The thought must arise occasionally: "There's got to be an easier way!" Fortunately, there are *miniaturized multitest systems*.

Miniaturized systems have the following advantages over the macromethods you have used to study the physiological characteristics of your unknown: (1) minimum media preparation, (2) simplicity of performance, (3) reliability, (4) rapid results, and (5) uniform results. These advantages have resulted in widespread acceptance of these systems by microbiologists.

Since it is not possible to describe all of the systems that are available, only four have been selected here: two by Analytab Products and two by Becton-Dickinson. All four of these products are designed specifically to provide rapid identification of medically important organisms, often within 5 hours. Each method consists of a plastic tube or strip that contains many different media to be inoculated and incubated. To facilitate rapid identification, these systems utilize numerical coding systems that can be applied to charts or computer programs.

The four multitest systems described in this unit have been selected to provide several options. Exercises 42 and 43 pertain to the identification of gram-negative, *oxidase-negative* bacteria (Enterobacteriaceae). Exercise 44 (Oxi/Ferm Tube) is used for identifying gram-negative, *oxidase-positive* bacteria. Exercise 45 (Staph-Ident) is a rapid system for the differentiation of the staphylococci.

As convenient as these systems are, one must not assume that the conventional macromethods of part 8 are becoming obsolete. Macromethods must still be used for culture studies and confirmatory tests; confirmatory tests by macromethods are often necessary when a particular test on a miniaturized system is in question. Another point to keep in mind is that all of the miniaturized multitest systems have

been developed for the identification of *medically important* microorganisms. If one is trying to identify a saprophytic organism of the soil, water, or some other habitat, there is no substitute for the conventional methods.

If these systems are available to you in this laboratory, they may be used to confirm your conclusions that were drawn in part 8 or they may be used in conjunction with some of the exercises in part 11. Your instructor will indicate what applications will be made.

Enterobacteriaceae Identification:
The API 20E System

The **API 20E System** is a miniaturized version of conventional tests that is used for the identification of members of the family Enterobacteriaceae and other gram-negative bacteria. It was developed by Analytab Products, of Plainview, New York. This system utilizes a plastic strip (figure 42.1) with 20 separate compartments. Each compartment consists of a depression, or *cupule*, and a small *tube* that contains a specific dehydrated medium (see illustration 4, figure 42.2). The system has a capacity of 23 biochemical tests.

To inoculate each compartment, it is necessary to first make up a saline suspension of the unknown organism; then, with the aid of a Pasteur pipette, fill each compartment with the bacterial suspension. The cupule receives the suspension and allows it to flow into the tube of medium. The dehydrated medium is reconstituted by the saline. To provide anaerobic conditions for some of the compartments, it is necessary to add sterile mineral oil to them.

After incubation for 18–24 hours, the reactions are recorded, test reagents are added to some compartments, and test results are tabulated. Once the test results are tabulated, a *profile number* (7 or 9 digits) is computed. By finding the profile number in a code book, the *Analytical Profile Index*, one is able to determine the name of the organism. If no *Analytical Profile Index* is available, characterization can be done by using chart III in Appendix D.

Although this system is intended for the identification of nonenterics, as well as the Enterobacteriaceae, only the identification of the latter will be pursued in this experiment. Proceed as follows to use the API 20E System to identify your unknown enteric.

FIRST PERIOD

Two things will be accomplished during this period: (1) the oxidase test will be performed if it has not been previously performed, and (2) the API 20E test strip will be inoculated. All steps are illustrated in figure 42.2. Proceed as follows to use this system:

MATERIALS

- agar slant or plate culture of unknown
- test tube of 5 ml 0.85% sterile saline
- API 20E test strip
- API incubation tray and cover
- squeeze bottle of tap water
- test tube of 5 ml sterile mineral oil

All tests: positive

All tests: negative

FIGURE 42.1   Positive and negative test results on API 20E test Courtesy of Analytab Products, Plainview, NY

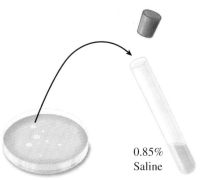

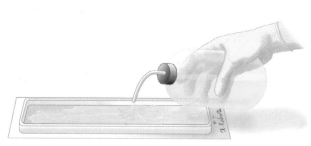

(1) Select one well-isolated colony to make a saline suspension of the unknown organism. Suspension should be well dispersed with a Vortex mixer.

(2) After labeling the end tab of a tray with your name and unknown number, dispense approximately 5 ml of tap water into bottom of tray.

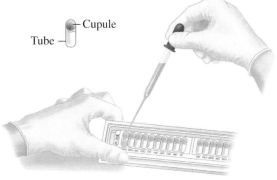

(3) Place an API 20E test strip into the bottom of the moistened tray. Be sure to seal the pouch from which the test strip was removed to prevent contamination of remaining strips.

(4) Dispense saline suspension of organisms into cupules of all twenty compartments. Slightly *underfill* ADH, LDC, ODC, H₂S, and URE. *Completely fill* cupules of CIT, VP, and GEL.

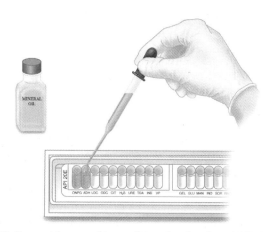

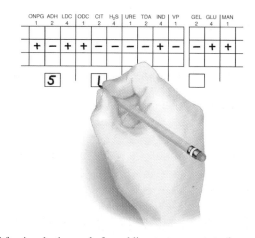

ONPG	ADH	LDC	ODC	CIT	H₂S	URE	TDA	IND	VP	GEL	GLU	MAN
1	2	4	1	2	4	1	2	4	1	2	4	1
+	−	+	+	−	−	−	−	−	+	−	−	+	+

5   L

(5) To provide anaerobic conditions for chambers ADH, LDC, ODC, H₂S, and URE, completely fill cupules of these chambers with sterile mineral oil. Use a fresh sterile Pasteur pipette.

(6) After incubation and after adding test reagents to four compartments, record all results and total numbers to arrive at 7-digit code. Consult the *Analytical Profile Index* to find the unknown.

FIGURE 42.2 **Procedure for preparing and inoculating the API 20E test strip**

- Pasteur pipettes (5 ml size)
- oxidase test reagent
- Whatman no. 2 filter paper
- empty petri dish
- Vortex mixer

1. If you haven't already done the **oxidase test** on your unknown, do so at this time. It must be established that your unknown is definitely oxidase-negative before using this system. Use the method that is described on page 259.

2. Prepare a **saline suspension** of your unknown by transferring organisms from the center of a well-established colony on an agar plate (or from a slant culture) to a tube of 0.85% saline solution. Disperse the organisms well throughout the saline.

3. Label the end strip of the API 20E tray with your name and unknown number. See illustration 2, figure 42.2.

4. Dispense about 5 ml of tap water into the tray with a squeeze bottle. Note that the bottom of the tray has numerous depressions to accept the water.

5. Remove an API 20E test strip from the sealed pouch and place it into the tray (see illustration 3). Be sure to reseal the pouch to protect the remaining strips.

6. Vortex mix the saline suspension to get uniform dispersal, and fill a sterile Pasteur pipette with the suspension. *Take care not to spill any of the organisms on the table or yourself. You may have a pathogen!*

7. Inoculate all the tubes on the test strip with the pipette by depositing the suspension into the cupules as you tilt the API tray (see illustration 4, figure 42.2).
   Important: Slightly *underfill* ADH, LDC, ODC, H₂S, and URE. (Note that the labels for these compartments are underlined on the strip.) Underfilling these compartments leaves room for oil to be added and facilitates interpretation of the results.

8. Since the media in |CIT|, |VP|, and |GEL| compartments require oxygen, *completely fill both the cupule and tube* of these compartments. Note that the labels on these three compartments are bracketed as shown here.

9. To provide anaerobic conditions for the ADH, LDC, ODC, H₂S, and URE compartments, dispense sterile **mineral oil** to the cupules of these compartments. Use another sterile Pasteur pipette for this step.

10. Place the lid on the incubation tray and incubate at 37° C for 18 to 24 hours. Refrigeration after incubation is not recommended.

SECOND PERIOD

(Evaluation of Tests)

During this period, all reactions will be recorded on the Laboratory Report, test reagents will be added to four compartments, and the seven-digit profile number will be determined so that the unknown can be looked up in the *API 20E Analytical Profile Index*. Proceed as follows:

MATERIALS

- incubation tray with API 20E test strip
- 10% ferric chloride
- Barritt's reagents A and B
- Kovacs' reagent
- nitrite test reagents A and B
- zinc dust or 20-mesh granular zinc
- hydrogen peroxide (1.5%)
- *API 20E Analytical Profile Index*
- Pasteur pipettes

1. Before any test reagents are added to any of the compartments, consult chart I, Appendix D, to determine the nature of positive reactions of each test, except TDA, VP, and IND.

2. Refer to chart II, Appendix D, for an explanation of the 20 symbols that are used on the plastic test strip.

3. Record the results of these tests on Laboratory Report 42.

4. **If GLU test is negative** (blue or blue-green), **and there are fewer than three positive reactions** before adding reagents, do not progress any further with this test as outlined here in this experiment. Organisms that are GLU-negative are nonenterics.

   For nonenterics, additional incubation time is required. If you wish to follow through on an organism of this type, consult your instructor for more information.

5. **If GLU test is positive** (yellow), **or there are more than three positive reactions**, proceed to add reagents as indicated in the following steps.

6. Add one drop of **10% ferric chloride** to the TDA tube. A positive reaction (brown-red), if it occurs, will occur immediately. A negative reaction color is yellow.

7. Add 1 drop each of **Barritt's A** and **B solutions** to the VP tube. Read the VP tube within 10 minutes. The pale pink color that occurs immediately has no significance. A positive reaction is dark pink or red and may take 10 minutes before it appears.

8. Add 1 drop of **Kovacs' reagent** to the IND tube. Look for a positive (red ring) reaction within 2 minutes.

   After several minutes, the acid in the reagent reacts with the plastic cupule to produce a color change from yellow to brownish-red, which is considered negative.

9. Examine the GLU tube closely for evidence of bubbles. Bubbles indicate the reduction of nitrate and the formation of N_2 gas. Note on the Laboratory Report that there is a place to record the presence of this gas.

10. Add 2 drops of each **nitrite test reagent** to the GLU tube. A positive (red) reaction should show up within 2 to 3 minutes if nitrates are reduced.

    If this test is negative, confirm negativity with **zinc dust** or 20-mesh granular zinc. A pink-orange color after 10 minutes confirms that nitrate reduction did not occur. A yellow color results if N_2 was produced.

11. Add 1 drop of **hydrogen peroxide** to each of the MAN, INO, and SOR cupules. If catalase is produced, gas bubbles will appear within 2 minutes. Best results will be obtained in tubes that have no gas from fermentation.

FINAL CONFIRMATION

After all test results have been recorded and the seven-digit profile number has been determined, according to the procedures outlined on the Laboratory Report, identify your unknown by looking up the profile number in the *API 20E Analytical Profile Index*.

CLEANUP

When finished with the test strip, be sure to place it in a container of disinfectant that has been designated for test strip disposal.

LABORATORY REPORT

Student: _____

Date: _____ Section: _____

EXERCISE 42 Enterobacteriaceae Identification: The API 20E System

A. Tabulation of Results

By referring to charts I and II, Appendix D, determine the results of each test and record these results as positive (+) or negative (−) in the table below. Note that the results of the oxidase test must be recorded in the last column on the right side of the table.

ONPG	ADH	LDC	ODC	CIT	H₂S	URE	TDA	IND	VP	GEL	GLU	MAN	INO	SOR	RHA	SAC	MEL	AMY	ARA	OXI
1	2	4	1	2	4	1	2	4	1	2	4	1	2	4	1	2	4	1	2	4

☐ ☐ ☐ ☐ ☐ ☐ ☐

NO₂	N2 GAS	MOT	MAC	OF-O	OF-F
1	2	4	1	2	4

☐ ☐ Additional Digits

B. Construction of Seven-Digit Profile

Note in the above table that each test has a value of 1, 2, or 4. To compute the seven-digit profile for your unknown, total up the positive values for each group.

Example:
5 144 572 = *E. coli*

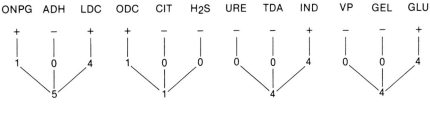

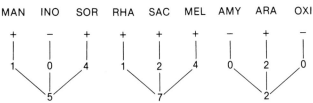

C. Using the API 20E Analytical Index or the API Characterization Chart

If the *API 20E Analytical Index* is available on the demonstration table, use it to identify your unknown, using the seven-digit profile number that has been computed. If no *Analytical Index* is available, use characterization chart III in Appendix D.

Name of Unknown: _____

D. **Additional Tabulation Blank**

If you need another form, use the one below:

api® 20E	Reference Number _____	Patient _____	Date _____
	Source/Site _____	Physician _____	Dept./Service _____

	ONPG 1	ADH 2	LDC 4	ODC 1	CIT 2	H₂S 4	URE 1	TDA 2	IND 4	VP 1	GEL 2	GLU 4	MAN 1	INO 2	SOR 4	RHA 1	SAC 2	MEL 4	AMY 1	ARA 2	OXI 4
5 h																					
24 h																					
48 h																					
Profile Number																					

	NO₂ 1	N₂ GAS 2	MOT 4	MAC 1	OF-O 2	OF-F 4	Additional Information
5 h							
24 h							Identification
48 h							
Additional Digits							00-42-012 E-3 (7/80)

E. **Questions**

1. What is the intended function of the API 20E system? _____

_____

_____

_____

2. In the "real world," who would use this system? _____

_____

_____

_____

3. What might be an explanation for the failure of this system to work with some of the bacterial cultures

we use? _____

_____

_____

Enterobacteriaceae Identification:
The Enterotube II System

The **Enterotube II** miniaturized multitest system was developed by Becton-Dickinson of Cockeysville, Maryland, for rapid identification of Enterobacteriaceae. It incorporates 12 different conventional media and 15 biochemical tests into a single ready-to-use tube that can be simultaneously inoculated in a moment's time with a minimum of equipment.

If you have an unknown gram-negative rod or coccobacillus that appears to be one of the Enterobacteriaceae, you may wish to try this system on it. Before applying this test, however, *make certain that your unknown is oxidase-negative*, since with only a few exceptions, all Enterobacteriaceae are oxidase-negative. If you have a gram-negative rod that is oxidase-positive, you might try the *Oxi/Ferm Tube II* instead, which is featured in Exercise 44.

Figure 43.1 illustrates an uninoculated tube (upper) and a tube with all positive reactions (lower). Figure 43.2 outlines the entire procedure for utilizing this system.

Each of the 12 compartments of an Enterotube II contains a different agar-based medium. Compartments that require aerobic conditions have openings for access to air. Those compartments that require anaerobic conditions have layers of paraffin wax over the media. Extending through all compartments of the entire tube is an inoculating wire. To inoculate the media, one simply picks up some organisms on the end of the wire and pulls the wire through each of the chambers in a single rotating action.

After incubation, the reactions in all the compartments are noted and the indole test is performed. The Voges-Proskauer test may also be performed as a confirmation test. Positive reactions are given numerical values, which are totaled to arrive at a five-digit code. Identification of the unknown is achieved by consulting a coding manual, the *Enterotube II Interpretation Guide*, which lists these numerical codes for the Enterobacteriaceae. Proceed as follows to use an Enterotube II in the identification of your unknown.

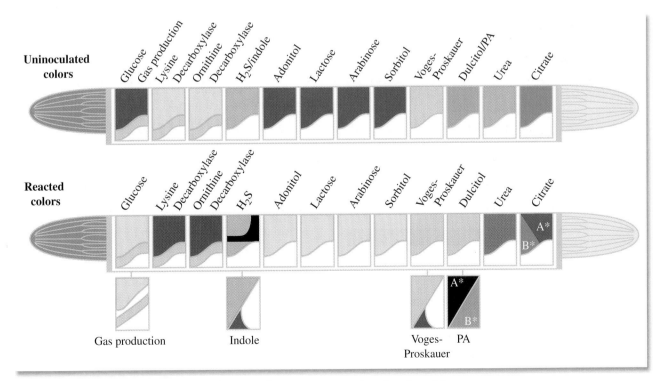

FIGURE 43.1   Enterotube II color differences between uninoculated and positive tests Courtesy of Becton-Dickinson, Cockeysville, Maryland.

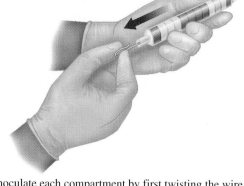

(1) Remove organisms from a well-isolated colony. Avoid touching the agar with the wire. To prevent damaging Enterotube II media, do not heat-sterilize the inoculating wire.

(2) Inoculate each compartment by first twisting the wire and then withdrawing it all the way out through the 12 compartments, using a turning movement.

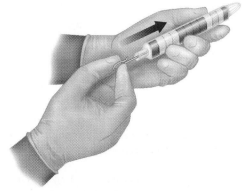

(3) Reinsert the wire (without sterilizing), using a turning motion through all 12 compartments until the notch on the wire is aligned with the opening of the tube.

(4) Break the wire at the notch by bending. The portion of the wire remaining in the tube maintains anaerobic conditions essential for true fermentation.

continued

FIGURE 43.2 The Enterotube II procedure

FIRST PERIOD

Inoculation and Incubation

The Enterotube II can be used to identify Enterobacteriaceae from colonies on agar that have been inoculated from urine, blood, sputum, and so on. The culture may be taken from media such as MacConkey, EMB, SS, Hektoen enteric, or trypticase soy agar.

MATERIALS

- culture plate of unknown
- 1 Enterotube II

1. Write your initials or unknown number on the white paper label on the side of the tube.
2. Unscrew both caps from the Enterotube II. The tip of the inoculating end is under the white cap.
3. *Without heat-sterilizing* the exposed inoculating wire, insert it into a well-isolated colony.
4. Inoculate each chamber by first twisting the wire and then withdrawing it through all 12 compart-

ments. Rotate the wire as you pull it through. See illustration 2, figure 43.2.
5. Again, *without sterilizing*, reinsert the wire, and with a turning motion, force it through all 12 compartments until the notch on the wire is aligned with the opening of the tube. (The notch is about $1\frac{5}{8}''$ from handle end of wire.) The tip of the wire should be visible in the citrate compartment. See illustration 3, figure 43.2.
6. Break the wire at the notch by bending, as shown in step 4, figure 43.2. The portion of the wire remaining in the tube maintains anaerobic conditions essential for fermentation of glucose, production of gas, and decarboxylation of lysine and ornithine.
7. With the retained portion of the needle, punch holes through the thin plastic coverings over the small depressions on the sides of the last eight compartments (adonitol, lactose, arabinose, sorbitol, Voges-Proskauer, dulcitol/PA, urea, and cit-

(5) Punch holes with broken-off part of wire through the thin plastic covering over depressions on sides of the last eight compartments (adonitol through citrate). Replace caps and incubate at 35° C for 18–24 hours.

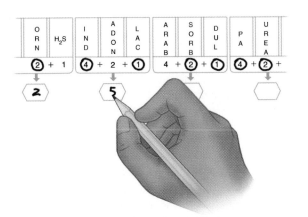

(6) After interpreting and recording positive results on the sides of the tube, perform the indole test by injecting 1 or 2 drops of Kovacs' reagent into the H_2S/indole compartment.

(7) Perform the Voges-Proskauer test, if needed for confirmation, by injecting the reagents into the H_2S/indole compartment.

After encircling the numbers of the positive tests on the Laboratory Report, total up the numbers of each bracketed series to determine the 5-digit code number. Refer to the *Enterotube II Interpretation Guide* for identification of the unknown by using the code number.

FIGURE 43.2 (continued)

rate). These holes will enable aerobic growth in these eight compartments.

8. Replace the caps at both ends.
9. Incubate at 35° to 37° C for 18 to 24 hours with the Enterotube II lying on its flat surface. *When incubating several tubes together, allow space between them to allow for air circulation.*

SECOND PERIOD

Reading Results

Reading the results on the Enterotube may be done in one of two ways: (1) by simply comparing the results with information on chart IV, Appendix D, or (2) by finding the five-digit code number you compute for your unknown in the *Enterotube II Interpretation Guide*. Of the two methods, the latter is much preferred. The chart in the appendix should be used *only* if the *Interpretation Guide* is not available.

Whether or not the *Interpretation Guide* is available, these three steps will be performed during this

period to complete this experiment: (1) positive test results must *first* be recorded on the Laboratory Report, (2) the indole test, a presumptive test, is performed on compartment 4, and (3) confirmatory tests, if needed, are performed. The Voges-Proskauer test falls in the latter category. Proceed as follows:

MATERIALS

- Enterotube II, inoculated and incubated
- Kovacs' reagent
- 10% KOH with 0.3% creatine solution
- 5% alpha-naphthol in absolute ethyl alcohol
- syringes with needles, or disposable Pasteur pipettes
- test-tube rack
- Enterotube II Results Pad (optional)
- coding manual: *Enterotube II Interpretation Guide*

1. Compare the colors of each compartment of your Enterotube II with the lower tube illustrated in figure 43.1.

TABLE 43.1 Biochemical Reactions of Enterotube II

Symbol	Uninoculated color	Reacted color	Type of reaction
GLU-GAS			**Glucose (GLU)** The end products of bacterial fermentation of glucose are either acid or acid and gas. The shift in pH due to the production of acid is indicated by a color change from red (alkaline) to yellow (acidic). Any degree of yellow should be interpreted as a positive reaction; orange should be considered negative. **Gas Production (GAS)** Complete separation of the wax overlay from the surface of the glucose medium occurs when gas is produced. The amount of separation between the medium and overlay will vary with strain of bacteria.
LYS			**Lysine Decarboxylase** Bacterial decarboxylation of lysine, which results in the formation of the alkaline end product cadaverine, is indicated by a change in the color of the indicator from pale yellow (acidic) to purple (alkaline). Any degree of purple should be interpreted as a positive reaction. The medium remains yellow if decarboxylation of lysine does not occur.
ORN			**Ornithine Decarboxylase** Bacterial decarboxylation of ornithine causes the alkaline end product putrescine to be produced. The acidic (yellow) nature of the medium is converted to purple as alkalinity occurs. Any degree of purple should be interpreted as a positive reaction. The medium remains yellow if decarboxylation of ornithine does not occur.
H₂S/IND			**H$_2$S Production** Hydrogen sulfide, liberated by bacteria that reduce sulfur-containing compounds such as peptones and sodium thiosulfate, reacts with the iron salts in the medium to form a black precipitate of ferric sulfide usually along the line of inoculation. Some **Proteus** and **Providencia** strains may produce a diffuse brown coloration in this medium, which should not be confused with true H$_2$S production. **Indole Formation** The production of indole from the metabolism of tryptophan by the bacterial enzyme tryptophanase is detected by the development of a pink to red color after the addition of Kovac's reagent.

continued

2. With a pencil, mark a small plus (+) or minus (−) near each compartment symbol on the white label on the side of the tube.
3. Consult table 43.1 for information as to the significance of each compartment label.
4. Record the results of the tests on the Laboratory Report. *All results must be recorded before doing the indole test.*
5. Record results on the Laboratory Report. **Important**: If at this point you discover that your unknown is GLU-negative, proceed no further with the Enterotube II because your unknown is not one of the Enterobacteriaceae. Your unknown may be *Acinetobacter* sp. or *Pseudomonas maltophilia*. If an Oxi/Ferm Tube is available, try it, using the procedure outlined in the next exercise.
6. **Indole Test**: Perform the indole test as follows:
   a. Place the Enterotube II into a test-tube rack with the GLU-GAS compartment pointing upward.
   b. Inject 1 or 2 drops of Kovacs' reagent onto the surface of the medium in the H₂S/indole compartment. This may be done with a syringe and needle through the thin Mylar plastic film that covers the flat surface, or with a disposable Pasteur pipette through a small hole made in the Mylar film with a hot inoculating needle.
   c. A positive test is indicated by the development of a **red color** on the surface of the medium or Mylar film within 10 seconds.
7. **Voges-Proskauer Test**: Since this test is used as a confirmatory test, it should be performed *only* when called for in the *Enterotube II Interpretation Guide*. If it is called for, perform the test in the following manner:
   a. Use a syringe or Pasteur pipette to inject 2 drops of potassium hydroxide containing creatine into the V-P section.

TABLE 43.1   Biochemical Reactions of Enterotube II (continued)

Symbol	Uninoculated color	Reacted color	Type of reaction
ADON			**Adonitol** Bacterial fermentation of adonitol, which results in the formation of acidic end products, is indicated by a change in color of the indicator present in the medium from red (alkaline) to yellow (acidic). Any sign of yellow should be interpreted as a positive reaction; orange should be considered negative.
LAC			**Lactose** Bacterial fermentation of lactose, which results in the formation of acidic end products, is indicated by a change in color of the indicator present in the medium from red (alkaline) to yellow (acidic). Any sign of yellow should be interpreted as a positive reaction; orange should be considered negative.
ARAB			**Arabinose** Bacterial fermentation of arabinose, which results in the formation of acidic end products, is indicated by a change in color from red (alkaline) to yellow (acidic). Any sign of yellow should be interpreted as a positive reaction; orange should be considered negative.
SORB			**Sorbitol** Bacterial fermentation of sorbitol, which results in the formation of acidic end products, is indicated by a change in color from red (alkaline) to yellow (acidic). Any sign of yellow should be interpreted as a positive reaction; orange should be considered negative.
V.P.			**Voges-Proskauer** Acetylmethylcarbinol (acetoin) is an intermediate in the production of butylene glycol from glucose fermentation. The presence of acetoin is indicated by the development of a red color within 20 minutes. Most positive reactions are evident within 10 minutes.
DUL-PA			**Dulcitol** Bacterial fermentation of dulcitol, which results in the formation of acidic end products, is indicated by a change in color of the indicator present in the medium from green (alkaline) to yellow or pale yellow (acidic). **Phenylalanine Deaminase** This test detects the formation of pyruvic acid from the deamination of phenylalanine. The pyruvic acid formed reacts with a ferric salt in the medium to produce a characteristic black to smoky gray color.
UREA			**Urea** The production of urease by some bacteria hydrolyzes urea in this medium to produce ammonia, which causes a shift in pH from yellow (acidic) to reddish-purple (alkaline). This test is strongly positive for **Proteus** in 6 hours and weakly positive for **Klebsiella** and some **Enterobacter** species in 24 hours.
CIT			**Citrate** Organisms that are able to utilize the citrate in this medium as their sole source of carbon produce alkaline metabolites that change the color of the indicator from green (acidic) to deep blue (alkaline). Any degree of blue should be considered positive.

Source: Courtesy of Becton-Dickinson, Cockeysville, Maryland.

b. Inject 3 drops of 5% alpha-naphthol.
c. A positive test is indicated by a **red color** within 10 minutes.
8. Record the indole and V-P results on the Laboratory Report.

LABORATORY REPORT

Determine the name of your unknown by following the instructions in Laboratory Report 43. Note that two methods of making the final determination are given.

LABORATORY REPORT

Student: _____

Date: _____ Section: _____

EXERCISE 43   Enterobacteriaceae Identification: The Enterotube II System

A. Tabulation of Results

Record the results of each test in the following table with a plus (+) or minus (−).

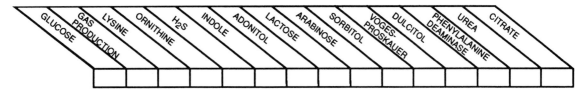

B. Identification by Chart Method

If no *Interpretation Guide* is available, apply the above results to chart IV, Appendix D, to find the name of your unknown. Note that the spacing of the above table matches the size of the spaces on chart IV. If this page is removed from the manual, folded, and placed on chart IV, the results on the above table can be moved down the chart to make a quick comparison of your results with the expected results for each organism.

C. Using the Enterotube II Interpretation Guide

If the *Interpretation Guide* is available, determine the five-digit code number by circling the numbers (4, 2, or 1) under each test that is positive, and then totaling these numbers within each group to form a digit for that group. Note that there are two tally charts in this Laboratory Report for your use.

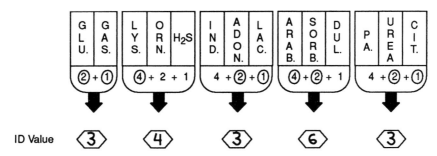

The "ID Value" 34363 can be found by thumbing the pages of the *Interpretation Guide*. The listing is as follows:

ID Value	Organism	Atypical Test Results
34363	*Klebsiella pneumoniae*	None

Conclusion: Organism was correctly identified as *Klebsiella pneumoniae*. In this case, the identification was made independent of the V-P test.

D. Tally Charts

ENTEROTUBE® II*

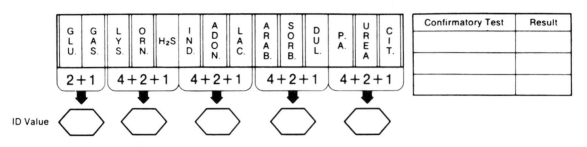

G L U.	G A S.	L Y S.	O R N.	H₂S	I N D.	A D O N.	L A C.	A R A B.	S O R B.	D U L.	P. A.	U R E A	C I T.

| 2 + 1 | 4 + 2 + 1 | 4 + 2 + 1 | 4 + 2 + 1 | 4 + 2 + 1 |

Confirmatory Test	Result

ID Value

Culture Number, Case Number or Patient Name   Date   Organism Identified

*VP utilized as confirmatory test only.

ENTEROTUBE® II*

G L U.	G A S.	L Y S.	O R N.	H₂S	I N D.	A D O N.	L A C.	A R A B.	S O R B.	D U L.	P. A.	U R E A	C I T.

| 2 + 1 | 4 + 2 + 1 | 4 + 2 + 1 | 4 + 2 + 1 | 4 + 2 + 1 |

Confirmatory Test	Result

ID Value

Culture Number, Case Number or Patient Name   Date   Organism Identified

*VP utilized as confirmatory test only.

E. Questions

1. What is the intended function of the Enterotube II System? _____

_____

_____

_____

2. In the "real world," who would use this system? _____

_____

_____

3. What might be an explanation for the failure of this system to work with some of the bacterial cultures

we use? _____

_____

_____

O/F Gram-Negative Rods Identification:
The Oxi/Ferm Tube II System

The Oxi/Ferm Tube II, produced by Becton-Dickinson, takes care of the identification of the oxidase-positive, gram-negative bacteria that cannot be identified by using the Enterotube II system. The two multitest systems were developed to work together. If an unknown gram-negative rod is oxidase-negative, the Enterotube II is used. If the organism is oxidase-positive, the Oxi/Ferm Tube II must be used. Whenever an oxidase-negative gram-negative rod turns out to be glucose-negative on the Enterotube II test, one must move on to use the Oxi/Ferm Tube II.

The Oxi/Ferm Tube II system is intended for the identification of nonfastidious species of oxidative-fermentative gram-negative rods from clinical specimens. This includes the following genera: *Aeromonas, Plesiomonas, Vibrio, Achromobacter, Alcaligenes, Bordetella, Moraxella,* and *Pasteurella.* Some other gram-negative bacteria can also be identified with additional biochemical tests. The system incorporates 12 different conventional media that can be inoculated simultaneously in a moment's time with a minimum of equipment. A total of 14 physiological tests are performed.

Like the Enterotube II system, the Oxi/Ferm Tube II has an inoculating wire that extends through all 12 compartments of the entire tube. To inoculate the media, one simply picks up some organisms on the end of the wire and pulls the wire through each of the chambers in a rotating action.

After incubation, the results are recorded and Kovacs' reagent is injected into one of the compartments to perform the indole test. Positive reactions are given numerical values that are totaled to arrive at a five-digit code. By looking up the code in an Oxi/Ferm *Biocode Manual,* one can quickly determine the name of the unknown and any tests that might be needed to confirm the identification.

Figure 44.1 illustrates an uninoculated tube and a tube with all positive reactions. Figure 44.2 illustrates

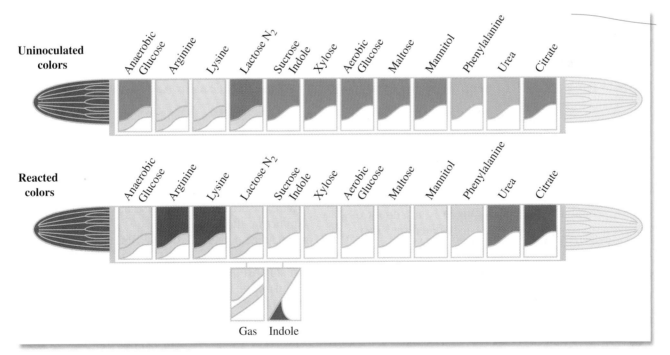

FIGURE 44.1 Oxi/Ferm Tube II color differences between uninoculated and positive tests

(1) Remove organisms from a well-isolated colony. Avoid touching the agar with the wire. To prevent damaging Enterotube II media, do not heat-sterilize the inoculating wire.

(2) Inoculate each compartment by first twisting the wire and then withdrawing it all the way out through the 12 compartments, using a turning movement.

(3) Reinsert the wire (without sterilizing), using a turning motion through all 12 compartments until the notch on the wire is aligned with the opening of the tube.

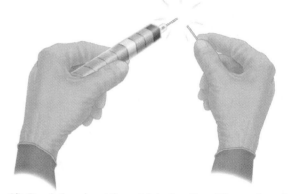

(4) Break the wire at the notch by bending. The portion of the wire remaining in the tube maintains anaerobic conditions essential for true fermentation.

continued

FIGURE 44.2   The Oxi/Ferm Tube II procedure

the entire procedure for utilizing this system. A minimum of two periods is required to use this system. Proceed as follows:

Inoculation and Incubation

The Oxi/Ferm Tube II must be inoculated with a large inoculum from a well-isolated colony. Culture purity is of paramount importance. If there is any doubt of purity, a TSA plate should be inoculated and incubated at 35° C for 24 hours, followed by 24 hours incubation at room temperature. If no growth occurs on TSA, but growth does occur on blood agar, the organism has special growth requirements. *Such organisms are too fastidious and cannot be identified with the Oxi/Ferm Tube II.*

MATERIALS

- culture plate of unknown

- 1 Oxi/Ferm Tube II
- 1 plate of trypticase soy agar (TSA) (for purity check, if needed)

1. Write your initials or unknown number on the side of the tube.
2. Unscrew both caps from the Oxi/Ferm Tube II. The tip of the inoculating end is under the white cap.
3. *Without heat-sterilizing* the exposed inoculating wire, insert it into a well-isolated colony. Do not puncture the agar.
4. Inoculate each chamber by first twisting the wire and then withdrawing it through all 12 compartments. Rotate the wire as you pull it through. See illustration 2, figure 44.2.
5. If a purity check of the culture is necessary, streak a petri plate of TSA with the inoculating wire that has just been pulled through the tube. **Do not flame.**
6. Again, *without sterilizing*, reinsert the wire, and with a turning motion, force it through all 12 compartments until the notch on the wire is aligned

(5) Punch holes with broken-off part of wire through the thin plastic covering over depressions on sides of the last eight compartments (sucrose/indole through citrate). Replace caps and incubate at 35° C for 18–24 hours.

(6) After interpreting and recording positive results on the sides of the tube, perform the indole test by injecting 1 or 2 drops of Kovacs' reagent into the sucrose/indole compartment.

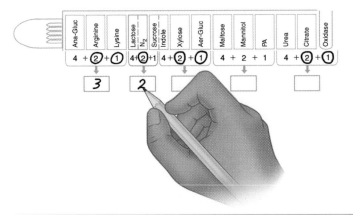

(7) After encircling the numbers of the positive tests on the Laboratory Report, total up the numbers of each bracketed series to determine the 5-digit code number. Refer to the *Biocode Manual* for identification of the unknown by using the code number.

FIGURE 44.2 (continued)

with the opening of the tube. (The notch is about $1\frac{5}{8}''$ from the handle end of the wire.) The tip of the wire should be visible in the citrate compartment. See illustration 3, figure 44.2.

7. Break the wire at the notch by bending, as noted in step 4, figure 44.2. The portion of the wire remaining in the tube maintains anaerobic conditions essential for true fermentation.

8. With the retained portion of the needle, punch holes through the thin plastic coverings over the small depressions on the sides of the last eight compartments (sucrose/indole, xylose, aerobic glucose, maltose, mannitol, phenylalanine, urea, and citrate). These holes will enable aerobic growth in these eight compartments.

9. Replace both caps on the tube.

10. Incubate at 35° to 37° C for 24 hours, with the tube lying on its flat surface or upright. At the end of 24 hours, inspect the tube to check results and continue incubation for another 24 hours. The 24-hour check may be needed for doing confir-

matory tests as required in the *Biocode Manual*. Occasionally, an Oxi/Ferm Tube II should be incubated longer than 48 hours.

SECOND PERIOD

EVALUATION OF TESTS

During this period, you will record the results of the various tests on your Oxi/Ferm Tube II, do an indole test, tabulate your results, use the *Biocode Manual*, and perform any confirmatory tests called for. Proceed as follows:

MATERIALS

- Oxi/Ferm Tube II, inoculated and incubated
- Kovacs' reagent
- syringes with needles, or disposable Pasteur pipettes
- Becton-Dickinson *Biocode Manual* (a booklet)

1. Compare the colors of each compartment of your Oxi/Ferm Tube II with the lower tube illustrated in figure 44.1.
2. With a pencil, mark a small plus (+) or minus (−) near each compartment symbol on the white label on the side of the tube.
3. Consult table 44.1 for information as to the significance of each compartment label.
4. Record the results of all the tests on the Laboratory Report. *All results must be recorded before doing the indole test.*

TABLE 44.1 Biochemical Reactions of the Oxi/Ferm Tube II

Reaction	Negative	Positive	Special remarks
Anaerobic Glucose			Positive fermentation is shown by change in color from green (neutral) to yellow (acid). Most oxidative-fermentative, gram-negative rods are negative.
Arginine Dihydrolase			Decarboxylation of arginine results in the formation of alkaline end products that changes bromcresol purple from yellow (acid) to purple (alkaline). Grey is negative.
Lysine			Decarboxylation of lysine results in the formation of alkaline end products that changes bromcresol purple from yellow (acid) to purple (alkaline). Grey is negative.
Lactose			Fermentation of lactose changes the color of the medium from red (neutral) to yellow (acid). Most O/F gram-negative rods are negative.
N_2 Gas-production			Gas production causes separation of wax overlay from medium. Occasionally, the gas will also cause separation of the agar from the compartment wall.
Sucrose			Bacterial oxidation of sucrose causes a change in color from green (neutral) to yellow (acid).
Indole			The bacterial enzyme tryptophanase metabolizes tryptophan to produce indole. Detection is by adding Kovacs' reagent to the compartment 48 hours after incubation.
Xylose			Bacterial oxidation of xylose causes a change in color from green (neutral) to yellow (acid).
Aerobic Glucose			Bacterial oxidation of glucose causes a change in color from green (neutral) to yellow (acid).
Maltose			Bacterial oxidation of maltose causes a change in color from green (neutral) to yellow (acid).
Mannitol			Bacterial oxidation of this carbohydrate is evidenced by a change in color from green (neutral) to yellow (acid).
Phenylalanine			Pyruvic acid is formed by deamination of phenylalanine. The pyruvic acid reacts with a ferric salt to produce a brownish tinge.
Urea			The production of ammonia by the action of urease on urea increases the alkalinity of the medium. The phenol red in this medium changes from beige (acid) to pink or purple. Pale pink should be considered negative.
Citrate			Organisms that grow on this medium are able to utilize citrate as their sole source of carbon. Utilization of citrate raises the alkalinity of the medium. The color changes from green (neutral) to blue (alkaline).

5. **Indole Test** (illustration 6, figure 44.2): Do an indole test by injecting 2 or 3 drops of Kovacs' reagent through the flat, plastic surface into the sucrose/indole compartment. Release the reagent onto the inside flat surface and allow it to drop down onto the agar.

If a Pasteur pipette is used instead of a syringe needle, it will be necessary to form a small hole in the Mylar film with a hot inoculating needle to admit the tip of the Pasteur pipette.

A positive test is indicated by the development of a **red color** on the surface of the medium or Mylar film within 10 seconds.

6. Record the results of the indole test on the Laboratory Report.

LABORATORY REPORT

Follow the instructions in Laboratory Report 44 for determining the five-digit code. Use the *Biocode Manual* booklet for identifying your unknown.

LABORATORY REPORT

Student: _____

Date: _____ Section: _____

EXERCISE 44   O/F Gram-Negative Rods Identification: The Oxi/Ferm Tube II System

A. Tabulation of Results and Code Determination

Once you have marked the positive reactions on the side of the tube and circled the numbers that are assigned to each of the positive chambers, as indicated in the example below, add the numbers in each bracketed group to get the five-digit code.

The final step is to look up the code number in the *Oxi/Ferm Tube II Biocode Manual* to determine the genus and species. If confirmatory tests are necessary, the manual will tell you which ones to perform.

In the example below, the code number is 32303. If you look up this number in the *Biocode Manual* you will find on page 25 that the organism is *Pseudomonas aeruginosa*.

Use this procedure to identify your unknown by applying your results to the blank diagrams provided.

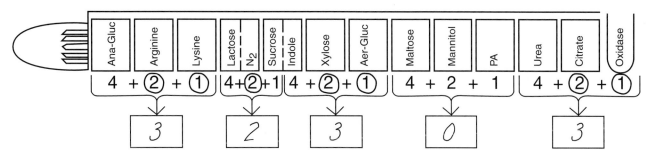

B. Results Pads

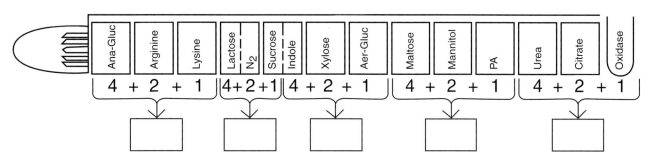

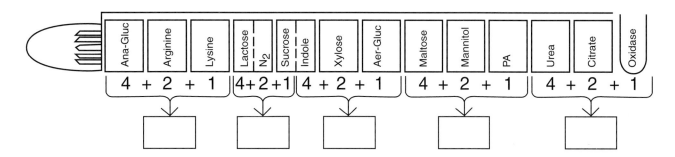

C. Questions

1. What is the intended function of the Oxi/Ferm Tube II System? _____
   _____
   _____
   _____

2. In the "real world," who would use this system? _____
   _____
   _____

3. What might be an explanation for the failure of this system to work with some of the bacterial cultures
   we use? _____
   _____
   _____

Staphylococcus Identification:
The API Staph-Ident System

The **API Staph-Ident System,** produced by Analytab Products of Plainview, New York, was developed to provide a rapid (5-hour) method for identifying 13 of the most clinically important species of staphylococci. This system consists of 10 microcupules that contain dehydrated substrates and/or nutrient media. Except for the coagulase test, all the tests that are needed for the identification of staphylococci are included on the strip.

Figure 45.1 illustrates two inoculated strips: the lower one just after inoculation and the upper one with all positive reactions. Note that the appearance of each microcupule undergoes a pronounced color change when a positive reaction occurs.

Figure 45.2 illustrates the overall procedure. The first step is to make a saline suspension of the organism from an isolated colony. A Staph-Ident strip is then placed in a tray that has a small amount of water added to it to provide humidity during incubation. Next, a sterile Pasteur pipette is used to dispense 2 to 3 drops of the bacterial suspension to each microcupule. The inoculated tray is then covered and incubated aerobically at 35° to 37° C for 5 hours. After incubation, a few drops of Staph-Ident reagent are added to the tenth microcupule and the results are read immediately. Finally, a four-digit profile is computed that is used to determine the species from a chart in Appendix D.

As simple as this system might seem, there are a few limitations that one must keep in mind. Final species determination by a competent microbiologist must take into consideration other factors such as the source of the specimen, the catalase reaction, colony characteristics, and antimicrobial susceptibility pattern. Very often there are confirmatory tests that must also be made.

If you have been working with an unknown that appears to be one of the staphylococci, use this system to confirm your conclusions. If you have already done the coagulase test and have learned that your organism is coagulase-negative, this system will enable you to identify one of the numerous coagulase-negative species that are not identifiable by the procedures in Exercise 52.

FIRST PERIOD

(INOCULATIONS AND COAGULASE TEST)

Before setting up this experiment, take into consideration that it must be completed at the end of 5 hours. Holding the test strips overnight is not recommended.

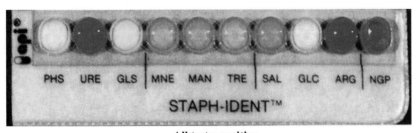

All tests: positive

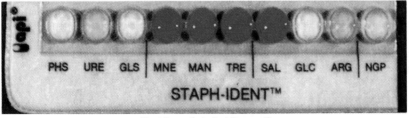

Just inoculated: All negative

FIGURE 45.1 Positive and negative results on API Staph-Ident test strips Courtesy of Analytab Products, Plainview, NY

(1) Use several loopfuls of organisms to make saline suspension of unknown. Turbidity of suspension should match McFarland No. 3 barium sulfate standard.

0.85% saline

(2) After labeling the end tab of a tray with your name and unknown number, dispense approximately 5 ml of tap water into bottom of tray.

(3) Place a STAPH-IDENT test strip into the bottom of the moistened tray. Take care not to contaminate the microcupules with fingers when handling test strip.

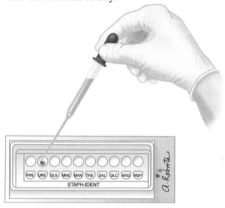

(4) With a Pasteur pipette dispense 2 to 3 drops of the bacterial suspension into each of the 10 microcupules. Cover the tray with the lid and incubate at 35°–37°C for 5 hours.

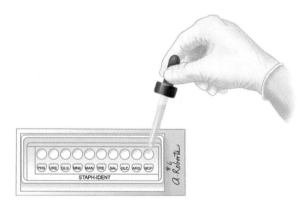

(5) After incubation, record results of first 9 microcupules and add 1–2 drops of STAPH-IDENT reagent to tenth microcupule as shown. A plum-purple color is positive. Record result.

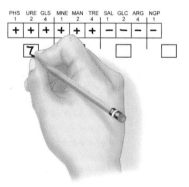

PHS	URE	GLS	MNE	MAN	TRE	SAL	GLC	ARG	NGP
1	2	4	1	2	4	1	2	4	1
+	+	+	+	+	+	−	−	−	−

(6) Once all results are recorded on Laboratory Report, total up positive values in each group to determine 4-digit profile. Consult chart VII, appendix D, to find unknown.

FIGURE 45.2   The API Staph-Ident procedure

MATERIALS

- API Staph-Ident test strip
- API incubation tray and cover
- blood agar plate culture of unknown (must not have been incubated over 30 hours)
- blood agar plate (if needed for purity check)
- serological tube of 2 ml sterile saline
- test-tube rack
- sterile swabs (optional in step 2 below)
- squeeze bottle of tap water
- tubes containing McFarland No. 3 (BaSO₄) standard (see Appendix B)
- sterile Pasteur pipette (5 ml size)

1. If the **coagulase test** has not been performed, refer to Exercise 52, page 318, for the procedure and perform it on your unknown.
2. Prepare a saline suspension of your unknown by transferring organisms to a tube of sterile saline from one or more colonies with a loop or sterile swab. Turbidity of the suspension should match a tube of No. 3 McFarland barium sulfate standard. **Important:** Do not allow the bacterial suspension to go unused for any great length of time. Suspensions older than 15 minutes become less effective.
3. Label the end strip of the tray with your name and unknown number. See illustration 2, figure 45.2.
4. Dispense about 5 ml of tap water into the bottom of the tray with a squeeze bottle. Note that the bottom of the tray has numerous depressions to accept the water.
5. Remove the API test strip from its sealed envelope and place the strip in the bottom of the tray.
6. After shaking the saline suspension to disperse the organisms, fill a sterile Pasteur pipette with the bacterial suspension.
7. Inoculate each of the microcupules with 2 or 3 drops of the suspension. If a purity check is necessary, use the excess suspension to inoculate another blood agar plate.
8. Place the plastic lid on the tray and incubate the strip aerobically for 5 hours at 35° to 37° C.

SECOND PERIOD

(Five Hours Later)

During this period, the results will be recorded on the Laboratory Report, the profile number will be de-

termined, and the unknown will be identified by looking up the number on the *Staph-Ident Profile Register* (or chart VII, Appendix D).

MATERIALS

- API Staph-Ident test strip (incubated 5 hours)
- 1 bottle of Staph-Ident reagent (room temperature)
- *Staph-Ident Profile Register*

1. After 5 hours incubation, refer to chart V, Appendix D, to interpret and record the results of the first nine microcupules (PHS through ARG).
2. Record the results in the Profile Determination Table in the Laboratory Report. Chart VI, Appendix D, reveals the biochemistry involved in these tests.
3. Add 1 or 2 drops of **Staph-Ident reagent** to the NGP microcupule. Allow 30 seconds for the color change to occur.

    A positive test results in a change of color to plum-purple. Record the results of this test. (figure 45.3)
4. Construct the profile number according to the instructions on the Laboratory Report and determine the name of your unknown.
    Use chart VII, Appendix D.

DISPOSAL

Once all the information has been recorded be sure to place the entire incubation unit in a receptacle that is to be autoclaved.

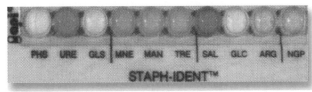

FIGURE 45.3 **Test results of a strip inoculated with** *S. aureus* (Courtesy of Analytab Products)

LABORATORY REPORT

Student: _____

Date: _____ Section: _____

EXERCISE 45   Staphylococcus Identification: The API Staph-Ident System

A. Tabulation of Results

By referring to charts V and VI, Appendix D, determine the results of each test, and record these results as positive (+) or negative (−) in the Profile Determination Table below. Note that two more of these tables have been printed on the next page for tabulation of additional organisms.

PHS 1	URE 2	GLS 4	MNE 1	MAN 2	TRE 4	SAL 1	GLC 2	ARG 4	NGP 1

RESULTS

PROFILE NUMBER ☐   ☐   ☐   ☐

GRAM STAIN ☐   COAGULASE ☐   Additional Information   Identification

MORPHOLOGY ☐   CATALASE ☐

B. Construction of Four-Digit Profile

Note in the above table that each test has a value of 1, 2, or 4. To compute the four-digit profile for your unknown, total up the positive values for each group.

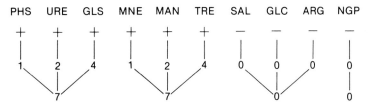

C. Final Determination

Refer to the Staph-Ident Profile Register (chart VII, Appendix D) to find the organism that matches your profile number. Write the name of your unknown in the space below and list any additional tests that are needed for final confirmation. If the materials are available for these tests, perform them.

Name of Unknown: _____

Additional Tests: _____

	PHS 1	URE 2	GLS 4	MNE 1	MAN 2	TRE 4	SAL 1	GLC 2	ARG 4	NGP 1
RESULTS										

PROFILE NUMBER ☐ ☐ ☐ ☐

GRAM STAIN ☐   COAGULASE ☐

MORPHOLOGY ☐   CATALASE ☐

Additional Information

Identification

	PHS 1	URE 2	GLS 4	MNE 1	MAN 2	TRE 4	SAL 1	GLC 2	ARG 4	NGP 1
RESULTS										

PROFILE NUMBER ☐ ☐ ☐ ☐

GRAM STAIN ☐   COAGULASE ☐

MORPHOLOGY ☐   CATALASE ☐

Additional Information

Identification

D. Questions

1. What is the intended function of the API Staph-Ident System? _____

   _____

   _____

   _____

2. In the "real world," who would use this system? _____

   _____

   _____

3. What might be an explanation for the failure of this system to work with some of the bacterial cultures

   we use? _____

   _____

   _____

APPLIED MICROBIOLOGY

Applied microbiology encompasses many aspects of modern microbiology. We use microorganisms to produce many of the foods we eat such as cheese, yogurt, bread, sauerkraut, and a whole list of fermented beverages. Microorganisms are important in industrial applications where they are involved in producing antibiotics, pharmaceuticals, and even solvents and starting materials for the manufacture of plastics. Their presence and numbers in our foods and drinking water determine if it is safe to consume these substances as they could cause us harm and disease. In the following exercises, you will explore some of the applications of microbiology by determining bacterial numbers and/or kinds in food and water. You will also study the process of alcohol fermentation as an example of food production.

Bacterial Counts of Foods

The standard plate count, as well as the multiple tube test, can be used on foods much in the same manner that they are used on milk and water to determine total counts and the presence of coliforms. To get the organisms in suspension, however, a food blender is necessary.

In this exercise, samples of ground meat, dried fruit, and frozen food will be tested for total numbers of bacteria. This will not be a coliform count. The instructor will indicate the specific kinds of foods to be tested and make individual assignments. Figure 46.1 illustrates the general procedure.

MATERIALS

per student:
- 3 petri plates
- 1 bottle (45 ml) of Plate Count agar or Standard Methods agar
- 1 99 ml sterile water blank
- 2 1.1 ml dilution pipettes

per class:
- food blender
- sterile blender jars (one for each type of food)

- sterile weighing paper
- 180 ml sterile water blanks (one for each type of food)
- samples of ground meat, dried fruit, and frozen vegetables, thawed 2 hours

1. Using aseptic techniques, weigh out on sterile weighing paper 20 grams of food to be tested.
2. Add the food and 180 ml of sterile water to a sterile mechanical blender jar. Blend the mixture for 5 minutes. This suspension will provide a 1:10 dilution.
3. With a 1.1 ml dilution pipette dispense from the blender 0.1 ml to plate 1 and 1.0 ml to the water blank. See figure 46.1
4. Shake the water blank 25 times in an arc for 7 seconds with your elbow on the table as done in Exercise 20 (Enumeration of Bacteria).
5. Using a fresh pipette, dispense 0.1 ml to plate III and 1.0 ml to plate II.
6. Pour agar (50° C) into the three plates and incubate them at 35° C for 24 hours.
7. Count the colonies on the best plate and record the results in Laboratory Report 46.

FIGURE 46.1 Dilution procedure for bacterial counts of food

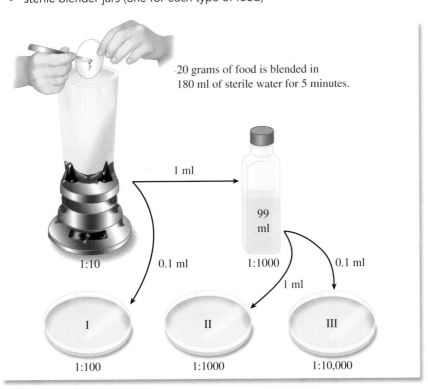

20 grams of food is blended in 180 ml of sterile water for 5 minutes.

1 ml

99 ml

1:10   0.1 ml   1:1000   0.1 ml

1 ml

I   II   III

1:100   1:1000   1:10,000

LABORATORY REPORT

Student: _____

Date: _____ Section: _____

EXERCISE 46   Bacterial Counts of Foods

A. Tabulation of Results

Record your count and the bacterial counts of various other foods made by other students.

TYPE OF FOOD	PLATE COUNT	DILUTION	ORGANISMS PER ML

B. Questions

1. Why is there such great variability in organisms per ml between different kinds of food? _____

_____

_____

_____

2. What dangers and undesirable results may occur from ground meats of high bacterial counts?

_____

_____

3. What bacterial pathogens might be present in frozen foods? _____

_____

_____

4. What harm can result from repeated thawing and freezing of foods? _____

_____

_____

5. What precautions are taken to prevent the spoilage of foods? _____

_____

_____

6. Which methods in question 5 are most effective? _____

_____

_____

Least effective? _____

Prior to the modern age of public health, water was a major means for the spread of infectious diseases such as cholera, dysentery, and typhoid fever. A physician, John Snow, showed in the 1840s that a cholera epidemic in London was the result of cesspool overflow into the Thames River from a tenement where cholera patients lived. When water for drinking was drawn by inhabitants near the cesspool discharge, the contaminated water and pump became the source for the spread of the disease to people in the area. Snow's solution was simply to remove the handle to the pump, and the epidemic abated. Water safety is still a primary concern of municipalities in today's world and it has become complex. Because of good public health measures, most of us are confident that the water we draw from our faucets is safe and will not cause us disease.

From a microbiological standpoint, it is not the numbers of bacteria that are present in water that is of primary concern to us but rather the kinds of bacteria. Water found in rivers, lakes, and streams can contain a variety of bacteria that may only be harmless saprophytes, which do not cause disease in humans. However, it is important that water not contain the intestinal pathogens that cause typhoid, cholera, and dysentery. In modern cities, treated sewage is discharged into receiving waters of lakes, rivers, and streams, and this constitutes a major sanitary problem because those same bodies of water are the sources of our drinking water. As a result, we have developed methods to treat water to eliminate the potential for disease, and we do microbiological tests to determine if water is potable and safe for consumption.

At first glance, it might seem reasonable to directly examine water for the presence of the pathogens *Vibrio cholerae, Salmonella typhi,* and *Shigella dysenteriae.* However, this not the case because it would be tedious and difficult to specifically test for each of the pathogens. Furthermore, these bacteria are often fastidious, and they might be overgrown by other bacteria in the water if we tried to culture and test for them. It is much easier to demonstrate the presence of some indicator bacterium, such as *Escherichia coli,* which is routinely found in the human intestine but is not found in the soil or water. The presence of these bacteria in water would then indicate the likelihood of fecal contamination and the potential for serious disease.

E. coli is a good indicator of fecal contamination and a good test organism. This is for several reasons: (1) it occurs primarily in the intestines of humans and some warm-blooded animals and it is not found routinely in soil or water; (2) the organism can be easily identified by microbiological tests; (3) it is not as fastidious as the intestinal pathogens, and hence it survives a little longer in water samples. By definition, organisms such as *E. coli* and *Enterobacter aerogenes* are designated as **coliforms**, which are gram-negative, facultative anaerobic, non–endospore forming rods that ferment lactose to produce acid and gas in 48 hours at 35°C. Lactose fermentation with the formation of acid and gas provides the basis for determining the total coliform count of water samples in the United States and therefore designates water purity. The presence of other bacteria, such as *Streptococcus faecalis,* which is a gram-positive enterococcus that inhabits the human intestine, can also indicate fecal contamination, but testing for this bacterium is not routinely done in the United States.

Three different tests are done to determine the coliform count (figure 47.1): presumptive, confirmed, and completed. Each test is based on one or more of the characteristics of a coliform. A description of each test follows.

Presumptive Test   In the presumptive test, 9 or 12 tubes of lactose broth are inoculated with measured amounts of water to see if the water contains any loctose-fermenting bacteria that produce gas. If, after incubation, gas is seen in any of the lactose broths, it is *presumed* that coliforms are present in the water sample. This test is also used to determine the most probable number (MPN) of coliforms present per 100 ml of water.

Confirmed Test   In this test, plates of Levine EMB agar or Endo agar are inoculated from positive (gas-producing) tubes to see if the organisms that are producing the gas are gram-negative (another coliform characteristic). Both of these media inhibit the growth of gram-positive bacteria and cause colonies of coliforms to be distinguishable from noncoliforms. On EMB agar, coliforms produce small colonies with dark centers (nucleated colonies). On Endo agar,

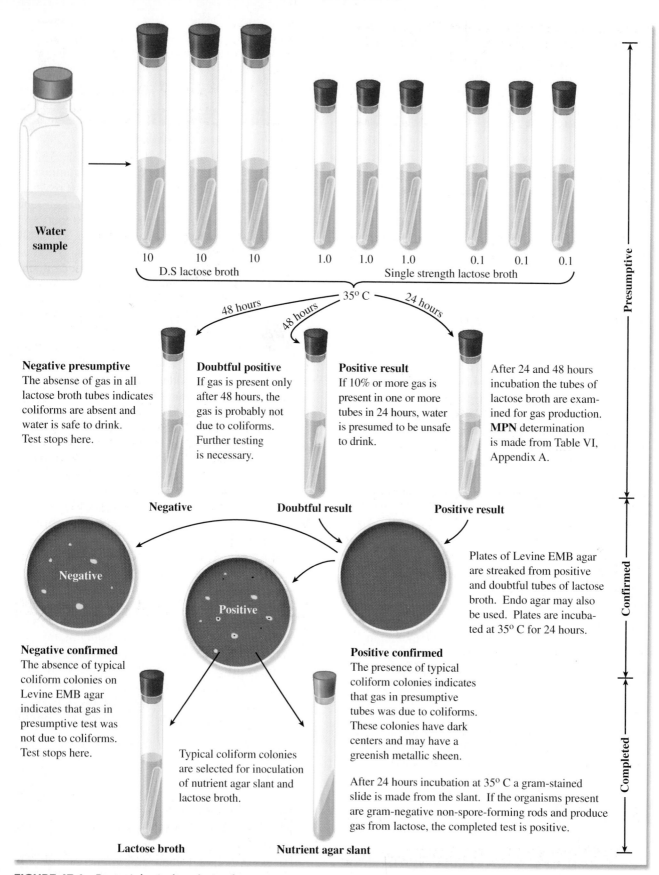

FIGURE 47.1 Bacteriological analysis of water

coliforms produce reddish colonies. The presence of coliformlike colonies confirms the presence of a lactose-fermenting, gram-negative bacterium.

Completed Test In the completed test, our concern is to determine if the isolate from the agar plates truly matches our definition of a coliform. Our media for this test include a nutrient agar slant and a Durham tube of lactose broth. If gas is produced in the lactose tube and a slide from the agar slant reveals that we have a gram-negative, non-spore-forming rod, we can be certain that we have a coliform.

The completion of these three tests with positive results establishes that coliforms are present; however, there is no certainty that *E. coli* is the coliform present. The organism might be *E. aerogenes*. Of the two, *E. coli* is the better sewage indicator since *E. aerogenes* can be of nonsewage origin. To differentiate these two species, one must perform the **IMViC tests,** which are described on page 243 in Exercise 40.

In this exercise, water will be tested from local ponds, streams, swimming pools, and other sources supplied by students and instructor. Enough known positive samples will be evenly distributed throughout the laboratory so that all students will be able to see positive test results. All three tests in figure 47.1 will be performed. If time permits, the IMViC tests may also be performed.

THE PRESUMPTIVE TEST

As stated earlier, the presumptive test is used to determine if gas-producing lactose fermenters are present in a water sample. If clear surface water is being tested, nine tubes of lactose broth will be used as shown in figure 47.1. For turbid surface water, an additional three tubes of single-strength lactose broth will be inoculated.

In addition to determining the presence or absence of coliforms, we can also use this series of lactose broth tubes to determine the **most probable number** (MPN) of coliforms present in 100 ml of water. See table VI, appendix A, to determine this value from the number of positive lactose tubes.

Before setting up your test, determine whether your water sample is clear or turbid. Note that a separate set of instructions is provided for each type of water.

Clear Surface Water

If the water sample is relatively clear, proceed as follows:

MATERIALS
- 3 Durham tubes of DSLB
- 6 Durham tubes of SSLB
- 1 10 ml pipette

- 1 1 ml pipette
  Note: DSLB designates double-strength lactose broth. It contains twice as much lactose as SSLB (single-strength lactose broth).

1. Set up 3 DSLB and 6 SSLB tubes as illustrated in figure 47.1. Label each tube according to the amount of water that is to be dispensed to it:*10 ml, 1.0 ml,* and *0.1 ml,* respectively.
2. Mix the bottle of water to be tested by shaking 25 times.
3. With a 10 ml pipette, transfer 10 ml of water to each of the DSLB tubes.
4. With a 1.0 ml pipette, transfer 1 ml of water to each of the middle set of tubes, and 0.1 ml to each of the last three SSLB tubes.
5. Incubate the tubes at 35° C for 24 hours.
6. Examine the tubes and record the number of tubes in each set that have 10% gas or more.
7. Determine the MPN by referring to table VI, Appendix A. Consider the following:
   Example: If you had gas in the first three tubes and gas only in one tube of the second series, but none in the last three tubes, your test would be read as 3–1–0. Table VI indicates that the MPN for this reading would be 43. This means that this particular sample of water would have approximately 43 organisms per 100 ml with 95% probability of there being between 7 and 210 organisms. *Keep in mind that the MPN figure of 43 is only a statistical probability figure.*
8. Record the data in Laboratory Report 47.

Turbid Surface Water

If your water sample appears to have considerable pollution, do as follows:

MATERIALS
- 3 Durham tubes of DSLB
- 9 Durham tubes of SSLB
- 1 10 ml pipette
- 2 1 ml pipettes
- 1 water blank (99 ml of sterile water)
  Note: See comment in previous materials list concerning DSLB and SSLB.

1. Set up 3 DSLB and 9 SSLB tubes in a test-tube rack, with the DSLB tubes on the left.
2. Label the 3 DSLB tubes *10 ml,* the next 3 SSLB tubes *1.0 ml*, the next 3 SSLB tubes *0.1 ml*, and the last 3 tubes *0.01 ml.*
3. Mix the bottle of water to be tested by shaking 25 times.
4. With a 10 ml pipette, transfer 10 ml of water to each of the DSLB tubes.
5. With a 1.0 ml pipette, transfer 1 ml to each of the next 3 tubes, and 0.1 ml to each of the third set of tubes.

6. With the same 1 ml pipette, transfer 1 ml of water to the 99 ml blank of sterile water and shake 25 times.
7. *With a fresh 1 ml pipette*, transfer 1.0 ml of water from the blank to the remaining tubes of SSLB. This is equivalent to adding 0.01 ml of full-strength water sample.
8. Incubate the tubes at 35° C for 24 hours.
9. Examine the tubes and record the number of tubes in each set that have 10% gas or more.
10. Determine the MPN by referring to table VI, Appendix A. This table is set up for only 9 tubes. To apply a 12-tube reading to it, do as follows:

   a. Select the three consecutive sets of tubes that have at least one tube with no gas.
   b. If the first set of tubes (10 ml tubes) are not used, multiply the MPN by 10.
   Example: Your tube reading was 3–3–3–1. What is the MPN?
      The first set of tubes (10 ml) is ignored and the figures 3–3–1 are applied to the table. The MPN for this series is 460. Multiplying this by 10, the MPN becomes 4600.
   Example: Your tube reading was 3–1–2–0. What is the MPN?
      The first three numbers are (3–1–2) applied to the table. The MPN is 210. Since the last set of tubes is ignored, 210 is the MPN.

THE CONFIRMED TEST

Once it has been established that gas-producing lactose fermenters are present in the water, it is *presumed* to be unsafe. However, gas formation may be due to noncoliform bacteria. Some of these organisms, such as *Clostridium perfringens*, are gram-positive. To confirm the presence of gram-negative lactose fermenters, the next step is to inoculate media such as Levine eosin-methylene blue agar or Endo agar from positive presumptive tubes.

Levine EMB agar contains methylene blue, which inhibits gram-positive bacteria. Gram-negative lactose fermenters (coliforms) that grow on this medium will produce "nucleated colonies" (dark centers). Colonies of *E. coli* and *E. aerogenes* can be differentiated on the basis of size and the presence of a greenish metallic sheen. *E. coli* colonies on this medium are small and have this metallic sheen, whereas *E. aerogenes* colonies usually lack the sheen and are larger. Differentiation in this manner is not completely reliable, however. It should be remembered that *E. coli* is the more reliable sewage indicator since it is not normally present in soil, while *E. aerogenes* has been isolated from soil and grains.

Endo agar contains a fuchsin sulfite indicator that makes identification of lactose fermenters rela-

tively easy. Coliform colonies and the surrounding medium appear red on Endo agar. Nonfermenters of lactose, on the other hand, are colorless and do not affect the color of the medium.

In addition to these two media, there are several other media that can be used for the confirmed test. Brilliant green bile lactose broth, Eijkman's medium, and EC medium are just a few examples that can be used.

To demonstrate the confirmation of a positive presumptive in this exercise, the class will use Levine EMB agar and Endo agar. One half of the class will use one medium; the other half will use the other medium. Plates will be exchanged for comparisons.

MATERIALS

- 1 petri plate of Levine EMB agar (odd-numbered students)
- 1 petri plate of Endo agar (even-numbered students)

1. Select one positive lactose broth tube from the presumptive test and streak a plate of medium according to your assignment. Use a streak method that will produce good isolation of colonies. If all your tubes were negative, borrow a positive tube from another student.
2. Incubate the plate for 24 hours at 35° C.
3. Look for typical coliform colonies on both kinds of media. Record your results on Laboratory Report 47. If no coliform colonies are present, the water is considered bacteriologically safe to drink.
   Note: In actual practice, confirmation of all presumptive tubes would be necessary to ensure accuracy of results.

THE COMPLETED TEST

A final check of the colonies that appear on the confirmatory media is made by inoculating a nutrient agar slant and a Durham tube of lactose broth. After incubation for 24 hours at 35° C, the lactose broth is examined for gas production. A Gram-stained slide is made from the slant, and the slide is examined under oil immersion optics.

If the organism proves to be a gram-negative, non-spore-forming rod that ferments lactose, we know that coliforms were present in the tested water sample. If time permits, complete these last tests and record the results in Laboratory Report 47.

THE IMVIC TESTS

Review the discussion of the IMViC tests on page 243. The significance of these tests should be much more apparent at this time. Your instructor will indicate whether these tests should also be performed if you have a positive completed test.

The Membrane Filter Method

The most probable number method for determining coliform bacteria in water samples is complicated and requires several days to complete. Furthermore, more than one kind of culture medium is needed for each phase of the test to finally establish the presence of coliforms in a water sample. A more rapid method is the **membrane filter method**, also recognized by the United States Public Health Service as a reliable procedure for determining coliforms. In this test, known volumes of a water sample are filtered through membrane filters that have pores 0.45 μm in diameter. Most bacteria, including coliforms, are larger than the pore diameters, and hence bacteria are retained on the membrane filter. Once the water sample has been filtered, the filter disk containing bacterial cells is placed in a petri dish with an absorbent pad saturated with Endo broth. The plate is then incubated at 35° C for 22 to 24 hours during which time individual cells on the filter multiply forming colonies.

Any coliforms that are present on the filter will ferment the lactose in the Endo broth producing acids. The acids produced from fermentation interact with basic fuschin, a dye in the medium, causing coliform colonies to have a characteristic metallic sheen. Noncoliform bacteria will not produce the metallic sheen. Gram-positive bacteria are inhibited from growing because of the presence of bile salts and sodium lauryl sulfate, which inhibit these bacteria. Colonies are easily counted on the filter disk, and the total coliform count is determined based on the volume of water filtered.

Figure 48.1 illustrates the procedure we will use in this experiment.

MATERIALS

- vacuum pump or water faucet aspirators
- membrane filter assemblies (sterile)
- side-arm flask, 1000 ml size, and rubber hose
- sterile graduates (100 ml or 250 ml size)
- sterile, plastic petri dishes, 50 mm dia (Millipore #PD10 047 00)
- sterile membrane filter disks (Millipore #HAWG 047 AO)
- sterile absorbent disks (packed with filters)
- sterile water
- 5 ml pipettes
- bottles of *m* Endo MF broth (50 ml)* water samples

1. Prepare a small plastic petri dish as follows:
   a. With a flamed forceps, transfer a sterile absorbent pad to a sterile plastic petri dish.
   b. Using a 5 ml pipette, transfer 2.0 ml of *m* Endo MF broth to the absorbent pad.
2. Assemble a membrane filtering unit as follows:
   a. *Aseptically* insert the filter holder base into the neck of a 1-liter side-arm flask.
   b. With a flamed forceps, place a sterile membrane filter disk, grid side up, on the filter holder base.
   c. Place the filter funnel on top of the membrane filter disk and secure it to the base with the clamp.
3. Attach the rubber hose to a vacuum source (pump or water aspirator) and pour the appropriate amount of water into the funnel.

   The amount of water used will depend on water quality. No less than 50 ml should be used. Waters with few bacteria and low turbidity permit samples of 200 ml or more. Your instructor will advise you as to the amount of water that you should use. Use a sterile graduate for measuring the water.
4. Rinse the inner sides of the funnel with 20 ml of sterile water.
5. Disconnect the vacuum source, remove the funnel, and carefully transfer the filter disk with sterile forceps to the petri dish of *m* Endo MF broth. *Keep grid side up.*
6. Incubate at 35° C for 22 to 24 hours. *Don't invert.*
7. After incubation, remove the filter from the dish and dry for 1 hour on absorbent paper.
8. Count the colonies on the disk with low-power magnification, using reflected light. Ignore all colonies that lack the golden metallic sheen. If desired, the disk may be held flat by mounting between two 2″ × 3″ microscope slides after drying. Record your count on the first portion of Laboratory Report 48.

See Appendix C for special preparation method.

(1) Sterile absorbent pad is aseptically placed in the bottom of a sterile plastic Petri dish.

(2) Absorbent pad is saturated with 2.0 ml of *m* Endo MF broth.

(3) Sterile membrane filter disk is placed on filter holder base with grid side up.

(4) Water sample is poured into assembled funnel, utilizing vacuum. A rinse of 20 ml of sterile water follows.

(5) Filter disk is carefully removed with sterile forceps after disassembling the funnel.

(6) Membrane filter disk is placed on medium-soaked absorbent pad with grid side up. Incubate at 35° C 24 hours.

FIGURE 48.1 Membrane filter routine

LABORATORY REPORT

Student: _____

Date: _____   Section: _____

EXERCISE 48   The Membrane Filter Method

A. Tabulation

A table similar to the one below will be provided for you, either on the chalkboard or as a photocopy. Record your coliform count on it. Once all data are available, complete this table.

SAMPLE	SOURCE	COLIFORM COUNT	AMOUNT OF WATER FILTERED	MPN*
A				
B				
C				
D				
E				
F				
G				
H				

$$\text{*MPN} = \frac{\text{Coliform Count} \times 100}{\text{Amount of Water Filtered}}$$

B. Questions

1. Give two limitations of the membrane filter technique:

   _____

   _____

2. Even if the membrane filter removed all bacteria from water being tested, is the water that passes through

   sterile? _____ Explain: _____

   _____

3. List some other applications of membrane filter technology in microbiology. _____

   _____

   _____

   _____

Reductase Test

49

EXERCISE

Milk that contains large numbers of actively growing bacteria will have a lowered oxidation-reduction potential due to the exhaustion of dissolved oxygen by microorganisms. The fact that methylene blue loses its color (becomes reduced) in such an environment is the basis for the **reductase test.** In this test, 1 ml of methylene blue (1:25,000) is added to 10 ml of milk. The tube is sealed with a rubber stopper and slowly inverted three times to mix. It is placed in a water bath at 35° C and examined at intervals up to 6 hours. The time it takes for the methylene blue to become colorless is the **methylene blue reduction time** (MBRT). The shorter the MBRT, the lower the quality of milk. An MBRT of 6 hours is very good. Milk with an MBRT of 30 minutes is of very poor quality (figure 49.1).

The validity of this test is based on the assumption that all bacteria in milk lower the oxidation-reduction potential at 35° C. Large numbers of psychrophiles, thermophiles, and thermodurics, which do not grow at this temperature, would not produce a positive test. Raw milk, however, will contain primarily *Streptococcus lactis* and *Escherichia coli*, which are strong reducers; thus, this test is suitable for screening raw milk at receiving stations. Its principal

value is that less technical training of personnel is required for its performance.

In this exercise, samples of low- and high-quality raw milk will be tested.

MATERIALS

- 2 sterile test tubes with rubber stoppers for each student
- raw milk samples of low and high quality (samples A and B)
- water bath set at 35° C
- methylene blue (1:25,000)
- 10 ml pipettes
- 1 ml pipettes
- gummed labels

1. Attach gummed labels with your name and type of milk to two test tubes. Each student will test a good-quality as well as a poor-quality milk.
2. Using separate 10 ml pipettes for each type of milk, transfer 10 ml to each test tube. To the milk in the tubes add 1 ml of methylene blue with a 1 ml pipette. Insert rubber stoppers and gently invert three times to mix. Record your name and the

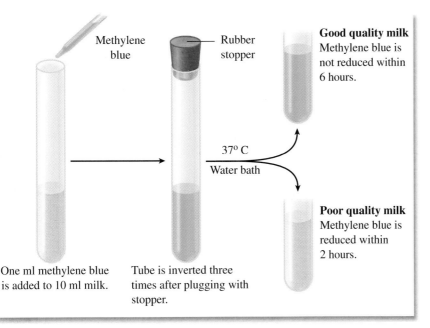

FIGURE 49.1   Procedure for testing raw milk with reductase test

Good quality milk
Methylene blue is not reduced within 6 hours.

Poor quality milk
Methylene blue is reduced within 2 hours.

Methylene blue · Rubber stopper

37° C Water bath

One ml methylene blue is added to 10 ml milk.

Tube is inverted three times after plugging with stopper.

301

time on the labels and place the tubes in the water bath, which is set at 35° C.

3. After 5 minutes incubation, remove the tubes from the bath and invert once to mix. This is the last time they should be mixed.

4. Carefully remove the tubes from the water bath 30 minutes later and every half hour until the end of the laboratory period. *When at least four-fifths of the tube has turned white*, the end point of reduction has taken place. Record this time in Laboratory Report 49. The classification of milk quality is as follows:

Class 1: Excellent, not decolorized in 8 hours

Class 2: Good, decolorized in less than 8 hours but not less than 6 hours

Class 3: Fair, decolorized in less than 6 hours, but not less than 2 hours

Class 4: Poor, decolorized in less than 2 hours

LABORATORY REPORT

Student: _____

Date: _____   Section: _____

EXERCISE 49   Reductase Test

1. How would you grade the two samples of milk that you tested? Give the MBRT for each one.

   Sample A: _____

   Sample B: _____

2. Is milk with a short reduction time necessarily unsafe to drink? _____

   Explain: _____

   _____

   _____

   _____

3. What other dye can be substituted for methylene blue in this test? _____

   _____

4. What advantage do you see in this method over the direct count method? _____

   _____

   _____

   _____

   _____

5. What kinds of organisms may be plentiful in a milk sample, yet give a negative reductase test?

   _____

   _____

   _____

   _____

Microbial Spoilage of Canned Food

Spoilage of heat-processed, commercially canned foods is confined almost entirely to the action of bacteria that produce heat-resistant endospores. Canning of foods normally involves heat exposure for long periods of time at temperatures that are adequate to kill spores of most bacteria. Particular concern is given to the processing of low-acid foods in which *Clostridium botulinum* can thrive to produce botulism food poisoning.

Spoilage occurs when the heat processing fails to meet accepted standards. This can occur for several reasons: (1) lack of knowledge on the part of the processor (usually the case in home canning); (2) carelessness in handling the raw materials before canning, resulting in an unacceptably high level of contamination that ordinary heat processing may be inadequate to control; (3) equipment malfunction that results in undetected underprocessing; and (4) defective containers that permit the entrance of organisms after the heat process.

Our concern here will be with the most common types of food spoilage caused by heat-resistant, spore-forming bacteria. There are three types: flat sour, T.A. spoilage, and stinker spoilage.

Flat sour pertains to spoilage in which acids are formed with no gas production; result: sour food in cans that have flat ends. **T.A. spoilage** is caused by thermophilic anaerobes that produce acid and gases (CO_2 and H_2, but not H_2S) in low-acid foods. Cans swell to various degrees, sometimes bursting. **Stinker spoilage** is due to spore-formers that produce hydrogen sulfide and blackening of the can and contents. Blackening is due to the reaction of H_2S with the iron in the can to form iron sulfide.

In this experiment, you will have an opportunity to become familiar with some of the morphological and physiological characteristics of organisms that cause canned food spoilage, including both aerobic and anaerobic endospore formers of *Bacillus* and *Clostridium*, as well as a non-spore-forming bacterium.

Working as a single group, the entire class will inoculate 10 cans of vegetables (corn and peas) with five different organisms. Figure 50.1 illustrates the procedure. Note that the cans will be sealed with solder after inoculation and incubated at different temperatures. After incubation the cans will be opened so that stained microscope slides can be made to determine Gram reaction and presence of endospores. Your instructor will assign individual students or groups of students to inoculate one or more of the 10 cans. One can of corn and one can of peas will be inoculated with each of the organisms. Proceed as follows:

FIRST PERIOD
(Inoculations)

MATERIALS

- 5 small cans of corn
- 5 small cans of peas
- cultures of *B. stearothermophilus*, *B. coagulans*, *C. sporogenes*, *C. thermosaccharolyticum*, and *E. coli*
- ice picks or awls
- hammer
- solder and soldering iron
- plastic bags
- gummed labels and rubber bands

1. Label the can or cans with the name of the organism that has been assigned to you. Use white gummed labels. In addition, place a similar label on one of the plastic bags to be used after sealing of the cans.
2. With an ice pick or awl, punch a small hole through a flat area in the top of each can. This can be done easily with the heel of your hand or a hammer, if available.
3. Pour off a small amount of the liquid from the can to leave an air space under the lid.
4. Use an inoculating needle to inoculate each can of corn or peas with the organism indicated on the label.
5. Take the cans up to the demonstration table where the instructor will seal the hole with solder.
6. After sealing, place each can in two plastic bags. Each bag must be closed separately with rubber bands, and the outer bag must have a label on it.

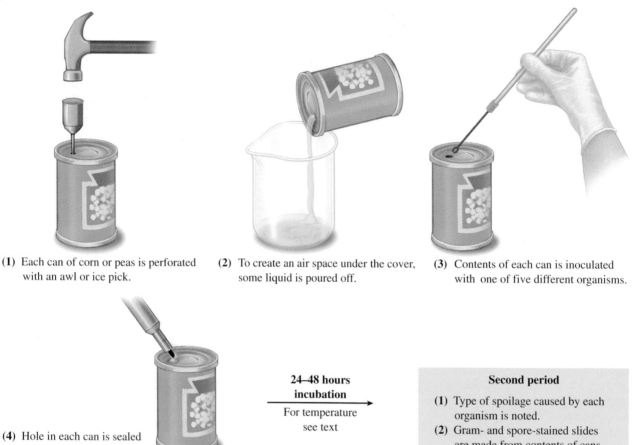

(1) Each can of corn or peas is perforated with an awl or ice pick.

(2) To create an air space under the cover, some liquid is poured off.

(3) Contents of each can is inoculated with one of five different organisms.

(4) Hole in each can is sealed by soldering over it.

24–48 hours incubation

For temperature see text

Second period

(1) Type of spoilage caused by each organism is noted.
(2) Gram- and spore-stained slides are made from contents of cans.

FIGURE 50.1   Canned food inoculation procedure

7. Incubation will be as follows until the next period:
   - **55° C**—*C. thermosaccharolyticum* and *B. stearothermophilus*
   - **37° C**—*C. sporogenes* and *B. coagulans*
   - **30° C**—*E. coli*

   Note: If cans begin to swell during incubation, they should be placed in refrigerator.

SECOND PERIOD
(Interpretation)

After incubation, place the cans under a hood to open them. The odors of some of the cans will be very strong due to H_2S production.

MATERIALS

- can opener, punch type
- small plastic beakers
- Parafilm
- Gram-staining kit
- spore-staining kit

1. Open each can carefully with a punch-type can opener. If the can is swollen, hold an inverted plastic funnel over the can during perforation to minimize the effects of any explosive release of contents.
2. Remove about 10 ml of the liquid through the opening, pouring it into a small plastic beaker. Cover with Parafilm. This fluid will be used for making stained slides.
3. Return the cans of food to the plastic bags, reclose them, and dispose in a proper trash bin.
4. Prepare Gram-stained and endospore-stained slides from your canned food extract as well as from the extracts of all the other cans. Examine under brightfield oil immersion.
5. Record your observations on the report sheet on the demonstration table. It will be duplicated and a copy will be made available to each student.

LABORATORY REPORT

Complete the first portion of Laboratory Report 50.

LABORATORY REPORT

Student: _____

Date: _____   Section: _____

EXERCISE 50   Microbial Spoilage of Canned Food

A. Results

Record your observations of the effects of each organism on the cans of vegetables. Share results with other students.

ORGANISM	PEAS		CORN	
	Gas Production + or −	Odor	Gas Production + or −	Odor
E. coli				
B. coagulans				
B. stearothermophilus				
C. sporogenes				
C. thermosaccharolyticum				

B. Microscopy

After making Gram-stained and spore-stained slides of all organisms from the canned food extracts, sketch in representatives of each species:

E. coli	B. coagulans	B. stearothermophilus	C. sporogenes	C. thermosac-charolyticum

C. Questions

1. Which organisms, if any, caused flat sour spoilage? _____

   _____

2. Which organisms, if any, caused T.A. spoilage? _____

   _____

3. Which organisms, if any, caused stinker spoilage? _____

   _____

4. Does flat sour cause a health problem? _____

5. Describe how typical spoilage resulting in botulism occurs. _____

   _____

   _____

Microbiology of Alcohol Fermentation

Fermented food and beverages are as old as civilization. Historical evidence indicates that beer and wine making were well established as long ago as 2000 B.C. An Assyrian tablet states that Noah took beer aboard the ark.

Beer, wine, vinegar, buttermilk, cottage cheese, sauerkraut, pickles, and yogurt are some of the products of fermentation. Most of these foods and beverages are produced by different strains of yeasts (*Saccharomyces*) or bacteria (*Lactobacillus, Acetobacter, etc.*).

Fermentation is actually a means of food preservation because the acids formed and the reduced environment (anaerobiasis) hold back the growth of many spoilage microbes.

Wine is essentially fermented fruit juice in which alcoholic fermentation is carried out by *Saccharomyces cerevisiae* var. *ellipsoideus*. Although we usually associate wine with fermented grape juice, it may also be made from various berries, dandelions, rhubarb, and so on. Three conditions are necessary:

simple sugar, yeast, and anaerobic conditions. The reaction is as follows:

$$C_6H_{12}O_6 \xrightarrow{\text{yeast}} 2C_2H_5OH + 2CO_2$$

Commercially, wine is produced in two forms: red and white. To produce red wines, the distillers use red grapes with the skins left on during the initial stage of the fermentation process. For white wines, either red or white grapes can be used, but the skins are discarded. White and red wines are fermented at 13° C (55° F) and 24° C (75° F), respectively.

In this exercise, we will set up a grape juice fermentation experiment to learn about some of the characteristics of sugar fermentation to alcohol. Note in figure 51.1 that a balloon will be attached over the mouth of the fermentation flask to exclude oxygen uptake and to trap gases that might be produced. To detect the presence of hydrogen sulfide production, we will tape a lead acetate test strip inside the neck of the flask. The pH of the substrate will also be monitored before and after the reaction to note any changes that occur.

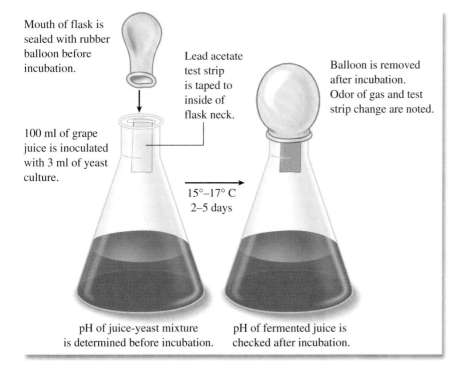

Mouth of flask is sealed with rubber balloon before incubation.

Lead acetate test strip is taped to inside of flask neck.

100 ml of grape juice is inoculated with 3 ml of yeast culture.

15°–17° C 2–5 days

Balloon is removed after incubation. Odor of gas and test strip change are noted.

pH of juice-yeast mixture is determined before incubation.

pH of fermented juice is checked after incubation.

FIGURE 51.1 Alcohol fermentation setup

FIRST PERIOD

MATERIALS

- 100 ml grape juice (no preservative)
- bottle of juice culture of wine yeast
- 125 ml Erlenmeyer flask
- 1 10 ml pipette
- balloon
- hydrogen sulfide (lead acetate) test paper
- tape
- pH meter

1. Label an Erlenmeyer flask with your initials and date.
2. Add about 100 ml of grape juice to the flask (fermenter).
3. Determine the pH of the juice with a pH meter and record the pH in Laboratory Report 51.
4. Agitate the container of yeast juice culture to suspend the culture, remove 5 ml with a pipette, and add it to the flask.
5. Attach a short strip of tape to a piece of lead-acetate test paper (3 cm long), and attach it to the inside surface of the neck of the flask. Make certain that neither the tape nor the test strip protrudes from the flask.
6. Cover the flask opening with a balloon.
7. Incubate at 15°–17° C for 2–5 days.

SECOND PERIOD

MATERIALS

- pH meter

1. Remove the balloon and note the aroma of the flask contents. Describe the odor in Laboratory Report 51.
2. Determine the pH and record it in the Laboratory Report.
3. Record any change in color of the lead-acetate-test strip in the Laboratory Report. If any H_2S is produced, the paper will darken due to the formation of lead sulfide as hydrogen sulfide reacts with the lead acetate.
4. Wash out the flask and return it to the drain rack.

LABORATORY REPORT

Complete Laboratory Report 51 by answering all the questions.

LABORATORY REPORT

Student: _____

Date: _____ Section: _____

EXERCISE 51   Microbiology of Alcohol Fermentation

A. Results

Record here your observations of the fermented product:

Aroma: _____

pH: _____

H₂S production: _____

B. Questions

1. Why must the fermentor be sealed? _____

    _____

    Why with a balloon? _____

2. What compound in the grape juice is being fermented? _____

3. Why would production of hydrogen sulfide by the yeast be of importance? _____

    _____

    _____

4. Why are we concerned about the pH of the fruit juice and the wine? _____

    _____

    _____

5. What happens to wine if *Acetobacter* takes over? _____

    _____

6. What process can be used to prevent the action of *Acetobacter* in the production of wine and beer?

    _____

MEDICAL MICROBIOLOGY AND IMMUNOLOGY

Although many of the exercises up to this point in this manual pertain in some way to medical microbiology, they also have applications that are nonmedical. The exercises of this unit, however, are primarily medical or dental in nature.

Medical (clinical) microbiology is primarily concerned with the isolation and identification of pathogenic organisms. The techniques for studying each type of organism are different. A complete coverage of this field of microbiology is very extensive, encompassing the Mycobacteriaceae, Brucellaceae, Enterobacteriaceae, Corynebacteriaceae, Micrococcaceae, and others. It is not possible to explore all of these groups in such a short period of time: however, this course would be incomplete if it did not include some routine procedures that are used in the identification of several common pathogens.

Exercise 60 differs from the other 9 exercises in that it pertains to the spread of disease (epidemiology) rather than to specific microorganisms. Its primary function is to provide an understanding of some of the tools used by public health epidemiologists to determine the sources of infection in the disease transmission cycle.

Since the most frequently encountered pathogenic bacteria are the gram-positive pyogenic cocci and the intestinal organisms, Exercises 52, 53, and 54 have been devoted to the study of those bacteria. The exercise that provides the greatest amount of depth is Exercise 53 (The Streptococci). To provide assistance in the identification of streptococci, it has been necessary to provide supplementary information in Appendix E.

Three exercises (55, 56, and 57) are related to various applications of the agglutination reaction to serological testing. Two of these exercises pertain to slide tests and one of them is a tube test. It is anticipated that the instructor will select those tests from this group that fit time and budget limitations.

Exercises 58 and 59 cover some of the basic hematological tests that might be included in a microbiology laboratory.

The Staphylococci:
Isolation and Identification

Often in conjunction with streptococci, the staphylococci cause abscesses, boils, carbuncles, osteomyelitis, and fatal septicemias. Collectively, the staphylococci and streptococci are referred to as the pyogenic (pus-forming) gram-positive cocci. Originally isolated from pus in wounds, the staphylococci were subsequently demonstrated to be normal inhabitants of the nasal membranes, the hair follicles, the skin, and the perineum of healthy individuals. The fact that 90% of hospital personnel are carriers of staphylococci portends serious epidemiological problems, especially since most strains are penicillin resistant.

The **staphylococci** are gram-positive, spherical bacteria that divide in more than one plane to form irregular clusters of cells (figure 52.1). They are listed in section 12, volume 2, of *Bergey's Manual of Systematic Bacteriology.* The genus *Staphylococcus* is grouped with three other genera in family Micrococcaceae:

SECTION 12 GRAM-POSITIVE COCCI
Family I Micrococcaceae
    Genus I *Micrococcus*
    Genus II *Stomatococcus*
    Genus III *Planococcus*
    Genus IV ***Staphylococcus***

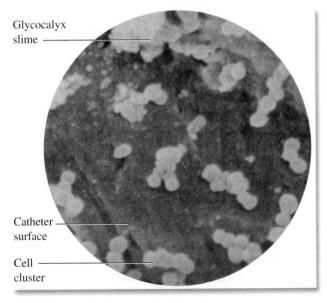

Glycocalyx slime

Catheter surface

Cell cluster

FIGURE 52.1 **Staphylococci** © Science VU/Charles W. Stratton/Visuals Unlimited

Family II Deinococcaceae
    Genus I *Deinococcus*
    Genus II ***Streptococcus***

Although the staphylococci make up a coherent phylogenetic group, they have very little in common with the streptococci except for their basic similarities of being gram-positive, non-spore-forming cocci. Note that *Bergey's Manual* puts these two genera into separate families due to their inherent differences.

Of the 19 species of staphylococci listed in *Bergey's Manual,* the most important ones are *S. aureus, S. epidermidis,* and *S. saprophyticus.* The single most significant characteristic that separates these species is the ability or inability of these organisms to coagulate plasma: only *S. aureus* has this ability; the other two are coagulase-negative.

Although *S. aureus* has, historically, been considered to be the only significant pathogen of the three, the others do cause infections. Some cerebrospinal fluid infections (2), prosthetic joint infections (3), and vascular graft infections (1) have been shown to be due to coagulase-negative staphylococci. Numbers in parentheses designate references at the end of this exercise.

Our concern in this exercise will pertain exclusively to the differentiation of only three species of staphylococci. If other species are encountered, the student may wish to use the API Staph-Ident miniaturized test strip system (Exercise 45).

In this experiment, we will attempt to isolate staphylococci from (1) the nose, (2) a fomite, and (3) an "unknown control." The unknown control will be a mixture containing staphylococci, streptococci, and some other contaminants. If the nasal membranes and fomite prove to be negative, the unknown control will yield positive results, provided all inoculations and tests are performed correctly.

Since *S. aureus* is by far the most significant pathogen in this group, most of our concern here will be with this organism. It is for this reason that the characteristics of only this pathogen will be outlined next.

Staphylococcus aureus cells are 0.8 to 1.0 μm in diameter and may occur singly, in pairs, or as clusters. Colonies of *S. aureus* on trypticase soy agar or blood agar are opaque, 1 to 3 mm in diameter, and yellow, orange, or white. They are salt tolerant and grow well on media containing 10% sodium chloride. Virtually

TABLE 52.1 Differentiation of Three Species of Staphylococci

	S. aureus	S. epidermidis	S. saprophyticus
Alpha toxin	+	–	–
Mannitol (acid only)	+	–	(+)
Coagulase	+	–	–
Biotin for growth	–	+	NS
Novobiocin	S	S	R

Note: NS = not significant; S = sensitive; R = resistant; (+) = mostly positive

all strains are coagulase positive. Mannitol is fermented aerobically to produce acid. Alpha toxin is produced that causes a wide zone of clear (beta-type) hemolysis on blood agar; in rabbits it causes local necrosis and death.

The other two species lack alpha toxin (do not exhibit hemolysis) and are coagulase negative. Mannitol is fermented to produce acid (no gas) by all strains of *S. aureus* and most strains of *S. saprophyticus*. Table 52.1 lists the principal characteristics that differentiate these three species of staphylococcus.

To determine the incidence of carriers in our classroom, as well as the incidence of the organism on common fomites, we will follow the procedure illustrated in figure 52.2. Results of class findings will be tabulated on the chalkboard so that all members of the class can record data required in Laboratory Report 52. The characteristics we will look for in our isolates will be (1) beta-type hemolysis (alpha toxin), (2) mannitol fermentation, and (3) coagulase production. Organisms found to be positive for these three characteristics will be *presumed* to be *S. aureus*. Final confirmation will be made with additional tests. Proceed as follows:

FIRST PERIOD

(Specimen Collection)

Note in figure 52.2 that swabs that have been applied to the nasal membranes and fomites will be placed in tubes of enrichment medium containing 10% NaCl (*m*-staphylococcus broth). Since your unknown control will lack a swab, initial inoculations from this culture will have to be done with a loop.

MATERIALS

- 1 tube containing numbered unknown control
- tubes of *m*-staphylococcus broth
- 2 sterile cotton swabs

1. Label the three tubes of *m*-staphylococcus broth NOSE, FOMITE, and the number of your unknown control.
2. Inoculate the appropriate tube of *m*-staphylococcus broth with one or two loopfuls of your unknown control.
3. After moistening one of the swabs by immersing partially into the "nose" tube of broth, swab the nasal membrane just inside your nostril. A small amount of moisture on the swab will enhance the pickup of organisms. Place this swab into the "nose" tube.
4. Swab the surface of a fomite with the other swab that has been similarly moistened and deposit this swab in the "fomite" tube.
   The fomite you select may be a coin, drinking glass, telephone mouthpiece, or any other item that you might think of.
5. Incubate these tubes of broth for 4 to 24 hours at 37° C.

SECOND PERIOD

(Primary Isolation Procedure)

Two kinds of media will be streaked for primary isolation: mannitol salt agar and staphylococcus medium 110.

Mannitol salt agar (MSA) contains mannitol, 7.5% sodium chloride, and phenol red indicator. The NaCl inhibits organisms other than staphylococci. If the mannitol is fermented to produce acid, the phenol red in the medium changes color from red to yellow.

Staphylococcus medium 110 (SM110) also contains NaCl and mannitol, but it lacks phenol red. Its advantage over MSA is that it favors colony pigmentation by different strains of *S. aureus*. Since this medium lacks phenol red, no color change takes place as mannitol is fermented.

MATERIALS

- 3 culture tubes from last period
- 2 petri plates of MSA
- 2 petri plates of SM110

1. Label the bottoms of the MSA and SM110 plates as shown in figure 52.2. Note that to minimize the number of plates required, it will be necessary to make half-plate inoculations for the nose and fomite. The unknown control will be inoculated on separate plates.
2. Quadrant streak the MSA and SM110 plates with the unknown control.
3. Inoculate a portion of the nose side of each plate with the swab from the nose tube; then, with a sterile loop, streak out the organisms on the remainder of the agar on that half of each plate. The swabbed areas will provide massive growth; the streaked-out areas should yield good colony isolation.

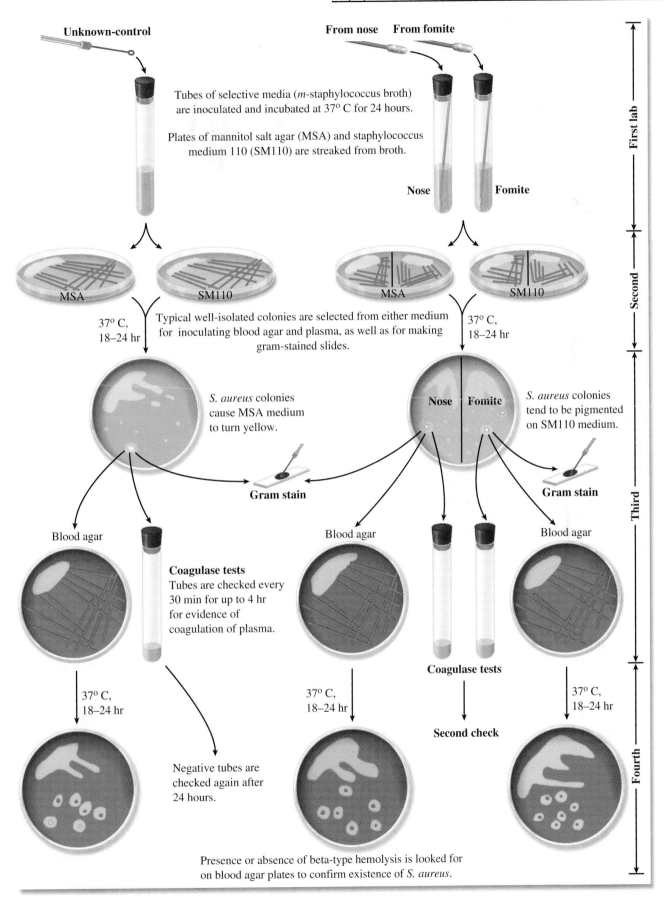

Unknown-control

From nose   From fomite

Tubes of selective media (*m*-staphylococcus broth) are inoculated and incubated at 37° C for 24 hours.

Plates of mannitol salt agar (MSA) and staphylococcus medium 110 (SM110) are streaked from broth.

Nose   **Fomite**

MSA   SM110   MSA   SM110

37° C, 18–24 hr   Typical well-isolated colonies are selected from either medium for inoculating blood agar and plasma, as well as for making gram-stained slides.   37° C, 18–24 hr

S. aureus colonies cause MSA medium to turn yellow.

Nose   Fomite

S. aureus colonies tend to be pigmented on SM110 medium.

Gram stain   **Gram stain**

Blood agar   Blood agar   Blood agar

Coagulase tests
Tubes are checked every 30 min for up to 4 hr for evidence of coagulation of plasma.

37° C, 18–24 hr   37° C, 18–24 hr   **Coagulase tests**   37° C, 18–24 hr

Second check

Negative tubes are checked again after 24 hours.

Presence or absence of beta-type hemolysis is looked for on blood agar plates to confirm existence of *S. aureus*.

First lab — Second — Third — Fourth

FIGURE 52.2   **Procedure for presumptive identification of staphylococci**

4. Repeat step 3 to inoculate the other half of each agar plate with the swab from the fomite tube.
5. Incubate the plates aerobically at 37° C for 24 to 36 hours.

THIRD PERIOD

(Plate Evaluations and Coagulase/DNase Tests)

During this period, we will perform the following tasks: (1) evaluate the plates from the previous period, (2) inoculate blood agar plates, (3) make gram-stained slides, and (4) perform coagulase and/or DNase tests on organisms from selected colonies. Proceed as follows:

MATERIALS

* MSA and SM110 plates from previous period
* 2 blood agar plates
* serological tubes containing 0.5 ml of 1:4 saline dilution of rabbit or human plasma (one tube for each isolate)
* petri plates of DNase agar
* gram-staining kit

Evaluation of Plates

1. Examine the mannitol salt agar plates. Has the phenol red in the medium surrounding any of the colonies turned yellow?

   If this color change exists, it can be presumed that you have isolated a strain of *S. aureus*. Record your results in Laboratory Report 52 and on the chalkboard. (Your instructor may wish to substitute a copy of the chart from the Laboratory Report to be filled out at the demonstration table.)
2. Examine the plates of SM110. The presence of growth here indicates that the organisms are salt-tolerant. Note color of the colonies (white, yellow, or orange).
3. Record your observations of these plates in Laboratory Report 52 and on the chalkboard.

Blood Agar Inoculations

1. Label the bottom of one blood agar plate with your unknown-control number, and streak out the organisms from a staphlike colony.
2. Select staphylococcus-like colonies from the MSA and SM110 plates from the nose and fomites for streaking out on another blood agar plate. Use half-plate streaking methods, if necessary.
3. Incubate the blood agar plates at 37° C for 18 to 24 hours. *Don't leave plates in incubator longer than 24 hours.* Overincubation will cause blood degeneration.

Coagulase Tests

The fact that 97% of the strains of *S. aureus* have proven to be coagulase positive and that the other two species are *always* coagulase negative makes the coagulase test an excellent definitive test for confirming identification of *S. aureus.*

The procedure is simple. It involves inoculating a small tube of plasma with several loopfuls of the organism and incubating it in a 37° C water bath for several hours. If the plasma coagulates, the organism is coagulase positive. Coagulation may occur in 30 minutes or several hours later. *Any degree of coagulation, from a loose clot suspended in plasma to a solid immovable clot, is considered to be a positive result, even if it takes 24 hours to occur.*

It should be emphasized that this test is valid only for gram-positive, staphylococcus-like bacteria, because some gram-negative rods, such as *Pseudomonas,* can cause a false-positive reaction. The mechanism of clotting in such organisms is not due to coagulase. Proceed as follows:

1. Label the plasma tubes NOSE, FOMITE, or UNKNOWN, depending on which of your plates have staphlike colonies.
2. With a wire loop, inoculate the appropriate tube of plasma with organisms from one or more colonies on SM110 or MSA. Use several loopfuls. Success is more rapid with a heavy inoculation. If positive colonies are present on both nose and fomite sides, be sure to inoculate a separate tube for each side.
3. Place the tubes in a 37° C water bath.
4. Check for solidification of the plasma every 30 minutes for the remainder of the period. Note in figure 52.3 that solidification may be complete, as in the lower tube, or show up as a semisolid ball, as seen in the middle tube.

   Any cultures that are negative at the end of the period will be left in the water bath. At 24 hours your instructor will remove them from the water bath and place them in the refrigerator, so that you can evaluate them in the next laboratory period.
5. Record your results in Laboratory Report 52.

DNase Test

The fact that coagulase-positive bacteria are also able to hydrolyze DNA makes the DNase test a reliable means of confirming *S. aureus* identification. The following procedure can be used to determine if a staphlike organism can hydrolyze DNA.

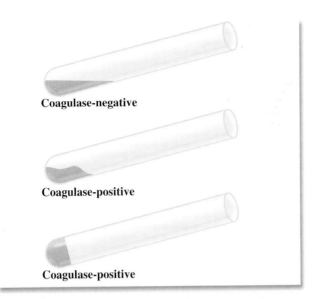

Coagulase-negative

Coagulase-positive

Coagulase-positive

FIGURE 52.3   Coagulase test results: one negative and two positive tests

1. Heavily streak the organism on a plate of DNase test agar. One plate can be used for several test cultures by making short streaks about 1 inch long.
2. Incubate for 18–24 hours at 35° C.

Gram-Stained Slides

While your tubes of plasma are incubating in the water bath, prepare Gram-stained slides from the same colonies that were used for the blood agar plates and coagulase tests.

   Examine the slides under oil immersion lens and draw the organisms in the appropriate areas of the Laboratory Report.

FOURTH PERIOD

(Confirmation)

During this period we will make final assessment of all tests and perform any other confirmatory tests that might be available to us.

MATERIALS

- coagulase tubes from previous tests
- blood agar plates from previous period
- DNase test agar plates from previous period
- 0.1N HCl

1. Examine any coagulase tubes that were carried over from the last laboratory period that were negative at the end of that period. Record your results in Laboratory Report 52.
2. Examine the colonies on your blood agar plates. Look for clear (beta-type) hemolysis around the colonies. The presence of alpha toxin is a definitive characteristic of *S. aureus*. Record your results in the Laboratory Report.
3. Look for zones of clearing near the streaks on the DNase agar plate. If none is seen, develop by flooding the plate with 0.1N HCl. The acid will render the hydrolyzed areas somewhat opaque.
4. Record your results on the chart on the chalkboard or chart on the demonstration table. If an instructor-supplied tabulation chart is used, the instructor will have copies made of it to be supplied to each student.

FURTHER TESTING

In addition to using the API Staph-Ident miniaturized test strip system (Exercise 45) to confirm your identification of staphylococci, you may wish to use the latex agglutination slide test described in Exercise 56. Your instructor will inform you as to the availability of these materials and the desirability of proceeding further.

LABORATORY REPORT

After recording your results on the chalkboard (or on the chart on the demonstration table), complete the chart in the Laboratory Report and answer all the questions.

The streptococci differ from the staphylococci in that they are arranged primarily in chains rather than in clusters (figure 53.1). In addition to causing many mixed infections with staphylococci, the streptococci can also, separately, cause diseases such as pneumonia, meningitis, endocarditis, pharyngitis, erysipelas, and glomerulonephritis.

Several species of streptococci are normal inhabitants of the pharynx. They can also be isolated from surfaces of the teeth, the saliva, skin, colon, rectum, and vagina.

The streptococci of greatest medical significance are *S. pyogenes, S. agalactiae,* and *S. pneumoniae.* Of lesser importance are *S. faecalis, S. faecium,* and *S. bovis.* Appendix E describes in greater detail the characteristics and significance of these and other streptococcal species.

The purpose of this exercise is twofold: (1) to learn about standard procedures for isolating streptococci from the pharynx and (2) to learn how to differentiate between the most significant medically important streptococci.

Figure 53.2 illustrates the overall procedure to be followed in the pursuit of the above two goals. Note that blood agar is used to separate the streptococci into two groups on the basis of the type of hemolysis they produce on blood agar. Those organisms that produce alpha hemolysis on blood agar can be differentiated by four tests. Those that produce beta-type hemolysis can be differentiated with the CAMP test and three other tests. The procedure outlined here is, primarily, designed to achieve *presumptive identification* of seven groups of streptococci. A few extra tests are usually required to confirm identification.

To broaden the application of these tests, your instructor may give you two or three unknown cultures of streptococci to be identified along with the pharyngeal isolates. *If unknowns are to be used, they will not be issued until physiological media are to be inoculated.*

FIRST PERIOD

(Making a Streak-Stab Agar Plate)

During this period, a plate of blood agar is swabbed and streaked in a special way to determine the type of hemolytic bacteria that are present in the pharynx. Before making such a streak plate, however, clinicians prefer to use a tube of enrichment broth (TSB) or a selective medium of TSB with a little crystal violet added to it (TSBCV). Media of this type are usually incubated at 37° C for 24 hours. This is particularly useful if the number of organisms might be low or if the swab cannot be applied to blood agar immediately. *Although this enrichment/selective step has been omitted here, it should be understood that the procedure is routine.*

Since swabbing one's own throat properly can be difficult, it will be necessary for you to work with your laboratory partner to swab each other's throats. Once your throat has been swabbed, you will proceed to use the swab to streak and stab your own agar plate according to a special procedure shown in figure 53.3.

MATERIALS

- 1 tongue depressor
- 1 sterile cotton swab
- inoculating loop
- 1 blood agar plate

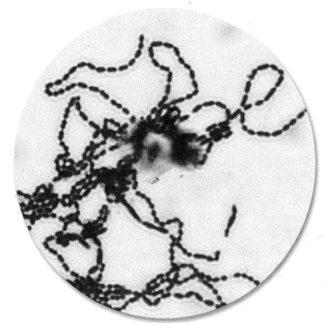

FIGURE 53.1 Streptococci

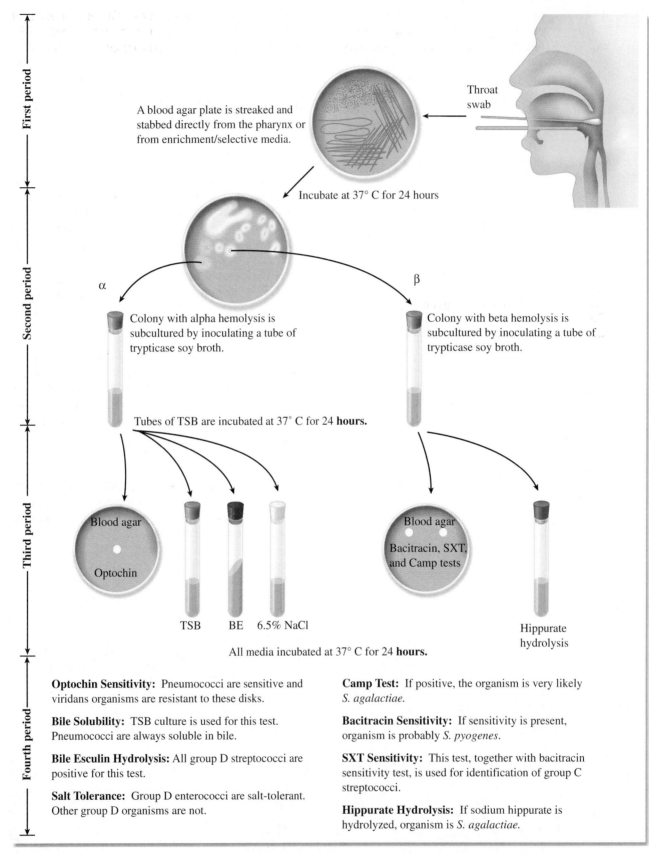

First period

Second period

Third period

Fourth period

Throat swab

A blood agar plate is streaked and stabbed directly from the pharynx or from enrichment/selective media.

Incubate at 37° C for 24 hours

α

β

Colony with alpha hemolysis is subcultured by inoculating a tube of trypticase soy broth.

Colony with beta hemolysis is subcultured by inoculating a tube of trypticase soy broth.

Tubes of TSB are incubated at 37° C for 24 **hours.**

Blood agar

Optochin

TSB   BE   6.5% NaCl

Blood agar

Bacitracin, SXT, and Camp tests

Hippurate hydrolysis

All media incubated at 37° C for 24 **hours.**

Optochin Sensitivity: Pneumococci are sensitive and viridans organisms are resistant to these disks.

Bile Solubility: TSB culture is used for this test. Pneumococci are always soluble in bile.

Bile Esculin Hydrolysis: All group D streptococci are positive for this test.

Salt Tolerance: Group D enterococci are salt-tolerant. Other group D organisms are not.

Camp Test: If positive, the organism is very likely *S. agalactiae.*

Bacitracin Sensitivity: If sensitivity is present, organism is probably *S. pyogenes.*

SXT Sensitivity: This test, together with bacitracin sensitivity test, is used for identification of group C streptococci.

Hippurate Hydrolysis: If sodium hippurate is hydrolyzed, organism is *S. agalactiae.*

FIGURE 53.2 Media inoculations for the presumptive identification of streptococci

1. With the subject's head tilted back and the tongue held down with the tongue depressor, rub the back surface of the pharynx up and down with the sterile swab.

   Also, *look for white patches* in the tonsillar area. Avoid touching the cheeks and tongue.

2. Since streptococcal hemolysis is most accurately analyzed when the colonies develop anaerobically beneath the surface of the agar, it will be necessary to use a streak-stab technique as shown in figure 53.3. The essential steps are as follows:
   - Roll the swab over an area approximating one-fifth of the surface. The entire surface of the swab should contact the agar.
   - With a wire loop, streak out three areas as shown to thin out the organisms.
   - Stab the loop into the agar to the bottom of the plate at an angle perpendicular to the surface to make a clean cut without ragged edges.
   - Be sure to make one set of stabs in an unstreaked area so that streptococcal hemolysis will be easier to interpret with a microscope.

   > **CAUTION** Dispose of swabs and tongue depressors in beaker of disinfectant.

3. Incubate the plate aerobically at 37° C for 24 hours. *Do not incubate longer than 24 hours.*

SECOND PERIOD

(Analysis and Subculturing)

During this period, two things must be accomplished: first, the type of hemolysis must be correctly determined and, second, well-isolated colonies must be selected for making subcultures. The importance of proper subculturing cannot be overemphasized: without a pure culture, future tests are certain to fail. Proceed as follows:

MATERIALS

- blood agar plate from previous period
- tubes of TSB (one for each different type of colony)
- dissecting microscope

1. Look for isolated colonies that have alpha or beta hemolysis surrounding them. Streptococcal colonies are characteristically very small.
2. Do any of the stabs appear to exhibit hemolysis? Examine these hemolytic zones near the stabs under 60 × magnification with a dissecting microscope.
3. Consult figure 53.4 to analyze the type of hemolysis. Note that the illustrations on the left side indicate what the colonies would look like if they were submerged under a layer of blood agar (two-layer pour plate). The illustrations on the right indicate the nature of hemolysis around stabs on streak-stab plates. Although this illustration is very diagrammatic, it reveals the microscopic differences between three kinds of hemolysis: alpha, alpha-prime, and beta.

   Only those stabs that are completely free of red blood cells in the hemolytic area are considered to be **beta hemolytic.** The chance of isolating a colony of this type from your own throat is very slim, for the beta hemolytic streptococci are the most serious pathogens.

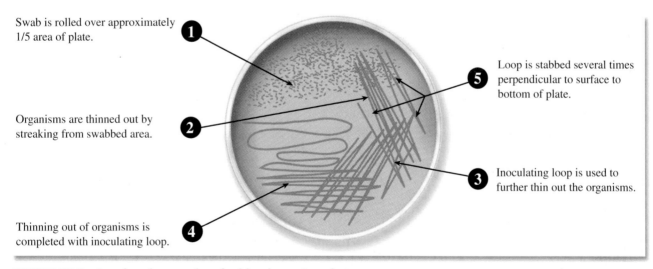

Swab is rolled over approximately 1/5 area of plate. **1**

Organisms are thinned out by streaking from swabbed area. **2**

Thinning out of organisms is completed with inoculating loop. **4**

5 Loop is stabbed several times perpendicular to surface to bottom of plate.

3 Inoculating loop is used to further thin out the organisms.

FIGURE 53.3   Streak-stab procedure for blood agar inoculations

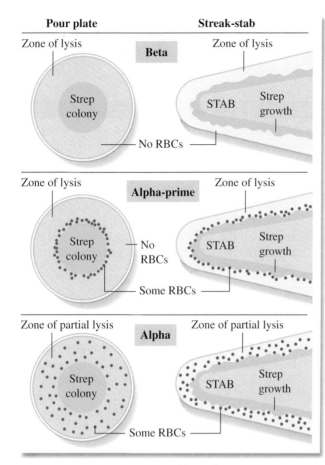

FIGURE 53.4 Comparison of hemolysis types as seen on pour plates and streak-stab plates

If some red blood cells are seen dispersed throughout the hemolytic zone, the organism is classified as **alpha-prime hemolytic.** Viridans streptococci often fall in this category.

4. Record your observations in Laboratory Report 53.
5. Select well-isolated colonies that exhibit hemolysis (alpha, beta, or both) for inoculating tubes of TSB. Be sure to label the tubes ALPHA or BETA. Whether or not the organism is alpha or beta is crucial in identification.

   Since the chances of isolating beta hemolytic streptococci from the pharynx are usually quite slim, notify your instructor if you think you have isolated one.
6. Incubate the tubes at 37° C for 24 hours.
7. **Important:** At some time prior to the next laboratory session, review the material in Appendix E that pertains to this exercise.

THIRD PERIOD
(Inoculations for Physiological Tests)

Presumptive identification of the various groups of streptococci is based on seven or eight physiological tests. Table 53.1 on page 329 reveals how they perform on these tests. Note that groups A, B, and C are all beta hemolytic; a few enterococci are also beta hemolytic. The remainder are all alpha hemolytic or nonhemolytic.

Since each of the physiological tests is specific for differentiating only two or three groups, it is not desirable to do all the tests on all unknowns. For economy and preciseness, only four tests that are mentioned for the third period in figure 53.2 should be performed on an isolate or unknown.

Before any inoculations are made, however, it is desirable to do a purity check on each TSB culture from the previous period. To accomplish this, it will be necessary to make a Gram-stained slide of each of the cultures.

If unknowns are to be issued, they will be given to you at this time. They will be tested along with your pharyngeal isolates. The only information that will be given to you about each unknown is its hemolytic type so that you will be able to determine what physiological tests to perform on each one. Proceed as follows:

Gram-Stained Slides (Purity Check)

MATERIALS
- TSB cultures from previous period
- Gram-staining kit

1. Make a Gram-stained slide from each of the pharyngeal isolates and examine them under oil immersion lens. Do they appear to be pure cultures?
2. Draw the organisms in the appropriate circles in Laboratory Report 53.

Beta-Type Inoculations

Use the following procedure to perform tests on each isolate that has beta-type hemolysis:

MATERIALS

for each isolate:
- 1 blood agar plate
- 1 tube of sodium hippurate broth
- 1 bacitracin differential disk
- 1 SXT sensitivity disk
- 1 broth culture of *S. aureus*
- dispenser or forceps for transferring disks

1. Label a blood agar plate and a tube of sodium hippurate broth with proper identification information of each isolate and unknown to be tested.
2. Follow the procedure outlined in figure 53.5 to inoculate each blood agar plate with the isolate (or unknown) and *S. aureus*.

   Note that a streak of the unknown is brought down perpendicular to the *S. aureus* streak, keeping the two organisms about 1 cm apart.

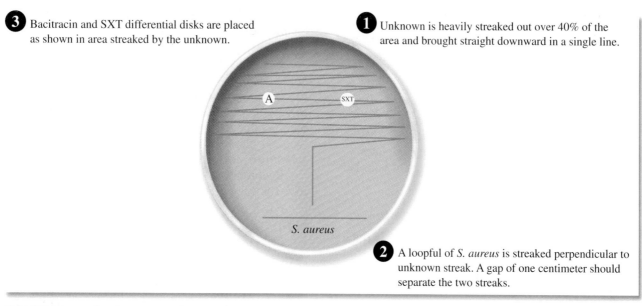

③ Bacitracin and SXT differential disks are placed as shown in area streaked by the unknown.

① Unknown is heavily streaked out over 40% of the area and brought straight downward in a single line.

S. aureus

② A loopful of *S. aureus* is streaked perpendicular to unknown streak. A gap of one centimeter should separate the two streaks.

FIGURE 53.5 Blood agar inoculation technique for the CAMP, bacitracin, and SXT tests

3. With forceps or dispenser, place one bacitracin differential disk and one SXT disk on the heavily streaked area at points shown in figure 53.5. Press down on each disk slightly.
4. Inoculate one tube of sodium hippurate broth for each isolate or unknown.
5. Incubate the blood agar plates at 37° C, aerobically, for 24 hours, and the hippurate broth tubes at 35° C, aerobically, for 24 hours. If the hippurate broths prove to be negative or weakly positive at 24 hours, they should be given more time to see if they change.

Alpha-Type Inoculations

As shown in figure 53.2, four inoculations will be made for each isolate or unknown that is alpha hemolytic.

MATERIALS

- 1 blood agar plate (for up to 4 unknowns)
- 1 6.5% sodium chloride broth
- 1 trypticase soy broth (TSB)
- 1 bile esculin (BE) slant
- 1 optochin (Taxo P) disk
- candle jar setup or CO_2 incubator

1. Mark the bottom of a blood agar plate to divide it into halves, thirds, or quarters, depending on the number of alpha hemolytic organisms to be tested. Label each space with the code number of each test organism.

2. Completely streak over each area of the blood agar plate with the appropriate test organism, and place one optochin (Taxo P) disk in the center of each area. Press down slightly on each disk to secure it to the medium.
3. Inoculate one tube each of TSB, BE, and 6.5% NaCl broth with each test organism.
4. Incubate all media at 35°–37° C as follows:
   Blood agar plates: 24 hours in a candle jar
   6.5% NaCl broths: 24, 48, and 72 hours
   Bile esculin slants: 48 hours
   Trypticase soy broths: 24 hours
   Note: While the blood agar plates should be incubated in a candle jar or CO_2 incubator, the remaining cultures can be incubated aerobically.

FOURTH PERIOD

(Evaluation of Physiological Tests)

Once all of the inoculated media have been incubated for 24 hours, you are ready to examine the plates and tubes and add test reagents to some of the cultures. Some of the tests will also have to be checked at 48 and 72 hours.

After you have assembled all the plates and tubes from the last period, examine the blood agar plates first that were double-streaked with the unknowns and *S. aureus*. Note that the second, third, and fourth tests listed in table 53.1 can be read from these plates. Proceed as follows:

CAMP Reaction

If you have an unknown that produces an enlarged arrowhead-shaped hemolytic zone at the juncture where the unknown meets the *S. aureus* streak, as seen in figure 53.6, the organism is *S. agalactiae*. This phenomenon is due to what is called the *CAMP factor.* The only problem that can arise from this test is that if the plate is incubated anaerobically, a positive CAMP reaction can occur on *S. pyogenes* inoculated plates.

Record the CAMP reactions for each of your isolates or unknowns in Laboratory Report 53.

Bacitracin Susceptibility

Any size zone of inhibition seen around the bacitracin disks should be considered to be a positive test result. Note in table 53.1 that *S. pyogenes* is positive for this characteristic.

This test has two limitations: (1) the disks must be of the *differential type,* not sensitivity type, and (2) the test should not be applied to alpha hemolytic streptococci. Reasons: Sensitivity disks have too high a concentration of the antibiotic, and many alpha hemolytic streptococci are sensitive to these disks.

Record the results of this test in the table under D of Laboratory Report 53.

SXT Sensitivity Test

The disks used in this test contain 1.25 mg of trimethoprim and 27.75 mg of sulfamethoxazole (SXT). The purpose of this test is to distinguish groups A and B from other beta hemolytic streptococci. Note in table 53.1 that both groups A and B are uniformly resistant to SXT.

FIGURE 53.6 Note positive SXT disk on right, negative bacitracin disk on left, and positive CAMP reaction (arrowhead). Organism: *S. agalactiae*

If a beta hemolytic streptococcus proves to be bacitracin resistant and SXT susceptible, it is classified as being a **non-group-A or -B beta hemolytic streptococcus.** This means that the organism is probably a species within group C. *Keep in mind that an occasional group A streptococcal strain is susceptible to both bacitracin and SXT disks.* One must always remember that exceptions to most tests do occur; that is why this identification procedure leads us only to *presumptive* conclusions.

Record any zone of inhibition (resistance) as positive for this test.

Hippurate Hydrolysis

Note in table 53.1 that hippurate hydrolysis and the CAMP test are grouped together as positive tests for *S. agalactiae*. If an organism is positive for both tests, or either one, one can assume with almost 100% certainty that the organism is *S. agalactiae*.

Proceed as follows to determine which of your isolates are able to hydrolyze sodium hippurate:

MATERIALS

- serological test tubes
- serological pipettes (1 ml size)
- ferric chloride reagent
- centrifuge

1. Centrifuge the culture for 3 to 5 minutes.
2. Pipette 0.2 ml of the supernatant and 0.8 ml of ferric chloride reagent into an empty serological test tube. Mix well.
3. Look for a **heavy precipitate** to form. If the precipitate forms and persists for 10 minutes or longer, the test is positive. If the culture proves to be weakly positive, incubate the culture for another 24 hours and repeat the test.
4. Record your results in Laboratory Report 53.

Bile Esculin (BE) Hydrolysis

This is the best physiological test that we have for the identification of group D streptococci. Both enterococcal and nonenterococcal species of group D are able to hydrolyze esculin in the agar slant, causing the slant to blacken.

A positive BE test tells us that we have a group D streptococcus; differentiation of the two types of group D streptococci depends on the salt-tolerance test.

Examine the BE agar slants, looking for **blackening of the slant,** as illustrated in figure 53.7. If less than half of the slant is blackened, or if no blackening occurs within 24 to 48 hours, the test is negative.

TABLE 53.1   Physiological Tests for Streptococcal Differentiation

	Type of hemolysis	Bacitracin susceptibility	CAMP reaction or hippurate hydrolysis	SXT sensitivity	Bile-esculin hydrolysis	Tolerance to 6.5% NaCl	Optochin susceptibility	Bile solubility
Group A *S. pyogenes*	beta	+	−	R	−	−	−	−	
Group B *S. agalactiae*	beta	− *	+	R	−	±	−	−	
Group C *S. equi S. equisimilis S. zooepidemicus*	beta	− *	−	S	−	−	−	−	
Group D** (enterococci) *S. faecalis S. faecium* etc.	alpha beta none	−	−	R	+	+	−	−	
Group D** (nonenterococci) *S. bovis* etc.	alpha none	−	−	R/S	+	−	−	−	
Viridans *S. mitis S. salivarius S. mutans* etc.	alpha none	− *	− *	S	−	−	−	−	
Pneumococci *S. pneumoniae*	alpha	±	−		−	−	+	+	

Note: R = resistant; S = sensitive; black = not significant.

*Exceptions occur occasionally.

**See comments on pp. 401 and 402 concerning correct genus.

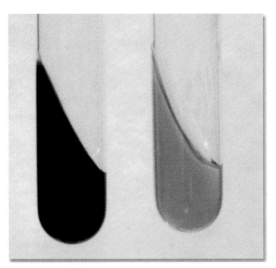

FIGURE 53.7   Positive bile esculin hydrolysis on left; negative on right

Salt Tolerance (6.5% NaCl)

All enterococci of group D produce heavy growth in 6.5% NaCl broth. As indicated in table 53.1, none of the nonenterococci, group D, grow in this medium. This test, then, provides us with a good method for differentiating the two types of group D streptococci.

A positive result shows up as turbidity within 72 hours. A color change of **purple to yellow** may also be present. If the tube is negative at 24 hours, incubate it and check it again at 48 and 72 hours. *If the organism is salt tolerant and BE positive, it is considered to be an enterococcus.* Parenthetically, it should be added here that approximately 80% of group B streptococci will grow in this medium.

Optochin Susceptibility

Optochin susceptibility is used for differentiation of the alpha hemolytic viridans streptococci from the pneumococci. The pneumococci are sensitive to these disks; the viridans organisms are resistant.

MATERIALS

- blood agar plates with optochin disks
- plastic metric ruler

1. Measure the diameters of zones of inhibition that surround each disk, evaluating whether the zones are large enough to be considered positive. The standards are as follows:
   - For 6 mm diameter disks, the zone must be at least 14 mm diameter to be considered positive.
   - For 10 mm diameter disks, the zone must be at least 16 mm diameter to be considered positive.
2. Record your results in Laboratory Report 53.

Bile Solubility

If an alpha hemolytic streptococcal organism is soluble in bile and positive on the optochin test, presumptive evidence indicates that the isolate is *S. pneumoniae*. Perform the bile solubility test on each of your alpha hemolytic isolates as follows:

MATERIALS

- 2 empty serological tubes (per test)
- dropping bottle of phenol red indicator
- dropping bottle of 0.05N NaOH

- TSB culture of unknown
- 2% bile solution (sodium desoxycholate)
- bottle of normal saline solution
- 2 serological pipettes (1 ml size)
- water bath (37°C)

1. Mark one empty serological tube BILE and the other SALINE. Into their respective tubes, pipette 0.5 ml of 2% bile and 0.5 ml of saline.
2. Shake the TSB unknown culture to suspend the organisms and pipette 0.5 ml of the culture into each tube.
3. Add 1 or 2 drops of phenol red indicator to each tube and adjust the pH to 7.0 by adding drops of 0.05N NaOH.
4. Place both tubes in a 37°C water bath and examine periodically for 2 hours. If the turbidity clears in the bile tube, it indicates that the cells have disintegrated and the organism is *S. pneumoniae*. Compare the tubes side by side.
5. Record your results in Laboratory Report 53.

FINAL CONFIRMATION

All the laboratory procedures performed so far lead us to presumptive identification. To confirm these conclusions, it is necessary to perform serological tests on each of the unknowns. If commercial kits are available for such tests, they should be used to complete the identification procedures.

LABORATORY REPORT

Complete Laboratory Report 53.

F. **Questions**

Record the answers for the following questions in the answer column.

1. What two physiological tests are significant in the identification of *S. agalactiae?*

2. If an alpha hemolytic streptococcus is able to hydrolyze bile esculin, what test can be used to tell whether the organism is an enterococcus?

3. What test is used for differentiating group A from group C streptococci if both organisms are bacitracin-susceptible?

4. What two tests are used to differentiate pneumococci from the viridians group?

5. What test is used for differentiating *S. pyogenes* from other beta hemolytic streptococci?

6. Hemolysis in streptococci can only be evaluated when the colonies develop_____ (*aerobically or anaerobically*) in blood agar.

7. Which streptococcal species is frequently present in the vagina of third-trimester pregnant women?

8. Only one beta hemolytic streptococcus is primarily of human origin. Which one is it?

9. Who developed the system of classifying streptococci into groups A, B, C, and so on?

10. Who is credited with grouping streptococci according to the type of hemolysis?

11. Which streptococcal species is seen primarily as paired cells (diplococci)?

12. Name two species of streptococci that are implicated in dental caries.

13. Where in the body can *S. bovis* be found?

14. After performing all physiological tests, what type of tests must be performed to confirm identification?

15. Which hemolysin produced by *S. pyogenes* is responsible for the beta-type hemolysis that is characteristic of this organism?

ANSWERS

1a. _____

b. _____

2. _____

3. _____

4a. _____

b. _____

5. _____

6. _____

7. _____

8. _____

9. _____

10. _____

11. _____

12a. _____

b. _____

13. _____

14. _____

15. _____

Gram-Negative Intestinal Pathogens

The enteric pathogens of prime medical concern are the **salmonella** and **shigella.** They cause enteric fevers, food poisoning, and bacillary dysentery. *Salmonella typhi,* which causes typhoid fever, is by far the most significant pathogen of the salmonella group. In addition to the typhoid organism, there are 10 other distinct salmonella species and over 2,200 serotypes. The shigella, which are the prime causes of human dysentery, comprise four species and many serotypes. *Serotypes* within genera are organisms of similar biochemical characteristics that can most easily be differentiated by serological typing.

Routine testing for the presence of these pathogens is a function of public health laboratories at various governmental levels. The isolation of these pathogenic enterics from feces is complicated by the fact that the colon contains a diverse population of bacteria. Species of such genera as *Escherichia, Proteus, Enterobacter, Pseudomonas,* and *Clostridium* exist in large numbers: hence it is necessary to use media that are differential and selective to favor the growth of the pathogens.

Figure 54.1 is a separation outline that is the basis for the series of tests that are used to demonstrate the presence of salmonella or shigella in a patient's blood, urine, or feces. Note that lactose fermentation separates the salmonella and shigella from most of the other Enterobacteriaceae. Final differentiation of the two enteric pathogens from *Proteus* relies on motility, hydrogen sulfide production, and urea hydrolysis. The differentiation information of the positive lactose fermenters on the left side of the separation outline is provided here mainly for comparative references that can be used for the identification of other unknown enterics.

The procedural diagram in figure 54.2 reveals how we will apply these facts in the identification of an unknown salmonella or shigella. The entire process will involve four laboratory periods.

In this experiment, you will be given a mixed culture containing a coliform, *Proteus,* and a salmonella or shigella. The pathogens will be of the less dangerous types, but their presence will, naturally, demand utmost caution in handling. Your problem will be to isolate the pathogen from the mixed culture and make a genus

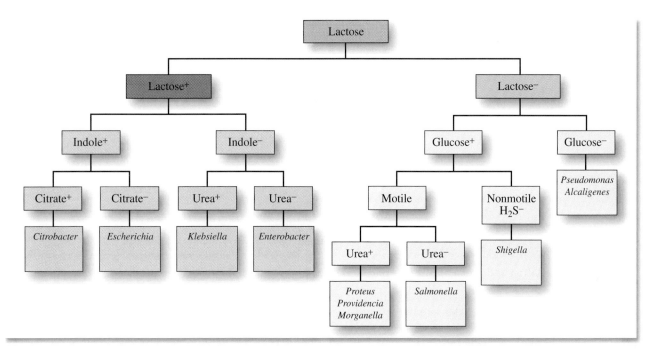

FIGURE 54.1 Separation outline of Enterobacteriaceae

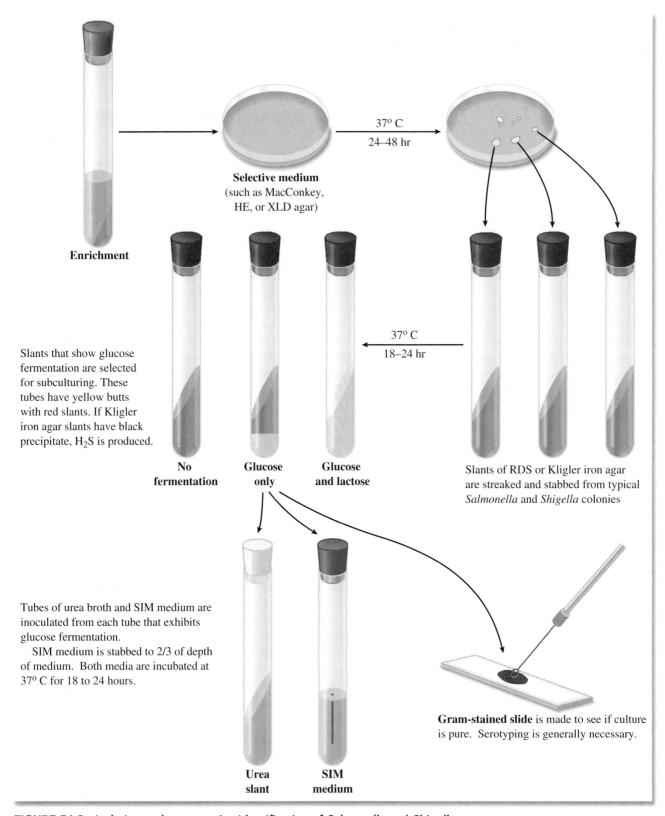

FIGURE 54.2 Isolation and presumptive identification of *Salmonella* and *Shigella*

identification. There are five steps that are used to prove the presence of these pathogens in a stool sample: (1) enrichment, (2) isolation, (3) fermentation tests, (4) final physiological tests, and (5) serotyping.

Enrichment

There are two enrichment media that are most frequently used to inhibit the nonpathogens and favor the growth of pathogenic enterics. They are selenite F and gram-negative (GN) broths. While most salmonella grow unrestricted in these two media, some of the shigella are inhibited to some extent in selenite F broth; thus, for shigella isolation, GN broth is preferred. In many cases, stool samples are plated directly on isolation media.

In actual practice, 1 to 5 grams of feces are placed in 10 ml of enrichment broth. In addition, plates of various kinds of selective media are inoculated directly. The broths are usually incubated for 4 to 6 hours.

Since we are not using stool samples in this exercise, the enrichment procedure is omitted. Instead, you will streak the isolation media directly from the unknown broth.

FIRST PERIOD

(Isolation)

There are several excellent selective differential media that have been developed for the isolation of these pathogens. Various inhibiting agents such as brilliant green, bismuth sulfite, sodium desoxycholate, and sodium citrate are included in them. For *Salmonella typhi,* bismuth sulfite agar appears to be the best medium. Colonies of *S. typhi* on this medium appear black due to the reduction of sulfite to sulfide.

Other widely used media are MacConkey agar, Hektoen Enteric agar (HE), and Xylose Lysine Desoxycholate (XLD) agar. These media may contain bile salts and/or sodium desoxycholate to inhibit gram-positive bacteria. To inhibit coliforms and other nonenterics, they may contain citrate. All of them contain lactose and a dye so that if an organism is a lactose fermenter, its colony will take on a color characteristic of the dye present.

Since the enrichment procedure is being omitted here, you will be issued an unknown broth culture with a pathogenic enteric. Your instructor will indicate which selective media will be used. Proceed as follows to inoculate the selective media with your unknown mixture:

MATERIALS

- unknown culture (mixture of a coliform, *Proteus,* and a salmonella or shigella)
- 1 or more petri plates of different selective media: MacConkey, Hektoen Enteric (HE), or Xylose Lysine Desoxycholate (XLD) agar

1. Label each plate with your name and unknown number.
2. With a loop, streak each plate with your unknown in a manner that will produce good isolation.
3. Incubate the plates at 37° C for 24 to 48 hours.

SECOND PERIOD

(Fermentation Tests)

As stated above, the fermentation characteristic that separates the SS pathogens from the coliforms is their *inability to ferment lactose.* Once we have isolated colonies on differential media that look like salmonella or shigella, the next step is to determine whether the isolates can ferment lactose. All media for this purpose contain at least two sugars, glucose and lactose. Some contain a third sugar, sucrose. They also contain phenol red to indicate when fermentation occurs. Russell Double Sugar (RDS) agar is one of the simpler media that works well. Kligler iron agar may also be used. It is similar to RDS with the addition of iron salts for detection of H_2S. Your instructor will indicate which one will be used.

Proceed as follows to inoculate three slants from colonies on the selective media that look like either salmonella or shigella. The reason for using three slants is that you may have difficulty distinguishing *Proteus* from the SS pathogens. By inoculating three tubes from different colonies, you will be increasing your chances of success.

MATERIALS

- 3 agar slants (RDS or Kligler iron)
- streak plates from first period

1. Label the three slants with your name and the number of your unknown.
2. Look for isolated colonies that look like salmonella or shigella organisms. The characteristics to look for on each medium are as follows:
   - **MacConkey agar**—*Salmonella, Shigella,* and other non-lactose-fermenting species produce smooth, colorless colonies. Coliforms that ferment lactose produce reddish, mucoid, or dark-centered colonies.
   - **Hektoen Enteric (HE) agar**—*Salmonella* and *Shigella* colonies are greenish-blue. Some species of *Salmonella* will have greenish-blue colonies with black centers due to H_2S production. Coliform colonies are salmon to orange and may have a bile precipitate.
   - **Xylose Lysine Desoxycholate (XLD) agar**— although most *Salmonella* produce red colonies with black centers, a few may produce red colonies that lack black centers. *Shigella* colonies are red. Coliform colonies are yellow. Some *Pseudomonas* produce false-positive red colonies.
3. With a straight wire, inoculate the three agar slants from separate SS-appearing colonies. Use

the streak-stab technique. When streaking the surface of the slant before stabbing, move the wire over the entire surface for good coverage.
4. Incubate the slants at 37° C for 18 to 24 hours. Longer incubation time may cause alkaline reversion. Even refrigeration beyond this time may cause reversion.

Alkaline reversion is a condition in which the medium turns yellow during the first part of the incubation period and then changes to red later due to increased alkalinity.

THIRD PERIOD
(Slant Evaluations and Final Inoculations)

During this period, you will inoculate tubes of SIM medium and urea broth with organisms from the slants of the previous period. Examination of the separation outline in figure 54.1 reveals that the final step in the differentiation of the SS pathogens is to determine whether a non-lactose-fermenter can do three things: (1) exhibit motility, (2) produce hydrogen sulfide, and (3) produce urease. You will also be making a Gram-stained slide to perform a purity check. If miniaturized multitest media are available, they can also be inoculated at this time.

MATERIALS
- RDS or Kligler's iron agar slants from previous period
- 1 tube of SIM medium for each positive slant
- 1 urea slant for each positive slant
- miniaturized multitest media such as API 20E or Enterotube II (optional)

1. Examine the slants from the previous period and **select those tubes that have a yellow butt with a red slant.** These tubes contain organisms that ferment only glucose (non-lactose-fermenters). If you used Kligler's iron agar, a black precipitate in the medium will indicate that the organism is a producer of H_2S.

 Note in figure 54.2 that slants that are completely yellow are able to ferment lactose as well as glucose. Tubes that are completely red are either nonfermenters or examples of alkaline reversion. Ignore those tubes.
2. With a loop, inoculate one urea slant from each slant that has a yellow butt and red slant (non-lactose-fermenter).
3. With a straight wire, stab one tube of SIM medium from each of the same agar slants. Stab in the center to two-thirds of depth of medium.

4. Incubate these tubes at 37° C for 18 to 24 hours.
5. Make Gram-stained slides from the same slants and confirm the presence of gram-negative rods.
6. If miniaturized multitest media are available, such as API 20E or Enterotube II, inoculate and incubate for evaluation in the next period. Consult Exercises 45 and 46 for instructions.
7. Refrigerate the positive RDS and Kligler iron slants for future use, if needed.

FOURTH PERIOD
(Final Evaluation)

During this last period, the tubes of SIM medium, urea broth, and any miniaturized multitest media from the last period will be evaluated. Serotyping can also be performed, if desired.

MATERIALS
- tubes of urea broth and SIM medium from previous period
- Kovacs' reagent and chloroform
- 5 ml pipettes
- miniaturized multitest media from previous period
- serological testing materials (optional)

1. Examine the tubes of SIM medium, checking for motility and H_2S production. If you see cloudiness spreading from the point of inoculation, the organism is motile. A black precipitate will be evidence of H_2S production.
2. Test for indole production by pipetting 2 ml of chloroform into each SIM tube and then adding 2 ml of Kovacs' reagent. A **pink to deep red color** will form in the chloroform layer if indole is produced.

 Salmonella are negative. Some *Shigella* may be positive. *Citrobacter* and *Escherichia* are positive.
3. Examine the urea slant tubes. If the medium has changed from yellow to **red** or **cerise color,** the organism is urease positive.
4. If a miniaturized multitest media was inoculated in the last period, complete it now.
5. If time and materials are available, confirm the identification of your unknown with serological typing. Refer to Exercise 55.

LABORATORY REPORT

Record the identity of your unknown in Laboratory Report 54 and answer all the questions.

LABORATORY REPORT

Student: _____

Date: _____ Section: _____

EXERCISE 54   Gram-Negative Intestinal Pathogens

A. **Unknown Identification**

1. What was the genus of your unknown?

   _____   _____

           Genus   No.

2. What problems, if any, did you encounter?

   _____

   _____

   _____

   _____

   _____

   _____

3. Now that you know the genus of your unknown, what steps would you follow to determine the species?

   _____

   _____

   _____

B. **General Questions**

1. Why are bile salts and sodium desoxycholate used in certain selective media in this exercise?

   _____

   _____

2. How can one identify coliforms on MacConkey agar? _____

   _____

   _____

3. How does one differentiate *Salmonella* from *Shigella* colonies on XLD agar_____

   _____

   _____

4. What characteristics do the salmonella and shigella have in common?

   _____

   _____

   _____

   _____

   _____

5. How do the salmonella and shigella differ?

_____

_____

_____

_____

6. What restrictions might be placed on a person who is a typhoid carrier?

_____

_____

_____

_____

_____

Slide Agglutination Test:
Serological Typing

Organisms of different species differ not only in morphology and physiology but also in protein makeup. The different proteins of bacterial cells that are able to stimulate antibody production when injected into an animal are **antigens.** The antigenic structure of each species of bacteria is unique to that species and, like the fingerprint of an individual, can be used to identify the organism. Many closely related microorganisms that are identical physiologically can be differentiated only by determining their antigenic nature.

The method of determining the presence of specific antigens in a microorganism is called **serological typing** (serotyping). It consists of adding a suspension of the organisms to **antiserum,** which contains antibodies that are specific for the known antigens. If the antigens are present, the antibodies in the antiserum will combine with the antigens, causing **agglutination,** or clumping, of the bacterial cells. Serotyping is particularly useful in the identification of various organisms that cause salmonella and shigella infections. In the identification of the various serotypes of these two genera, the use of antisera is generally performed after basic biochemical tests have been utilized as in Exercise 54.

In this exercise, you will be issued two unknown organisms, one of which is a salmonella. By following the procedure shown in figure 55.1, you will determine which one of the unknowns is salmonella. Note that you will use two test controls. A **negative test control** will be set up in depression A on the slide to see what the absence of agglutination looks like. The negative control is a mixture of antigen and saline (antibody is lacking). A **positive test control** will be performed in depression C with standardized antigen and antiserum to give you a typical reaction of agglutination.

MATERIALS

- 2 numbered unknowns per student (slant cultures of a salmonella and a coliform)
- salmonella O antigen, group B (Difco #2840-56)
- salmonella O antiserum, poly A-I (Difco #2264-47)
- depression slides or spot plates

- dropping bottle of phenolized saline solution (0.85% sodium chloride, 0.5% phenol)
- 2 serological tubes per student
- 1-ml pipettes

> **CAUTION** Keep in mind that *Salmonella typhimurium* is a pathogen and can cause gastroenteritis. Be careful!

1. Label three depressions on a spot plate or depression slide **A, B,** and **C,** as shown in figure 55.1.
2. Make a phenolized saline suspension of each unknown in separate serological tubes by suspending one or more loopfuls of organisms in 1 ml of phenolized saline. Mix the organisms sufficiently to ensure complete dispersion of clumps of bacteria. The mixture should be very turbid.
3. Transfer 1 loopful (0.05 ml) from the phenolized saline suspension of one tube to depressions A and B.

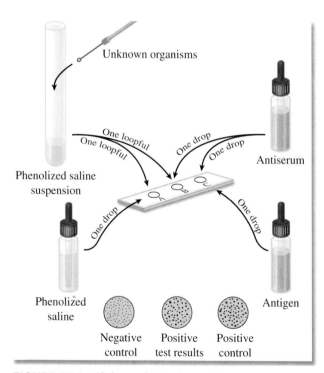

FIGURE 55.1   Slide agglutination technique

4. To depressions B and C add 1 drop of salmonella O polyvalent antiserum. To depression A, add 1 drop of phenolized saline, and to depression C, add 1 drop of salmonella O antigen, group B.
5. Mix the organisms in each depression with a clean wire loop. Do not go from one depression to the other without washing the loop first.
6. Compare the three mixtures. Agglutination should occur in depression C (positive control), but not in depression A (negative control). If agglutination occurs in depression B, the organism is salmonella.

7. Repeat this process on another slide for the other organism.

> **CAUTION** Deposit all slides and serological tubes in a container of disinfectant provided by the instructor.

LABORATORY REPORT

Record your results on the first portion of Laboratory Report 55, 56.

Slide Agglutination (Latex) Test:
For *S. aureus* Identification

Many manufacturers of reagents for slide agglutination tests utilize polystyrene latex particles as carriers for the antibody particles. By adsorbing reactive antibody units to these particles, an agglutination reaction results that occurs rapidly and is much easier to see than ordinary precipitin-type reactions that might be used to demonstrate the presence of a soluble antigen.

In this exercise, we will use reagents manufactured by Difco Laboratories to determine if a suspected staphylococcus organism produces coagulase and/or protein A. The test reagent (*Difco Staph Reagent*) is a suspension of yellow latex particles sensitized with antibodies for coagulase and protein A. Reagents are also included to provide positive and negative controls in the test. Instead of using depression slides or spot plates, Difco provides disposable cards with eight black circles printed on them for performing the test. As indicated in figure 56.1, only three circles are used when performing the test on one unknown. The additional circles are provided for testing five additional unknowns at the same time. The black background of the cards facilitates rapid interpretation by providing good contrast for the yellow clumps that form.

There are two versions of this test: direct and indirect. The procedure for the direct method is illustrated in figure 56.1. The indirect method differs in that saline is used to suspend the organism being tested.

It should be pointed out that the reliability correlation between this test for coagulase and the tube test (page 318) is very high. Studies reveal that a reliability correlation of over 97% exists. Proceed as follows to perform this test.

MATERIALS

- plate culture of staphylococcus-like organism (trypticase soy agar plus blood)
- Difco Staph Latex Test kit #3850-32-7, which consists of:
  - bottle of Bacto Staph Latex Reagent
  - bottle of Bacto Staph Positive Control
  - bottle of Bacto Staph Negative Control
  - bottle of Bacto Normal Saline Reagent
  - disposable test slides (black circle cards)
  - mixing sticks (minimum of 3)
- slide rotator

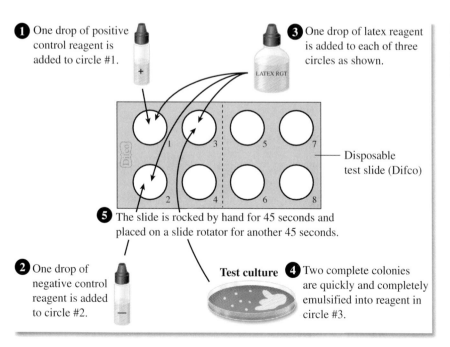

❶ One drop of positive control reagent is added to circle #1.

❸ One drop of latex reagent is added to each of three circles as shown.

LATEX RGT

Disposable test slide (Difco)

❺ The slide is rocked by hand for 45 seconds and placed on a slide rotator for another 45 seconds.

❷ One drop of negative control reagent is added to circle #2.

Test culture

❹ Two complete colonies are quickly and completely emulsified into reagent in circle #3.

FIGURE 56.1 Slide agglutination test (direct method) for the presence of coagulase and/or protein A

DIRECT METHOD

If the direct method is to be used, as illustrated in figure 56.1, follow this procedure:

1. Place 1 drop of Bacto Staph Positive Control reagent onto circle #1.
2. Place 1 drop of Bacto Staph Negative Control reagent on circle #2.
3. Place 1 drop of Bacto Staph Latex Reagent onto circles #1, #2, and #3.
4. Using a sterile inoculating needle or loop, quickly and completely emulsify *two isolated colonies* from the culture to be tested into the drop of Staph Latex Reagent in circle #3.

   Also, emulsify the Staph Latex Reagent in the positive and negative controls in circles #1 and #2 using separate mixing sticks supplied in the kit.

   All mixing in these three circles should be done quickly to minimize drying of the latex on the slide and to avoid extended reaction times for the first cultures emulsified.
5. Rock the slide by hand for 45 seconds.
6. Place the slide on a slide rotator capable of providing 110 to 120 rpm and rotate it for another 45 seconds.
7. Read the results immediately, according to the descriptions provided in the table at right. If agglutination occurs before 45 seconds, the results may be read at that time. *The slide should be read at normal reading distance under ambient light.*

INDIRECT METHOD

The only differences between the direct and indirect methods pertain to the amount of inoculum and the use of saline to emulsify the unknown being tested. Proceed as follows:

1. Place 1 drop of Bacto Staph Positive Control reagent onto test circle #1.
2. Place 1 drop of Bacto Staph Negative Control onto circle #2.
3. Place 1 drop of Bacto Normal Saline Reagent onto circle #3.

4. Using a sterile inoculating needle or loop, completely emulsify *four isolated colonies* from the culture to be tested into the circle containing the drop of saline (circle #3).
5. Add 1 drop of Bacto Staph Latex Reagent to each of the three circles.
6. Quickly mix the contents of each circle, using individual mixing sticks.
7. Rock the slide by hand for 45 seconds.
8. Place the slide on a slide rotator capable of providing 110 to 120 rpm and rotate it for another 45 seconds.
9. Read the results immediately according to the descriptions provided in the table below. If agglutination occurs before 45 seconds, the results may be read at that time. *The slide should be read at normal reading distance under ambient light.*

Positive Reactions	
4 +	Large to small clumps of aggregated yellow latex beads; clear background
3 +	Large to small clumps of aggregated yellow latex beads; slightly cloudy background
2 +	Medium to small but clearly visible clumps of aggregated yellow latex beads; moderately cloudy background
1 +	Fine clumps of aggregated yellow latex beads; cloudy background
Negative Reactions	
+	Smooth cloudy suspension; particulate grainy appearance that cannot be identified as agglutination
−	Smooth, cloudy suspension; free of agglutination or particles

LABORATORY REPORT

Record your results on the last portion of Laboratory Report 55, 56.

LABORATORY REPORT

Student: _____

Date: _____ Section: _____

EXERCISE 55   Slide Agglutination Test: Serological Typing

1. Record the unknown number that proved to be a salmonella. _____

2. Why was phenolized saline used instead of plain physiological saline? _____
    _____
    _____

3. If your results were negative for both cultures, what might be the explanation? _____
    _____
    _____

EXERCISE 56   Slide Agglutination (Latex) Test: For *S. aureus* Identification

1. If your test turned out to be positive for *S. aureus,* record the degree of positivity here: _____

2. What other test for *S. aureus* is highly correlated with this test? _____

3. What two kinds of antibodies are attached to the latex particles in the Difco latex reagent?

    a. _____   b. _____

4. What role, if any, do the staphylococci cells play in this reaction? _____
    _____
    _____

Infectious mononucleosis (IM) is a benign disease, occurring principally in individuals in the 13 to 25 year age group. It is caused by the Epstein-Barr virus (EBV), a herpesvirus, that is one of the most ubiquitous viruses in humans. Studies have shown that the virus can be isolated from saliva of patients with IM, as well as from some healthy, asymptomatic individuals. Between 80% and 90% of all adults possess antibodies for EBV.

The disease is characterized by a sudden onset of fever, sore throat, and pronounced enlargement of the cervical lymph nodes. There is also moderate leukocytosis with a marked increase in the number of lymphocytes (50% to 90%).

The serological test for IM takes advantage of an unusual property: the antibodies produced against the EBV coincidentally agglutinate sheep red blood cells. This is an example of a **heterophile antigen**— a substance isolated from a living form that stimulates the production of antibodies capable of reacting with tissues of other organisms. The antibodies are referred to as **heterophile antibodies.**

This test is performed by adding a suspension of sheep red blood cells to dilutions of inactivated patient's serum and incubating the tubes overnight in the refrigerator. Figure 57.1 illustrates the overall procedure. Agglutination titers of 320 or higher are considered significant. Titers of 40,960 have been obtained.

Proceed as follows to perform this test on a sample of test serum:

FIRST PERIOD

MATERIALS

- test-tube rack (Wasserman type) with 10 clean serological tubes
- bottle of saline solution (0.85% NaCl), clear or filtered
- 1 ml pipettes
- 5 ml pipettes
- 2% suspension of sheep red blood cells
- patient's serum (known to be positive)

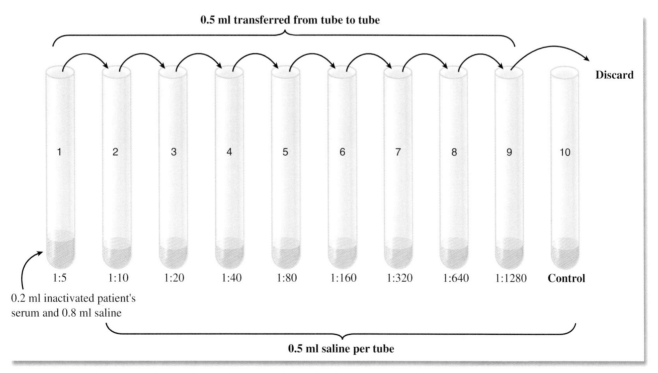

0.5 ml transferred from tube to tube

| 1 | 2 | 3 | 4 | 5 | 6 | 7 | 8 | 9 | 10 |

Discard

1:5   1:10   1:20   1:40   1:80   1:160   1:320   1:640   1:1280   **Control**

0.2 ml inactivated patient's serum and 0.8 ml saline

0.5 ml saline per tube

FIGURE 57.1   Procedure for setting up heterophile antibody test

1. Place the test serum in a 56° C water bath for 30 minutes to inactivate the complement.
2. Set up a row of 10 serological tubes in the front row of a test-tube rack and number them from 1 to 10 (left to right) with a marking pencil.
3. Into tube 1, pipette 0.8 ml of physiological saline.
4. Dispense 0.5 ml of physiological saline to tubes 2 through 10. Use a 5 ml pipette.
5. With a 1 ml pipette add 0.2 ml of the inactivated serum to tube 1. Mix the contents of this tube by drawing into the pipette and expelling about five times.
6. Transfer 0.5 ml from tube 1 to tube 2, mix five times, and transfer 0.5 ml from tube 2 to tube 3, are so on, through the ninth tube. *Discard 0.5 ml from the ninth tube after mixing.* Tube 10 is the **control.**
7. Add 0.2 ml of 2% sheep red blood cells to all tubes (1 through 10) and shake the tubes. Final dilutions of the serum are shown in figure 57.1.

8. Allow the rack of tubes to stand at room temperature for 1 hour, then transfer the tubes to a small wire basket, and place in a refrigerator to remain overnight.

SECOND PERIOD

Set up the tubes in a tube rack in order of dilution and compare each tube with the control by holding the tubes overhead and looking up at the bottoms of the tubes. Nonagglutinated cells will tumble to the bottom of the tube and form a small button (as in control tube). Agglutinated cells will form a more-amorphous "blanket."

The **titer** should be recorded as the reciprocal of the last tube in the series that shows positive agglutination.

LABORATORY REPORT

Complete Laboratory Report 57.

FIGURE 57.2 Agglutination is more readily seen when the tube is examined against a black surface

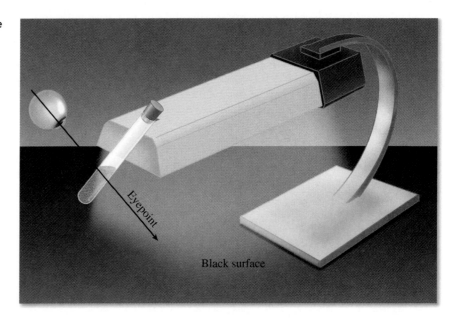

Eyepoint

Black surface

LABORATORY REPORT

Student: _____

Date: _____ Section: _____

EXERCISE 57 Tube Agglutination Test: The Heterophile Antibody Test

1. What was the titer of the serum that you tested? _____

2. For what disease is this diagnostic test used? _____

3. Below what titer would this test be considered to be negative? _____

4. What is the name of the virus that causes this disease? _____

5. What is unusual about a heterophile antibody? _____

   _____

6. Indicate the type of antigen (*soluble protein, red blood cells,* or *bacteria*) that is used for each of the following serological tests:

   Agglutination: _____

   Precipitation: _____

   Hemolysis: _____

7. For which one of the above tests is complement necessary? _____

   _____

In 1883, at the Pasteur Institute in Paris, Metchnikoff published a paper proposing the **phagocytic theory of immunity.** On the basis of his studies performed on transparent starfish larvae, he postulated that amoeboid cells in the tissue fluid and blood of all animals are the major guardians of health against bacterial infection. He designated the large phagocytic cells of the blood as *macrophages* and the smaller ones as *microphages.* Today, Metchnikoff's macrophages are known as monocytes and his microphages as neutrophils or polymorphonuclear leukocytes.

Figure 58.1 illustrates the five types of leukocytes that are normally seen in the blood. Blood platelets and erythrocytes also are shown to present a complete picture of all formed elements in the blood. When observed as living cells under the microscope, they appear as refractile, colorless structures. As shown here, however, they reflect the dyes that are imparted by Wright's stain.

In this exercise, we will do a study of the white blood cells in human blood. This study may be made from a prepared stained microscope slide or from a slide made from your own blood. By scanning an entire slide and counting the various types, you will have an opportunity to encounter most, if not all, types. The erythrocytes and blood platelets will be ignored.

Figures 58.1 and 58.2 will be used to identify the various types of cells. Figure 58.3 illustrates the procedure for preparing a slide stained with Wright's stain. The relative percentages of each type will be determined after a total of 100 white blood cells have been identified. This method of white blood cell enumeration is called a **differential WBC count.**

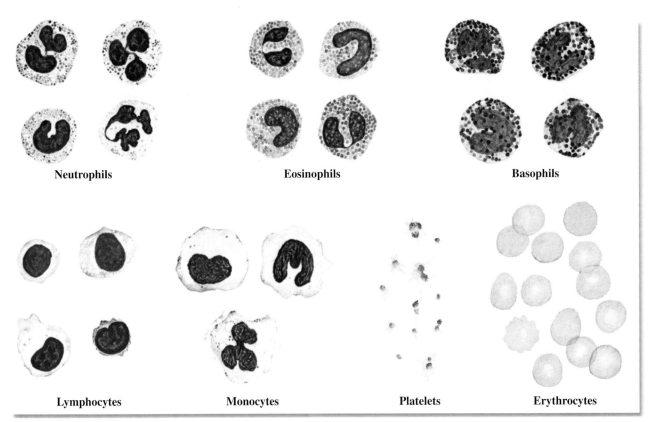

Neutrophils   Eosinophils   Basophils

Lymphocytes   Monocytes   Platelets   Erythrocytes

FIGURE 58.1   Formed elements of blood

Granulocytes

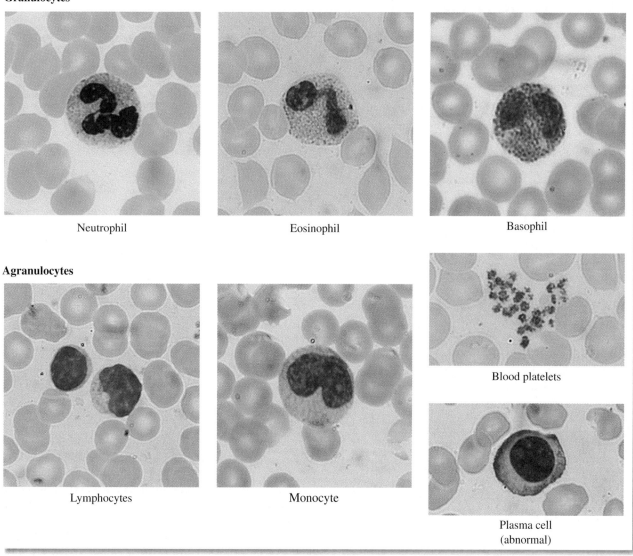

Agranulocytes

FIGURE 58.2   Photomicrographs of formed elements in blood

As you proceed with this count, it will become obvious that the neutrophils are most abundant (50%–70%). The next most prominent cells are the lymphocytes (20%–30%). Monocytes comprise about 2%–6%; eosinophils, 1%–5%; and basophils, less than 1%.

A normal white blood cell count is between 5,000 and 10,000 white cells per cubic millimeter. Elevated white blood cell counts are referred to as *leukocytosis;* counts of 30,000 or 40,000 represent marked leukocytosis. When counts fall considerably below 5,000, *leukopenia* is said to exist. Both conditions can have grave implications.

The value of a differential count is immeasurable in the diagnosis of infectious diseases. High neutrophil counts, or *neutrophilia,* often signal localized infections, such as appendicitis or abscesses in some other part of the body. *Neutropenia,* a condition in which there is a marked decrease in the numbers of neutrophils, occurs in typhoid fever, undulant fever, and influenza. *Eosinophilia* may indicate allergic conditions or invasions by parasitic roundworms such as *Trichinella spiralis,* the "pork worm." Counts of eosinophils may rise to as high as 50% in cases of trichinosis. High lymphocyte counts, or *lymphocytosis,* are present in whooping cough and some viral infections. Increased numbers of monocytes, or *monocytosis,* may indicate the presence of the Epstein-Barr virus, which causes infectious mononucleosis.

Note in the materials list that items needed for making a slide (option B) are listed separately. If a prepared slide (option A) is to be used, ignore the instructions under the heading "Preparation of Slide,"

(1) A small drop of blood is placed about 3/4 inch away from one end of slide. The drop should not exceed 1/8" diameter.

(2) The spreader slide is moved in direction of arrow, allowing drop of blood to spread along slide's back edge.

(3) The spreader slide is pushed along the slide, dragging the blood over the surface of the slide.

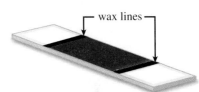

wax lines

(4) A china marking pencil is used to mark off both ends of the smear to retain the staining solution on the slide.

FIGURE 58.3   Smear preparation technique for making a stained blood slide

and proceed to the heading "Performing the Cell Count." Your instructor will indicate which option will be used. Proceed as follows:

PRECAUTIONS

When working with blood observe the following precautions:
1. Always disinfect the finger with alcohol prior to piercing it.
2. Use sterile disposable lancets only one time.
3. Dispose of used lancets by placing them into a beaker of disinfectant.
4. Avoid skin contact with blood of other students. Wear disposable latex gloves.
5. Disinfect finger with alcohol after blood has been taken.

MATERIALS

prepared blood slide (option A):
• stained with Wright's or Giemsa's stains

for staining a blood smear (option B):
• 2 or 3 clean microscope slides (should have polished edges)
• sterile disposable lancets
• disposable latex gloves
• sterile absorbent cotton, 70% alcohol
• Wright's stain, wax pencil, bibulous paper
• distilled water in dropping bottle

PREPARATION OF SLIDE

Figure 58.3 illustrates the procedure that will be used to make a stained slide of a blood smear. The most difficult step in making such a slide is getting a good spread of the blood, which is thick at one end and thin at the other end. If done properly, the smear will have a gradient of cellular density that will make it possible to choose an area that is ideal for study. The angle at which the spreading slide is held in making the smear will determine the thickness of the smear. It may be necessary for you to make more than one slide to get an ideal one.

1. Clean three or four slides with soap and water. Handle them with care to avoid getting their flat surfaces soiled by your fingers. Although only two slides may be used, it is often necessary to repeat the spreading process, thus the extra slides.
2. Scrub the middle finger with 70% alcohol and stick it with a lancet. Put a drop of blood on the slide $\frac{3}{4}$" from one end and spread with another slide in the manner illustrated in figure 58.3.
   Note that the blood is dragged over the slide, not pushed. Do not pull the slide over the smear a second time. If you don't get an even smear the first time, repeat the process on a fresh clean slide. To get a smear that will be the proper thickness, hold the spreading slide at an angle somewhat greater than 45°.

3. Draw a line on each side of the smear with a wax pencil to confine the stain that is to be added. (Note: This step is helpful for beginners, and usually omitted by professionals.)

4. Cover the film with Wright's stain, *counting the drops* as you add them. Stain for **4 minutes** and then add the same number of drops of distilled water to the stain and let stand for another **10 minutes.** Blow gently on the mixture every few minutes to keep the solutions mixed.

5. Gently wash off the slide under running water for 30 seconds and shake off the excess. Blot dry with bibulous paper.

PERFORMING THE CELL COUNT

Whether you are using a prepared slide or one that you have just stained, the procedure is essentially the same. Although the high-dry objective can be used for the count, the oil immersion lens is much better. Differentiation of some cells is difficult with high-dry optics. Proceed as follows:

1. Scan the slide with the low-power objective to find an area where cell distribution is best. A good area is one in which the cells are not jammed together or scattered too far apart.

2. Systematically scan the slide, following the pathway indicated in figure 58.4. As each leukocyte is encountered, identify it, using figures 58.1 and 58.2 for reference.

3. Tabulate your count on the Laboratory Report sheet according to the instructions there. It is best to remove the lab report sheet from the back of the manual for this identification and tabulation procedure.

LABORATORY REPORT

Place your results in the table in Part A of Laboratory Report 58.

FIGURE 58.4   Path to follow when seeking cells

LABORATORY REPORT

Student: _____

Date: _____   Section: _____

EXERCISE 58 White Blood Cell Study: The Differential WBC Count

A. Tabulation of Results

As you move the slide in the pattern indicated in figure 58.4, record all the different types of cells in the following table. Refer to figures 58.1 and 58.2 for cell identification. Use this method of tabulation: ꟷꟷ ꟷꟷ ꟷ. Identify and tabulate 100 leukocytes. Divide the total of each kind of cell by 100 to determine percentages.

NEUTROPHILS	LYMPHOCYTES	MONOCYTES	EOSINOPHILS	BASOPHILS
Total				
Percent				

B. Questions

1. Were your percentages for each type within the normal ranges? _____

2. What errors might one be likely to make when doing this count for the first time? _____

_____

_____

_____

3. Differentiate between the following:

   Cellular immunity: _____

   Humoral immunity: _____

4. Do cellular and humoral immunity work independently? _____ Explain: _____

_____

_____

_____

_____

_____

_____

Blood Grouping

Exercises 55 through 57 illustrate three uses of agglutination tests as related to (1) the identification of serological types, (2) species identification (*S. aureus*), and (3) disease identification (infectious mononucleosis and typhoid fever). The typing of blood is another example of a medical procedure that relies on this useful phenomenon.

The procedure for blood typing was developed by Karl Landsteiner around 1900. He is credited with having discovered that human blood types can be separated into four groups on the basis of two antigens that are present on the surface of red blood cells. These antigens are designated as A and B. The four groups (types) are A, B, AB, and O. The last group type O, which is characterized by the absence of A or B antigens, is the most common type in the United States (45% of the population). Type A is next in frequency, found in 39% of the population. The incidences of types B and AB are 12% and 4%, respectively.

Blood typing is performed with antisera containing high titers of anti-A and anti-B antibodies. The test may be performed by either slide or tube methods. In both instances, a drop of each kind of antiserum is added to separate samples of saline suspension of red blood cells. Figure 59.1 illustrates the slide technique. If agglutination occurs only in the suspension to which the anti-A serum was added, the blood is type A. If agglutination occurs only in the anti-B mixture, the blood is type B. Agglutination in both samples indicates that the blood is type AB. The absence of agglutination indicates that the blood is type O.

Between 1900 and 1940, a great deal of research was done to uncover the presence of other antigens in human red blood cells. Finally, in 1940, Landsteiner and Wiener reported that rabbit sera containing antibodies against the red blood cells of the rhesus monkey would agglutinate the red blood cells of 5% of white humans. This antigen in humans, which was first designated as the **Rh factor** (in due respect to the rhesus monkey), was later found to exist as six antigens: C, c, D, d, E, and e. Of these six antigens, the D factor is responsible for the Rh-positive condition and is found in 85% of Whites, 94% of Blacks, and 99% of Orientals.

Typing blood for the Rh factor can also be performed by both tube and slide methods, but there are certain differences in the two techniques. First of all, the antibodies in the typing sera are of the incomplete albumin variety, which *will not agglutinate human red cells when they are diluted with saline.* Therefore, it is necessary to use whole blood or dilute the cells with plasma. Another difference is that the test *must be performed at higher temperatures:* 37° C for tube test, 45° C for the slide test.

In this exercise, two separate slide methods are presented for typing blood. If only the Landsteiner ABO groups are to be determined, the first method may be preferable. If Rh typing is to be included, the second method, which utilizes a slide warmer, will be followed. The availability of materials will determine which method is to be used.

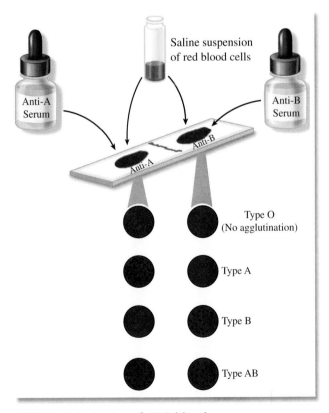

Saline suspension of red blood cells

Anti-A Serum

Anti-B Serum

Anti-A

Anti-B

Type O (No agglutination)

Type A

Type B

Type AB

FIGURE 59.1 Typing of ABO blood groups

Review the precautionary comments highlighted on page 353.

ABO BLOOD TYPING

MATERIALS

- small vial (10 mm dia × 50 mm long)
- disposable lancets (B-D Microlance, Serasharp, etc.)
- 70% alcohol and cotton
- china marking pencil
- microscope slides
- typing sera (anti-A and anti-B)
- applicators or toothpicks
- saline solution (0.85% NaCl)
- 1 ml pipettes
- disposable latex gloves

1. Mark a slide down the middle with a marking pencil, dividing the slide into two halves as shown in figure 59.1. Write "anti-A" on the left side and "anti-B" on the right side.
2. Pipette 1 ml of saline solution into a small vial or test tube.
3. Scrub the middle finger with a piece of cotton saturated with 70% alcohol and pierce it with a sterile disposable lancet. Allow 2 or 3 drops of blood to mix with the saline by holding the finger over the end of the vial and washing it with the saline by inverting the tube several times.
4. Place a drop of this red cell suspension on each side of the slide.

5. Add a drop of anti-A serum to the left side of the slide and a drop of anti-B serum to the right side.
   Do not contaminate the tips of the serum pipettes with the material on the slide.
6. After mixing each side of the slide with separate applicators or toothpicks, look for agglutination. The slide should be held about 6″ above an illuminated white background and rocked gently for 2 or 3 minutes. Record your results in Laboratory Report 59 as of 3 minutes.

COMBINED ABO AND RH TYPING

As stated, Rh typing must be performed with heat on blood that has not been diluted with saline. A warming box such as the one in figure 59.2 is essential in this procedure. In performing this test, two factors are of considerable importance: first, only a small amount of blood must be used (a drop of about 3 mm diameter on the slide) and, second, proper agitation must be executed. The agglutination that occurs in this antibody-antigen reaction results in finer clumps; therefore, closer examination is essential. If the agitation is not properly performed, agglutination may not be as apparent as it should be.

In this combined method, we will use whole blood for the ABO typing as well as for the Rh typing. Although this method works satisfactorily as a classroom demonstration for the ABO groups, it is *not as reliable* as the previous method in which saline and room temperature are used. *This method is not recommended for clinical situations.*

FIGURE 59.2 Blood typing with warming box

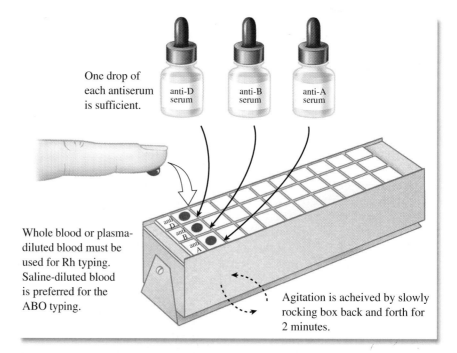

One drop of each antiserum is sufficient.

anti-D serum

anti-B serum

anti-A serum

Whole blood or plasma-diluted blood must be used for Rh typing. Saline-diluted blood is preferred for the ABO typing.

Agitation is acheived by slowly rocking box back and forth for 2 minutes.

MATERIALS

- slide warming box with a special marked slide
- anti-A, anti-B, and anti-D typing sera
- applicators or toothpicks
- 70% alcohol and cotton
- disposable sterile lancets

1. Scrub the middle finger with a piece of cotton saturated with 70% alcohol and pierce it with a sterile disposable lancet. Place a small drop of blood in each of three squares on the marked slides on the warming box.

   To get the proper proportion of serum to blood, do not use a drop larger than 3 mm diameter on the slide.

2. Add a drop of anti-D serum to the blood in the anti-D square, mix with a toothpick, and note the time. **Only 2 minutes should be allowed for agglutination.**
3. Add a drop of anti-B serum to the anti-B square and a drop of anti-A serum to the anti-A square. Mix the sera and blood in both squares with *separate* fresh toothpicks.
4. Agitate the mixtures on the slide by slowly rocking the box back and forth on its pivot. At the end of 2 minutes, examine the anti-D square carefully for agglutination. If no agglutination is apparent, consider the blood to be Rh-negative. By this time the ABO type can also be determined.
5. Record your results in Laboratory Report 59.

LABORATORY REPORT

Student: _____

Date: _____ Section: _____

EXERCISE 59   Blood Grouping

1. On the basis of this test, what is your blood type? _____

2. What antibodies are present in each type of blood?

   Type A: _____   Type B: _____   Type AB: _____   Type O: _____

3. Why does a person of type A blood go into a transfusion reaction when given type B blood? _____

   _____

   _____

   _____

4. Why can Rh-positive blood be given only once to a person who is Rh-negative? _____

   _____

   _____

   _____

A Synthetic Epidemic

A disease caused by microorganisms that enter the body and multiply in the tissues at the expense of the host is said to be an **infectious disease.** Infectious diseases that are transmissible to other persons are considered to be **communicable.** The transfer of communicable infectious agents between individuals can be accomplished by direct contact, such as in handshaking, kissing, and sexual intercourse, or they can be spread indirectly through food, water, objects, animals, and so on.

Epidemiology is the study of how, when, where, what, and who are involved in the spread and distribution of diseases in human populations. An epidemiologist is, in a sense, a medical detective who searches out the sources of infection so that the transmission cycle can be broken.

Whether an epidemic actually exists is determined by the epidemiologist by comparing the number of new cases with previous records. If the number of newly reported cases in a given period of time in a specific area is excessive, an **epidemic** is considered to be in progress. If the disease spreads to one or more continents, a **pandemic** is occurring.

In this experiment, we will have an opportunity to approximate, in several ways, the work of the epidemiologist. Each member of the class will take part in the spread of a "synthetic infection." The mode of transmission will be handshaking. For obvious safety reasons, the agent of transmission will not be a pathogen.

Two different approaches to this experiment are given: procedures A and B. In procedure A, a white powder is used. In Procedure B, two nonpathogens (*Micrococcus luteus* and *Serratia marcescens*) will be used. The advantage of procedure A is that it can be completed in one laboratory session. Procedure B, on the other hand, is more realistic in that viable organisms are used; however, it involves two periods. Your instructor will indicate which procedure is to be followed.

PROCEDURE A

In this experiment, each student will be given a numbered container of white powder. Only one member in the class will be given a powder that is to be considered the infectious agent. The other members will be issued a transmissible agent that is considered noninfectious. After each student has spread the powder on his or her hands, all members of the class will engage in two rounds of handshaking, directed by the instructor. A record of the handshaking contacts will be recorded on a chart similar to the one in the Laboratory Report. After each round of handshaking, the hands will be rubbed on blotting paper so that a chemical test can be applied to it to determine the presence or absence of the infectious agent.

Once all the data are compiled, an attempt will be made to determine two things: (1) the original source of the infection, and (2) who the carriers are. The type of data analysis used in this experiment is similar to the procedure that an epidemiologist would employ. Proceed as follows:

MATERIALS
- 1 numbered container of white powder*
- 1 piece of white blotting paper
- spray bottles of "developer solution"*

Preliminaries

1. After assembling your materials, write your name and unknown number at the top of your sheet of blotting paper. In addition, draw a line down the middle, top to bottom, and label the left side ROUND 1 and the right side ROUND 2.
2. Wash and dry your hands thoroughly.
3. Moisten the right hand with water and prepare it with the agent by thoroughly coating it with the white powder, especially on the palm surface. This step is similar to the contamination that would occur to one's hand if it were sneezed into during a cold.

   IMPORTANT: Once the hand has been prepared, do not rest it on the tabletop or allow it to touch any other object.

---

*Instructor: To prevent students from preguessing the outcome of this experiment, the compositions of the powders and developer solution are known only to the instructor. The Instructor's Handbook provides this information.

Round 1

1. On the cue of the instructor, you will begin the first round of handshaking. Your instructor will inform you when it is your turn to shake hands with someone. You may shake with anyone, but it is best not to shake your neighbor's hand. *Be sure to use only your treated hand, and avoid extracurricular glad-handing.*

2. In each round of handshaking, you will be selected by the instructor *only once* for handshaking; however, due to the randomness of selection by the handshakers, it is possible that you may be selected as the "shakee" several times.

3. After every member of the class has shaken someone's hand, you need to assess just who might have picked up the "microbe." To accomplish this, wipe your fingers and palm of the contaminated hand on the left side of your blotting paper. Press fairly hard, but don't tear the surface.

   IMPORTANT: Don't allow your hand to touch any other object. A second round of handshaking follows.

Round 2

1. On the cue of your instructor, shake hands with another person. Avoid contact with any other objects.

2. Once the second handshaking episode is finished, rub the fingers and palm of the contaminated hand on the right side of the blotting paper.

   CAUTION: Keep your contaminated hand off the left side of the blotting paper.

Chemical Identification

1. To determine who has been "infected," we will now spray the developer solution on the handprints of both rounds. One at a time, each student, with the help of the instructor, will spray his or her blotting paper with developer solution.

2. Color interpretation is as follows:
   • Blue:— positive for infectious agent
   • Brown or yellow:— negative

Tabulation of Results

1. Tabulate the results on the chalkboard, using a table similar to the one in the Laboratory Report.

2. Once all results have been recorded, proceed to determine the originator of the epidemic. The easiest way to determine this is to put together a flowchart of shaking.

3. Identify those persons that test positive. You will be working backward with the kind of information an epidemiologist has to work with (contacts and infections). Eventually, a pattern will emerge that shows which person started the epidemic.

4. Complete Laboratory Report 60.

PROCEDURE B

In this experiment, each student will be given a piece of hard candy that has had a drop of *Micrococcus luteus* or *Serratia marcescens* applied to it. Only one person in the class will receive candy with *S. marcescens*, the presumed pathogen. All others will receive *M. luteus*.

After each student has handled the piece of candy with a glove-covered right hand, he or she will shake hands (glove to glove) with another student as directed by the instructor. A record will be kept of who takes part in each contact. Two rounds of handshaking will take place. After each round, a plate of trypticase soy agar will be streaked.

After incubating the plates, a tabulation will be made for the presence or absence of *S. marcescens* on the plates. From the data collected, an attempt will be made to determine two things: (1) the original source of the infection and (2) who the carriers are. The type of data analysis used in this experiment is similar to the procedure that an epidemiologist would employ. Proceed as follows:

> **CAUTION** Although the pathogenicity of *S. marcescens* is considered to be relatively low, avoid allowing any skin contact during this experiment.

MATERIALS

- sterile rubber surgical gloves (1 per student)
- hard candy contaminated with *M. luteus*
- hard candy contaminated with *S. marcescens*
- sterile swabs (2 per student)
- TSA plates (1 per student)

Preliminaries

1. Draw a line down the middle of the bottom of a TSA plate, dividing it into two halves. Label one half ROUND 1 and the other ROUND 2.

2. Put a sterile rubber glove on your right hand. Avoid contaminating the palm surface.

3. Grasp the piece of candy in your gloved hand, rolling it around the surface of your palm. Discard the candy into a beaker of disinfectant set aside for disposal. You are now ready to do the first-round handshake.

Round 1

1. *On the cue of your instructor,* select someone to shake hands with. You may shake with anyone, but it is best not to shake hands with your neighbor.

2. In each round of handshaking, you will be selected by the instructor *only once* for handshaking; however, due to the randomness of selection

by the handshakers, it is possible that you may be selected as the "shakee" several times. The instructor or a recorder will record the initials of the shaker and shakee each time.

3. After you have shaken someone's hand, swab the surface of your palm and transfer the organisms to the side of your plate designated as ROUND 1. Discard this swab into the appropriate container for disposal.

Round 2

1. Again, on the cue of your instructor, select someone at random to shake hands with. Be sure not to contaminate your gloved hand by touching something else.
2. With a fresh swab, swab the palm of your hand and transfer the organisms to the side of your plate designated as ROUND 2. Make sure that your initials and the initials of the shakee are recorded by the instructor or recorder.

3. Incubate the TSA plate at room temperature for 48 hours.

Tabulation and Analysis

1. After 48 hours incubation look for typical red *S. marcescens* colonies on your petri plate. If such colonies are present, record them as positive on your Laboratory Report chart and on the chart on the chalkboard.
2. Fill out the chart on Laboratory Report 60 with all the information from the chart on the chalkboard.
3. Identify those persons that test positive. You will be working backward with the kind of information an epidemiologist has to work with (contacts and infections). Eventually a pattern will emerge that shows which person started the epidemic.

LABORATORY REPORT

Complete Laboratory Report 60.

Descriptive Chart

STUDENT: _____

LAB SECTION: _____

Habitat: _____ Culture No.: _____

Source: _____

Organism: _____

MORPHOLOGICAL CHARACTERISTICS

Cell Shape:

Arrangement:

Size:

Spores:

Gram's Stain:

Motility:

Capsules:

Special Stains:

CULTURAL CHARACTERISTICS

Colonies:

   Nutrient Agar:

   Blood Agar:

Agar Slant:

Nutrient Broth:

Gelatin Stab:

Oxygen Requirements:

Optimum Temp.:

PHYSIOLOGICAL CHARACTERISTICS

	TESTS	RESULTS
Fermentation	Glucose	
	Lactose	
	Sucrose	
	Mannitol	
Hydrolysis	Gelatin Liquefaction	
	Starch	
	Casein	
	Fat	
IMViC	Indole	
	Methyl Red	
	V-P (acetylmethylcarbinol)	
	Citrate Utilization	
	Nitrate Reduction	
	H_2S Production	
	Urease	
	Catalase	
	Oxidase	
	DNase	
	Phenylalanase	

	REACTION	TIME
Litmus Milk	Acid	_____
	Alkaline	_____
	Coagulation	_____
	Reduction	_____
	Peptonization	_____
	No Change	_____

Descriptive Chart

STUDENT: _____

LAB SECTION: _____

Habitat: _____ Culture No.: _____

Source: _____

Organism: _____

MORPHOLOGICAL CHARACTERISTICS

Cell Shape:

Arrangement:

Size:

Spores:

Gram's Stain:

Motility:

Capsules:

Special Stains:

CULTURAL CHARACTERISTICS

Colonies:

   Nutrient Agar:

   Blood Agar:

Agar Slant:

Nutrient Broth:

Gelatin Stab:

Oxygen Requirements:

Optimum Temp.:

PHYSIOLOGICAL CHARACTERISTICS

	TESTS	RESULTS
Fermentation	Glucose	
	Lactose	
	Sucrose	
	Mannitol	
Hydrolysis	Gelatin Liquefaction	
	Starch	
	Casein	
	Fat	
IMViC	Indole	
	Methyl Red	
	V-P (acetylmethylcarbinol)	
	Citrate Utilization	
	Nitrate Reduction	
	H_2S Production	
	Urease	
	Catalase	
	Oxidase	
	DNase	
	Phenylalanase	

	REACTION	TIME
Litmus Milk	Acid	_____
	Alkaline	_____
	Coagulation	_____
	Reduction	_____
	Peptonization	_____
	No Change	_____

Descriptive Chart

STUDENT: _____

LAB SECTION: _____

Habitat: _____ Culture No.: _____

Source: _____

Organism: _____

MORPHOLOGICAL CHARACTERISTICS

Cell Shape:

Arrangement:

Size:

Spores:

Gram's Stain:

Motility:

Capsules:

Special Stains:

CULTURAL CHARACTERISTICS

Colonies:

   Nutrient Agar:

   Blood Agar:

Agar Slant:

Nutrient Broth:

Gelatin Stab:

Oxygen Requirements:

Optimum Temp.:

PHYSIOLOGICAL CHARACTERISTICS

	TESTS	RESULTS
Fermentation	Glucose	
	Lactose	
	Sucrose	
	Mannitol	
Hydrolysis	Gelatin Liquefaction	
	Starch	
	Casein	
	Fat	
IMViC	Indole	
	Methyl Red	
	V-P (acetylmethylcarbinol)	
	Citrate Utilization	
	Nitrate Reduction	
	H$_2$S Production	
	Urease	
	Catalase	
	Oxidase	
	DNase	
	Phenylalanase	

	REACTION	TIME
Litmus Milk	Acid	_____
	Alkaline	_____
	Coagulation	_____
	Reduction	_____
	Peptonization	_____
	No Change	_____

Descriptive Chart

STUDENT: _____

LAB SECTION: _____

Habitat: _____ Culture No.: _____

Source: _____

Organism: _____

MORPHOLOGICAL CHARACTERISTICS

Cell Shape:

Arrangement:

Size:

Spores:

Gram's Stain:

Motility:

Capsules:

Special Stains:

CULTURAL CHARACTERISTICS

Colonies:

 Nutrient Agar:

 Blood Agar:

Agar Slant:

Nutrient Broth:

Gelatin Stab:

Oxygen Requirements:

Optimum Temp.:

PHYSIOLOGICAL CHARACTERISTICS

	TESTS	RESULTS
Fermentation	Glucose	
	Lactose	
	Sucrose	
	Mannitol	
Hydrolysis	Gelatin Liquefaction	
	Starch	
	Casein	
	Fat	
IMViC	Indole	
	Methyl Red	
	V-P (acetylmethylcarbinol)	
	Citrate Utilization	
	Nitrate Reduction	
	H_2S Production	
	Urease	
	Catalase	
	Oxidase	
	DNase	
	Phenylalanase	

	REACTION	TIME
Litmus Milk	Acid	_____
	Alkaline	_____
	Coagulation	_____
	Reduction	_____
	Peptonization	_____
	No Change	_____

Tables

TABLE I International Atomic Weights

Element	Symbol	Atomic Number	Atomic Weight
Aluminum	Al	13	26.97
Antimony	Sb	51	121.76
Arsenic	As	33	74.91
Barium	Ba	56	137.36
Beryllium	Be	4	9.013
Bismuth	Bi	83	209.00
Boron	B	5	10.82
Bromine	Br	35	79.916
Cadmium	Cd	48	112.41
Calcium	Ca	20	40.08
Carbon	C	6	12.010
Chlorine	Cl	17	35.457
Chromium	Cr	24	52.01
Cobalt	Co	27	58.94
Copper	Cu	29	63.54
Fluorine	F	9	19.00
Gold	Au	79	197.2
Hydrogen	H	1	1.0080
Iodine	I	53	126.92
Iron	Fe	26	55.85
Lead	Pb	82	207.21
Magnesium	Mg	12	24.32
Manganese	Mn	25	54.93
Mercury	Hg	80	200.61
Nickel	Ni	28	58.69
Nitrogen	N	7	14.008
Oxygen	O	8	16.0000
Palladium	Pd	46	106.7
Phosphorus	P	15	30.98
Platinum	Pt	78	195.23
Potassium	K	19	39.096
Radium	Ra	88	226.05
Selenium	Se	34	78.96
Silicon	Si	14	28.06
Silver	Ag	47	107.880
Sodium	Na	11	22.997
Strontium	Sr	38	87.63
Sulfur	S	16	32.066
Tin	Sn	50	118.70
Titanium	Ti	22	47.90
Tungsten	W	74	183.92
Uranium	U	92	238.07
Vanadium	V	23	50.95
Zinc	Zn	30	65.38
Zirconium	Zr	40	91.22

TABLE II Four-Place Logarithms

N	0	1	2	3	4	5	6	7	8	9
10	0000	0043	0086	0128	0170	0212	0253	0294	0334	0374
11	0414	0453	0492	0531	0569	0607	0645	0682	0719	0755
12	0792	0828	0864	0899	0934	0969	1004	1038	1072	1106
13	1139	1173	1206	1239	1271	1303	1335	1367	1399	1430
14	1461	1492	1523	1553	1584	1614	1644	1673	1703	1732
15	1761	1790	1818	1847	1875	1903	1931	1959	1987	2014
16	2041	2068	2095	2122	2148	2175	2201	2227	2253	2279
17	2304	2330	2355	2380	2405	2430	2455	2480	2504	2529
18	2553	2577	2601	2625	2648	2672	2695	2718	2742	2765
19	2788	2810	2833	2856	2878	2900	2923	2945	2967	2989
20	3010	3032	3054	3075	3096	3118	3139	3160	3181	3201
21	3222	3243	3263	3284	3304	3324	3345	3365	3385	3404
22	3424	3444	3464	3483	3502	3522	3541	3560	3579	3598
23	3617	3636	3655	3674	3692	3711	3729	3747	3766	3784
24	3802	3820	3838	3856	3874	3892	3909	3927	3945	3962
25	3979	3997	4014	4031	4048	4065	4082	4099	4116	4133
26	4150	4166	4183	4200	4216	4232	4249	4265	4281	4298
27	4314	4330	4346	4362	4378	4393	4409	4425	4440	4456
28	4472	4487	4502	4518	4533	4548	4564	4579	4594	4609
29	4624	4639	4654	4669	4683	4698	4713	4728	4742	4757
30	4771	4786	4800	4814	4829	4843	4857	4871	4886	4900
31	4914	4928	4942	4955	4969	4983	4997	5011	5024	5038
32	5051	5065	5079	5092	5105	5119	5132	5145	5159	5172
33	5185	5198	5211	5224	5237	5250	5263	5276	5289	5302
34	5315	5328	5340	5353	5366	5378	5391	5403	5416	5428
35	5441	5453	5465	5478	5490	5502	5514	5527	5539	5551
36	5563	5575	5587	5599	5611	5623	5635	5647	5658	5670
37	5682	5694	5705	5717	5729	5740	5752	5763	5775	5786
38	5798	5809	5821	5832	5843	5855	5866	5877	5888	5899
39	5911	5922	5933	5944	5955	5966	5977	5988	5999	6010
40	6021	6031	6042	6053	6064	6075	6085	6096	6107	6117
41	6128	6138	6149	6160	6170	6180	6191	6201	6212	6222
42	6232	6243	6253	6263	6274	6284	6294	6304	6314	6325
43	6335	6345	6355	6365	6375	6385	6395	6405	6415	6425
44	6435	6444	6454	6464	6474	6484	6493	6503	6513	6522
45	6532	6542	6551	6561	6571	6580	6590	6599	6609	6618
46	6628	6637	6646	6656	6665	6675	6684	6693	6702	6712
47	6721	6730	6739	6749	6758	6767	6776	6785	6794	6803
48	6812	6821	6830	6839	6848	6857	6866	6875	6884	6893
49	6902	6911	6920	6928	6937	6946	6955	6964	6972	6981
50	6990	6998	7007	7016	7024	7033	7042	7050	7059	7067
51	7076	7084	7093	7101	7110	7118	7126	7135	7143	7152
52	7160	7168	7177	7185	7193	7202	7210	7218	7226	7235
53	7243	7251	7259	7267	7275	7284	7292	7300	7308	7316
54	7324	7332	7340	7348	7356	7364	7372	7380	7388	7396
N	0	1	2	3	4	5	6	7	8	9

TABLE II   Four-Place Logarithms cont.

N	0	1	2	3	4	5	6	7	8	9
55	7404	7412	7419	7427	7435	7443	7451	7459	7466	7474
56	7482	7490	7497	7505	7513	7520	7528	7536	7543	7551
57	7559	7566	7574	7582	7589	7597	7604	7612	7619	7627
58	7634	7642	7649	7657	7664	7672	7679	7686	7694	7701
59	7709	7716	7723	7731	7738	7745	7752	7760	7767	7774
60	7782	7789	7796	7803	7810	7818	7825	7832	7839	7846
61	7853	7860	7868	7875	7882	7889	7896	7903	7910	7917
62	7924	7931	7938	7945	7952	7959	7966	7973	7980	7987
63	7993	8000	8007	8014	8021	8028	8035	8041	8048	8055
64	8062	8069	8075	8082	8089	8096	8102	8109	8116	8122
65	8129	8136	8142	8149	8156	8162	8169	8176	8182	8189
66	8195	8202	8209	8215	8222	8228	8235	8241	8248	8254
67	8261	8267	8274	8280	8287	8293	8299	8306	8312	8319
68	8325	8331	8338	8344	8351	8357	8363	8370	8376	8382
69	8388	8395	8401	8407	8414	8420	8426	8432	8439	8445
70	8451	8457	8463	8470	8476	8482	8488	8494	8500	8506
71	8513	8519	8525	8531	8537	8543	8549	8555	8561	8567
72	8573	8579	8585	8591	8597	8603	8609	8615	8621	8627
73	8633	8639	8645	8651	8657	8663	8669	8675	8681	8686
74	8692	8698	8704	8710	8716	8722	8727	8733	8739	8745
75	8751	8756	8762	8768	8774	8779	8785	8791	8797	8802
76	8808	8814	8820	8825	8831	8837	8842	8848	8854	8859
77	8865	8871	8876	8882	8887	8893	8899	8904	8910	8915
78	8921	8927	8932	8938	8943	8949	8954	8960	8965	8971
79	8976	8982	8987	8993	8998	9004	9009	9015	9020	9025
80	9031	9036	9042	9047	9053	9058	9063	9069	9074	9079
81	9085	9090	9096	9101	9106	9112	9117	9122	9128	9133
82	9138	9143	9149	9154	9159	9165	9170	9175	9180	9186
83	9191	9196	9201	9206	9212	9217	9222	9227	9232	9238
84	9243	9248	9253	9258	9263	9269	9274	9279	9284	9289
85	9294	9299	9304	9309	9315	9320	9325	9330	9335	9340
86	9345	9350	9355	9360	9365	9370	9375	9380	9385	9390
87	9395	9400	9405	9410	9415	9420	9425	9430	9435	9440
88	9445	9450	9455	9460	9465	9469	9474	9479	9484	9489
89	9494	9499	9504	9509	9513	9518	9523	9528	9533	9538
90	9542	9547	9552	9557	9562	9566	9571	9576	9581	9586
91	9590	9595	9600	9605	9609	9614	9619	9624	9628	9633
92	9638	9643	9647	9652	9657	9661	9666	9671	9675	9680
93	9685	9689	9694	9699	9703	9708	9713	9717	9722	9727
94	9731	9736	9741	9745	9750	9754	9759	9763	9768	9773
95	9777	9782	9786	9791	9795	9800	9805	9809	9814	9818
96	9823	9827	9832	9836	9841	9845	9850	9854	9859	9863
97	9868	9872	9877	9881	9886	9890	9894	9899	9903	9908
98	9912	9917	9921	9926	9930	9934	9939	9943	9948	9952
99	9956	9961	9965	9969	9974	9978	9983	9987	9991	9996
100	0000	0004	0009	0013	0017	0022	0026	0030	0035	0039
N	0	1	2	3	4	5	6	7	8	9

TABLE III Temperature Conversion Table Centigrade to Fahrenheit

°C	0	1	2	3	4	5	6	7	8	9
−50	**−58.0**	**−59.8**	**−61.6**	**−63.4**	**−65.2**	**−67.0**	**−68.8**	**−70.6**	**−72.4**	**−74.2**
−40	−40.0	−41.8	−43.6	−45.4	−47.2	−49.0	−50.8	−52.6	−54.4	−56.2
−30	−22.0	−23.8	−25.6	−27.4	−29.2	−31.0	−32.8	−34.6	−36.4	−38.2
−20	− 4.0	− 5.8	− 7.6	− 9.4	−11.2.	−13.0	−14.8	−16.6	−18.4	−20.2
−10	+14.0	+12.2	+10.4	+ 8.6	+ 6.8	+ 5.0	+ 3.2	+ 1.4	− 0.4	− 2.2
− 0	+32.0	+30.2	+28.4	+26.6	+24.8	+23.0	+21.2	+19.4	+17.6	+15.8
0	**32.0**	**33.8**	**35.6**	**37.4**	**39.2**	**41.0**	**42.8**	**44.6**	**46.4**	**48.2**
10	50.0	51.8	53.6	55.4	57.2	59.0	60.8	62.6	64.4	66.2
20	68.0	69.8	71.6	73.4	75.2	77.0	78.8	80.6	82.4	84.2
30	86.0	87.8	89.6	91.4	93.2	95.0	96.8	98.6	100.4	102.2
40	104.0	105.8	107.6	109.4	111.2	113.0	114.8	116.6	118.4	120.2
50	122.0	123.8	125.6	127.4	129.2	131.0	132.8	134.6	136.4	138.2
60	**140.0**	**141.8**	**143.6**	**145.4**	**147.2**	**149.0**	**150.8**	**152.6**	**154.4**	**156.2**
70	158.0	159.8	161.6	163.4	165.2	167.0	168.8	170.6	172.4	174.2
80	176.0	177.8	179.6	181.4	183.2	185.0	186.8	188.6	190.4	192.2
90	194.0	195.8	197.6	199.4	201.2	203.0	204.8	206.6	208.4	210.2
100	212.0	213.8	215.6	217.4	219.2	221.0	222.8	224.6	226.4	228.2
110	**230.0**	**231.8**	**233.6**	**235.4**	**237.2**	**239.0**	**240.8**	**242.6**	**244.4**	**246.2**
120	248.0	249.8	251.6	253.4	255.2	257.0	258.8	260.6	262.4	264.2
130	266.0	267.8	269.6	271.4	273.2	275.0	276.8	278.6	280.4	282.2
140	284.0	285.8	287.6	289.4	291.2	293.0	294.8	296.6	298.4	300.2
150	302.0	303.8	305.6	307.4	309.2	311.0	312.8	314.6	316.4	318.2
160	**320.0**	**321.8**	**323.6**	**325.4**	**327.2**	**329.0**	**330.8**	**332.6**	**334.4**	**336.2**
170	338.0	339.8	341.6	343.4	345.2	347.0	348.8	350.6	352.4	354.2
180	356.0	357.8	359.6	361.4	363.2	365.0	366.8	368.6	370.4	372.2
190	374.0	375.8	377.6	379.4	381.2	383.0	384.8	386.6	388.4	390.2
200	392.0	393.8	395.6	397.4	399.2	401.0	402.8	404.6	406.4	408.2
210	**410.0**	**411.8**	**413.6**	**415.4**	**417.2**	**419.0**	**420.8**	**422.6**	**424.4**	**426.2**
220	428.0	429.8	431.6	433.4	435.2	437.0	438.8	440.6	442.4	444.2
230	446.0	447.8	449.6	451.4	453.2	455.0	456.8	458.6	460.4	462.2
240	464.0	465.8	467.6	469.4	471.2	473.0	474.8	476.6	478.4	480.2
250	482.0	483.8	485.6	487.4	489.2	491.0	492.8	494.6	496.4	498.2

$$°F \ = \ °C \ \times \ 9/5 \ + \ 32 \qquad\qquad °C \ = \ °F \ - \ 32 \ \times \ 5/9$$

TABLE IV Autoclave Steam Pressures and Corresponding Temperatures

Steam Pressure lb/sq in	Temperature °C	°F	Steam Pressure lb/sq in	Temperature °C	°F	Steam Pressure lb/sq in	Temperature °C	°F
0	100.0	212.0						
1	101.9	215.4	11	116.4	241.5	21	126.9	260.4
2	103.6	218.5	12	117.6	243.7	22	127.8	262.0
3	105.3	221.5	13	118.8	245.8	23	128.7	263.7
4	106.9	224.4	14	119.9	247.8	24	129.6	265.3
5	108.4	227.1	15	121.0	249.8	25	130.4	266.7
6	109.8	229.6	16	122.0	251.6	26	131.3	268.3
7	111.3	232.3	17	123.0	253.4	27	132.1	269.8
8	112.6	234.7	18	124.1	255.4	28	132.9	271.2
9	113.9	237.0	19	125.0	257.0	29	133.7	272.7
10	115.2	239.4	20	126.0	258.8	30	134.5	274.1

Figures are for steam pressure only and the presence of any air in the autoclave invalidates temperature readings from the above table.

TABLE V Autoclave Temperatures as Related to the Presence of Air

Gauge Pressure, lb	Pure steam, complete air discharge °C	°F	Two-thirds air discharge, 20-in. vacuum °C	°F	One-half air discharge, 15-in. vacuum °C	°F	One-third air discharge, 10-in. vacuum °C	°F	No air discharge °C	°F
5	109	228	100	212	94	202	90	193	72	162
10	115	240	109	228	105	220	100	212	90	193
15	121	250	115	240	112	234	109	228	100	212
20	126	259	121	250	118	245	115	240	109	228
25	130	267	126	259	124	254	121	250	115	240
30	135	275	130	267	128	263	126	259	121	250

TABLE VI   MPN Determination from Multiple Tube Test

NUMBER OF TUBES GIVING POSITIVE REACTION OUT OF			MPN Index per 100 ml	95 PERCENT CONFIDENCE LIMITS	
3 of 10 ml each	3 of 1 ml each	3 of 0.1 ml each		Lower	Upper
0	0	1	3	<0.5	9
0	1	0	3	<0.5	13
1	0	0	4	<0.5	20
1	0	1	7	1	21
1	1	0	7	1	23
1	1	1	11	3	36
1	2	0	11	3	36
2	0	0	9	1	36
2	0	1	14	3	37
2	1	0	15	3	44
2	1	1	20	7	89
2	2	0	21	4	47
2	2	1	28	10	150
3	0	0	23	4	120
3	0	1	39	7	130
3	0	2	64	15	380
3	1	0	43	7	210
3	1	1	75	14	230
3	1	2	120	30	380
3	2	0	93	15	380
3	2	1	150	30	440
3	2	2	210	35	470
3	3	0	240	36	1,300
3	3	1	460	71	2,400
3	3	2	1,100	150	4,800

From *Standard Methods for the Examination of Water and Wastewater,* Twelfth edition (New York: The American Public Health Association, Inc.), p. 608.

TABLE VII Significance of Zones of Inhibition in Kirby-Bauer Method of Antimicrobic Sensitivity Testing (1995)

Antimicrobial Agent	Disk Potency	R Resistant mm	I Intermediate mm	S Sensitive mm
Amikacin	30 mcg	<14	15–16	>17
Amoxicillin/Clavulinic Acid	30 mcg			
Staphylococci		<19	14–17	>20
Other gram-positive organisms		<13	14–17	>18
Ampicillin	75 mcg			
Gram-negative enterics		<13	14–16	>17
Staphylococci		<28		>29
Enterococci		<16		>17
Streptococci (not *S. pneumoniae*)		<21	22–29	>30
Haemophilus spp.		<18	19–21	>22
Listeria monocytogenes		<19		>20
Azlocillin (*Pseudomonas aeruginosa*)	75 mcg	<17		>18
Carbenicillin (*P. aeruginosa*)	100 mcg	<13	14–16	>17
Other gram-negative organisms		<19	20–22	>23
Cefactor	30 mcg	<14	15–17	>18
Cephalothin	30 mcg	<14	15–17	>18
Chloramphenicol	30 mcg	<12	13–17	>18
S. pneumoniae		<20		>21
Clarithromycin	15 mcg	<13	14–17	>18
S. pneumoniae		<16	17–20	>21
Clindamycin	2 mcg	<14	15–20	>21
S. pneumoniae		<20		>21
Erythromycin	15 mcg	<13	14–22	>23
S. pneumoniae		<15	16–20	>21
Gentamicin	10 mcg	<12	13–14	>15
Impenem	10 mcg	<13	14–15	>16
Haemophilus spp.				>16
Kanamycin	30 mcg	<13	14–17	>18
Lomefloxacin	10 mcg	<18	19–21	>22
Loracarbef	30 mcg	<14	15–17	>18
Mezlocillin (*P. aeruginosa*)	75 mcg	<15		>16
Other gram-negative organisms		<17	18–20	>19
Minocycline	30 mcg	<14	15–18	>19
Moxalactam	30 mcg	<14	15–22	>23
Nafcillin	1 mcg	<10	11–12	>13
Nalidixic Acid	30 mcg	<13	14–18	>19
Netilmicin	30 mcg	<12	13–14	>15
Norfloxacin	10 mcg	<12	13–16	>17
Ofloxacin	5 mcg	<12	13–15	>16
Penicillin G (Staphylococci)	10 units	<28		>29
Enterococci		<14		>15
Streptococci (not *S. pneumoniae*)		<19	20–27	>28
Neisseria gonorrhoeae		<26	27–46	>47
L. monocytogenes		<19		>20
Piperacillin/Tazobactam	100/10 mcg			
Staphylococci		<17		>18
P. aeruginosa		<17		>18
Other gram-negative organisms		<14	15–19	>20
Rifampin	5 mcg	<16	17–19	>20
Haemophilus spp.		<16	17–19	>20
S. pneumoniae		<16	17–18	>19
Streptomycin	10 mcg	<11	12–14	>15
Sulfisoxazole	300 mcg	<12	13–16	>17
Tetracycline	30 mcg	<14	15–18	>19
S. pneumoniae		<17	18–21	>22
Tobramycin	10 mcg	<12	13–14	>15
Trimethoprim/Sulfamethoxazole	1.25/23.75	<10	11–15	>16
Vancomycin	30 mcg	<14	15–16	>17

TABLE VIII Indicators of Hydrogen Ion Concentration

Many of the following indicators are used in the media of certain exercises in this manual. This table indicates the pH range of each indicator and the color changes that occur. To determine the exact pH within a particular range one should use a set of standard colorimetric tubes that are available from the prep room. Consult your lab instructor.

Indicator	Full Acid Color	Full Alkaline Color	pH Range
Cresol Red	red	yellow	0.2 – 1.8
Metacresol Purple (acid range)	red	yellow	1.2 – 2.8
Thymol Blue	red	yellow	1.2 – 2.8
Bromphenol Blue	yellow	blue	3.0 – 4.6
Bromcresol Green	yellow	blue	3.8 – 5.4
Chlorcresol Green	yellow	blue	4.0 – 5.6
Methyl Red	red	yellow	4.4 – 6.4
Chlorphenol Red	yellow	red	4.8 – 6.4
Bromcresol Purple	yellow	purple	5.2 – 6.8
Bromthymol Blue	yellow	blue	6.0 – 7.6
Neutral Red	red	amber	6.8 – 8.0
Phenol Red	yellow	red	6.8 – 8.4
Cresol Red	yellow	red	7.2 – 8.8
Metacresol Purple (alkaline range)	yellow	purple	7.4 – 9.0
Thymol Blue (alkaline range)	yellow	blue	8.0 – 9.6
Cresolphthalein	colorless	red	8.2 – 9.8
Phenolphthalein	colorless	red	8.3 – 10.0

Indicators, Stains, Reagents

INDICATORS

All the indicators used in this manual can be made by (1) dissolving a measured amount of the indicator in 95% ethanol, (2) adding a measured amount of water, and (3) filtering with filter paper. The following chart provides the correct amounts of indicator, alcohol, and water for various indicator solutions.

Indicator Solution	Indicator (gm)	95% Ethanol (ml)	Distilled H_2O (ml)
Bromcresol green	0.4	500	500
Bromcresol purple	0.4	500	500
Bromthymol blue	0.4	500	500
Cresol red	0.2	500	500
Methyl red	0.2	500	500
Phenolphthalein	1.0	50	50
Phenol red	0.2	500	500
Thymol blue	0.4	500	500

STAINS AND REAGENTS

Acid-Alcohol (for Ziehl-Neelsen stain)

3 ml concentrated hydrochloric acid in 100 ml of 95% ethyl alcohol.

Alcohol, 70% (from 95%)

Alcohol, 95% . . . . .368.0 ml
Distilled water . . . . .132.0 ml

Barritt's Reagent (Voges-Proskauer test)

Solution A: 6 g alpha-naphthol in 100 ml 95% ethyl alcohol.

Solution B: 16 g potassium hydroxide in 100 ml water.

Note that no creatine is used in these reagents as is used in O'Meara's reagent for the V-P test.

Carbolfuchsin Stain (Ziehl's)

Solution A: Dissolve 0.3 g of basic fuchsin (90% dye content) in 10 ml 95% ethyl alcohol.

Solution B: Dissolve 5 g of phenol in 95 ml of water.

Mix solutions A and B.

Crystal Violet Stain (Hucker modification)

Solution A: Dissolve 2.0 g of crystal violet (85% dye content) in 20 ml of 95% ethyl alcohol.

Solution B: Dissolve 0.8 g ammonium oxalate in 80.0 ml distilled water.

Mix solutions A and B.

Diphenylamine Reagent (nitrate test)

Dissolve 0.7 g diphenylamine in a mixture of 60 ml of concentrated sulfuric acid and 28.8 ml of distilled water.

Cool and add slowly 11.3 ml of concentrated hydrochloric acid. After the solution has stood for 12 hours, some of the base separates, showing that the reagent is saturated.

Ferric Chloride Reagent (Ex. 53)

$FeCl_3 \cdot 6H_2O$ . . . . .12 gm
2% Aqueous HCl . .100 ml

Make up the 2% aq. HCl by adding 5.4 ml of concentrated HCl (37%) to 94.6 ml H_2O. Inoculate with two or three colonies of beta hemolytic streptococci, incubate at 35°C for 20 or more hours. Centrifuge the medium to pack the cells, and pipette 0.8 ml of the clear supernate into a Kahn tube. Add 0.2 ml of the ferric chloride reagent to the Kahn tube and mix well. If a heavy precipitate remains longer than 10 minutes, the test is positive.

Gram's Iodine (Lugol's)

Dissolve 2.0 g of potassium iodide in 300 ml of distilled water and then add 1.0 g iodine crystals.

Iodine, 5% Aqueous Solution (Ex. 34)

Dissolve 4 g of potassium iodide in 300 ml of distilled water and then add 2.0 g iodine crystals.

Kovacs' Reagent (indole test)

n-amyl alcohol .75.0 ml
Hydrochloric acid (conc.) . . . . .25.0 ml
ρ-dimethylamine-benzaldehyde . . . . .5.0 g

Lactophenol Cotton Blue Stain

Phenol crystals .20 g
Lactic acid . . . . .20 ml
Glycerol . . . . .40 ml
Cotton blue . . . . .0.05 g

Dissolve the phenol crystals in the other ingredients by heating the mixture gently under a hot water tap.

Malachite Green Solution (spore stain)

Dissolve 5.0 g malachite green oxalate in 100 ml distilled water.

McFarland Nephelometer Barium Sulfate Standards (Ex. 45)

Prepare 1% aqueous barium chloride and 1% aqueous sulfuric acid solutions.

Add the amounts indicated in table 1 to clean, dry ampoules. Ampoules should have the same diameter as the test tube to be used in subsequent density determinations.

Seal the ampoules and label them.

Methylene Blue (Loeffler's)

Solution A: Dissolve 0.3 g of methylene blue (90% dye content) in 30.0 ml ethyl alcohol (95%).

Solution B: Dissolve 0.01 g potassium hydroxide in 100.0 ml distilled water. Mix solutions A and B.

Naphthol, alpha

5% alpha-naphthol in 95% ethyl alcohol

Caution: Avoid all contact with human tissues. Alpha-naphthol is considered to be carcinogenic.

TABLE 1 Amounts for Standards

Tube	Barium Chloride 1% (ml)	Sulfuric Acid 1% (ml)	Corresponding Approx. Density of Bacteria (million/ml)
1	0.1	9.9	300
2	0.2	9.8	600
3	0.3	9.7	900
4	0.4	9.6	1200
5	0.5	9.5	1500
6	0.6	9.4	1800
7	0.7	9.3	2100
8	0.8	9.2	2400
9	0.9	9.1	2700
10	1.0	9.0	3000

Nigrosine Solution (Dorner's)

Nigrosine, water soluble . . . . .10 g
Distilled water .100 ml

Boil for 30 minutes. Add as a preservative 0.5 ml formaldehyde (40%). Filter twice through double filter paper and store under aseptic conditions.

Nitrate Test Reagent
(see Diphenylamine)

Nitrite Test Reagents

Solution A: Dissolve 8 g sulfanilic acid in 1000 ml 5N acetic acid (1 part glacial acetic acid to 2.5 parts water).

Solution B: Dissolve 5 g dimethyl-alpha-naphthylamine in 1000 ml 5N acetic acid. Do not mix solutions.

Caution: Although at this time it is not known for sure, there is a possibility that dimethyl-α-naphthylamine in solution B may be carcinogenic. For reasons of safety, avoid all contact with tissues.

Oxidase Test Reagent

Mix 1.0 g of dimethyl-ρ-phenylenediamine hydrochloride in 100 ml of distilled water.

Preferably, the reagent should be made up fresh, daily. It should not be stored longer than one week in the refrigerator. Tetramethyl-ρ-phenylenediamine dihydrochloride (1%) is even more sensitive, but is considerably more expensive and more difficult to obtain.

Phenolized Saline

Dissolve 8.5 g sodium chloride and 5.0 g phenol in 1 liter distilled water.

Physiological Saline

Dissolve 8.5 g sodium chloride in 1 liter distilled water.

Potassium permanganate
(for fluorochrome staining)

KMnO$_4$ . . . . .2.5 g

Distilled water . . . . .500.0 ml

Safranin (for gram staining)

Safranin O (2.5% sol'n in 95%
        ethyl alcohol) . . . . .10.0 ml

Distilled water . . . . .100.0 ml

Trommsdorf's Reagent (nitrite test)

Add slowly, with constant stirring, 100 ml of a 20% aqueous zinc chloride solution to a mixture of 4.0 g of starch in water. Continue heating until the starch is dissolved as much as possible, and the solution is nearly clear. Dilute with water and add 2 g of potassium iodide. Dilute to 1 liter, filter, and store in amber bottle.

Vaspar

Melt together 1 pound of Vaseline and 1 pound of paraffin. Store in small bottles for student use.

Voges-Proskauer Test Reagent
(see Barritt's)

White Blood Cell (WBC) Diluting Fluid

Hydrochloric acid   5 ml

Distilled water   495 ml

Add 2 small crystals of thymol as a preservative.

Conventional Media   The following media are used in the experiments of this manual. All of these media are available in dehydrated form from either Difco Laboratories, Detroit, Michigan, or Baltimore Biological Laboratory (BBL), a division of Becton, Dickinson & Co., Cockeysville, Maryland. Compositions, methods of preparation, and usage will be found in their manuals, which are supplied upon request at no cost. The source of each medium is designated as (B) for BBL and (D) for Difco.

Bile esculin (D)
Brewer's anaerobic agar (D)
Desoxycholate citrate agar (B,D)
Desoxycholate lactose agar (B,D)
DNase test agar (B,D)
Endo agar (B,D)
Eugonagar (B,D)
Fluid thioglycollate medium (B,D)
Heart infusion agar (D)
Hektoen Enteric Agar (B,D)
Kligler iron agar (B,D)
Lead acetate agar (D)
Levine EMB agar (B,D)
Lipase reagent (D)
Litmus milk (B,D)
Lowenstein-Jensen medium (B,D)
MacConkey Agar (B,D)
Mannitol salt agar (B,D)
MR-VP medium (D)
Mueller-Hinton medium (B,D)
Nitrate broth (D)
Nutrient agar (B,D)
Nutrient broth (B,D)

Nutrient gelatin (B,D)
Phenol red sucrose broth (B,D)
Phenylalanine agar (D)
Phenylethyl alcohol medium (B)
Russell double sugar agar (B,D)
Sabouraud's glucose (dextrose)
    agar (D)
Semisolid medium (B)
Simmons citrate agar (B,D)
Snyder test agar (D)
Sodium hippurate (D)
Spirit blue agar (D)
SS agar (B,D)
m-Staphylococcus broth (D)
Staphylococcus medium 110 (D)
Starch agar (D)
Trypticase soy agar (B)
Trypticase soy broth (B)
Tryptone glucose extract agar (B,D)
Urea (urease test) broth (B,D)
Veal infusion agar (B,D)
Xylose Lysine Desoxycholate Agar
    (B,D)

Special Media   The following media are not included in the manuals that are supplied by Difco and BBL; therefore, methods of preparation are presented here.

Blood Agar

Trypticase soy agar power . . . . .40 g
Distilled water .1000 ml
    Final pH of 7.3
Defibrinated sheep or rabbit blood . . . . . .50 ml

Liquefy and sterilize 1000 ml of trypticase soy agar in a large Erlenmeyer flask. While the TSA is being sterilized, warm up 50 ml of defibrinated blood to 50°C. After cooling the TSA to 50°C, aseptically transfer the blood to the flask and mix by gently rotating the flask (cold blood may cause lumpiness).

Pour 10–12 ml of the mixture into sterile petri plates. If bubbles form on the surface of the medium, flame the surface gently with a Bunsen burner before the medium solidifies. It is best to have an assistant to lift off the petri plate lids while pouring the medium into the plates. A full flask of blood agar is somewhat cumbersome to handle with one hand.

Bromthymol Blue Carbohydrate Broths

Make up stock indicator solution:

Bromthymol blue . . . . .8 g

95% ethyl alcohol . . . . .250 ml

Distilled water . . . . .250 ml

Indicator is dissolved first in alcohol and then water is added.

Make up broth:

Sugar base (lactose, sucrose, glucose, etc.) . .5 g

Tryptone . . . . .10 g

Yeast extract . . . . .5 g

Indicator solution . . . . .2 ml

Distilled water .1000 ml
  Final pH 7.0

Emmons' Culture Medium for Fungi

C. W. Emmons developed the following recipe as an improvement over Sabouraud's glucose agar for the cultivation of fungi. Its principal advantage is that a neutral pH does not inhibit certain molds that have difficulty growing on Sabouraud's agar (pH 5.6). Instead of relying on a low pH to inhibit bacteria, it contains chloramphenicol, which does not adversely affect the fungi.

Glucose . . . . .20 g

Neopeptone . . . . .10 g

Agar . . . . .20 g

Chloramphenicol . . . . .40 mg

Distilled water .1000 ml

After the glucose, peptone, and agar are dissolved, heat to boiling, add the chloramphenicol which has been suspended in 10 ml of 95% alcohol and remove quickly from the heat. Autoclave for only 10 minutes

Glucose Peptone Acid Agar

Glucose . . . . .10 g

Peptone . . . . .5 g

Monopotassium phosphate . . . . .1 g

Magnesium sulfate ($MgSO_4 \cdot 7H_2O$) . . . . .0.5 g

Agar . . . . .15 g

Water . . . . .1000 ml

While still liquid after sterilization, add sufficient sulfuric acid to bring the pH down to 4.0.

Glycerol Yeast Extract Agar

Glycerol . . . . .5 ml

Yeast extract . . . . .2 g

Dipotassium phosphate . . . . .1 g

Agar . . . . .15 g

Water . . . . .1000 ml

m Endo MF Broth (Ex. 45)

This medium is extremely hygroscopic in the dehydrated form and oxidizes quickly to cause deterioration of the medium after the bottle has been opened. Once a bottle has been opened it should be dated and discarded after one year. If the medium becomes hardened within that time it should be discarded. Storage of the bottle inside a larger bottle that contains silica gel will extend shelf life.

Failure of Exercise 45 can often be attributed to faulty preparation of the medium. It is best to make up the medium the day it is to be used. It should not be stored over 96 hours prior to use. The Millipore Corporation recommends the following method for preparing this medium. (These steps are not exactly as stated in the Millipore Application Manual AM302.)

1. Into a 250 ml screw-cap Erlenmeyer flask place the following:

   Distilled water . . . . .50 ml

   95% ethyl alcohol .2 ml

   Dehydrated medium (m Endo MF broth) . .4.8 g

   Shake the above mixture by swirling the flask until the medium is dissolved and then add another 50 ml of distilled water.
2. Cap the flask loosely and immerse it into a pan of boiling water. As soon as the medium begins to simmer, remove the flask from the water bath. Do not boil the medium any further.
3. Cool the medium to 45°C, and adjust the pH to between 7.1 and 7.3.
4. If the medium must be stored for a few days, place it in the refrigerator at 2°–10°C, with screw-cap tightened securely.

Milk Salt Agar (15% NaCl)

Prepare three separate beakers of the following ingredients:

1. Beaker containing 200 grams of sodium chloride.
2. Large beaker (2000 ml size) containing 50 grams of skim milk powder in 500 ml of distilled water.
3. Glycerol-peptone agar medium:

   $MgSO_4 \cdot 7H_2O$ .5.0 g

   $MgNO_3 \cdot 6H_2O$ .1.0 g

   $FeCl_3 \cdot 7H_2O$ .0.025 g

   Difco proteose-peptone #3 . . . . .5.0 g

   Glycerol . . . . .10.0 g

   Agar . . . . .30.0 g

   Distilled water .500.0 ml

   Sterilize the above three beakers separately. The milk solution should be sterilized at 113°–115° C (8 lb pressure) in autoclave for 20 minutes. The salt and glycerol-peptone agar can be sterilized at conventional pressure and temperature. After the milk solution has cooled to 55°C, add the sterile salt, which should also be cooled down to a moderate temperature. If the salt is too hot, coagulation may occur. Combine the milk-salt and glycerol-peptone agar solutions by gently swirling with a glass rod. Dispense aseptically into petri plates.

Nitrate Broth

Beef extract . . . . .3 g
Peptone . . . . .5 g
Potassium nitrate . . . . .1 g
Distilled water .1000 ml
     Final pH 7.0 at 25°C

Nitrate Agar

Beef extract . . . . .3 g
Peptone . . . . .5 g
Potassium nitrate . . . . .1 g
Agar . . . . .12 g
Distilled water .1000 ml
     Final pH 6.8 at 25°C

Phage Growth Medium (Ex. 22)

KH_2PO_4 . . . . .1.5 g
Na_2HPO_4 . . . . .3.0 g
NH_4Cl . . . . .1.0 g
$MgSO_4 \cdot 7H_2O$ . . . . .0.2 g
Glycerol . . . . .10.0 g
Acid-hydrolyzed casein . . . . .5.0 g
dl-Tryptophan . . . . .0.01 g
Gelatin . . . . .0.02 g
Tween-80 . . . . .0.2 g
Distilled water . . . . .1000.0 ml
Sterilize in autoclave at 121° C for 20 minutes.

Phage Lysing Medium (Ex. 22)

Add sufficient sodium cyanide (NaCN) to the above growth medium to bring the concentration up to 0.02M. For 1 liter of lysing medium this will amount to about 1 gram (actually 0.98 g) of NaCN. When an equal amount of this lysing medium is added to the growth medium during the last 6 hours of incubation, the concentration of NaCN in the combined medium is 0.01M.

Russell Double Sugar Agar (Ex. 54)

Beef extract . . . . .1 g
Proteose Peptone No.3 (Difco) . . . . .12 g
Lactose . . . . .10 g
Dextrose . . . . .1 g
Sodium chloride . . . . .5 g
Agar . . . . .15 g

Phenol red (Difco) . . . . .0.025 g
Distilled water .1000 ml
     Final pH 7.5 at 25°C
Dissolve ingredients in water, and bring to boiling. Cool to 50°–60°C, and dispense about 8 ml per tube (16 mm dia tubes). Slant tubes to cool. Butt depth should be about $\frac{1}{2}$″.

Skim Milk Agar

Skim milk powder 100g
Agar 15g
Distilled water .1000 ml
Dissolve the 15 g of agar into 700 ml of distilled water by boiling. Pour into a large flask and sterilize at 121°C, 15 lb pressure.

In a separate container, dissolve the 100 g of skim milk powder into 300 ml of water heated to 50° C. Sterilize this milk solution at 113°–115° C (8 lb pressure) for 20 minutes.

After the two solutions have been sterilized, cool to 55° C and combine in one flask, swirling gently to avoid bubbles. Dispense into sterile petri plates.

Soft Nutrient Agar (for bacteriophage)

Dehydrated nutrient broth . . . . .8 g
Agar . . . . .7 g
Distilled water .1000 ml
Sterilize in autoclave at 121°C for 20 minutes.

Spirit Blue Agar (Ex. 36)

This medium is used to detect lipase production by bacteria. Lipolytic bacteria cause the medium to change from pale lavender to deep blue.
Spirit blue agar (Difco) . . . . .35 g
Lipase reagent (Difco) . . . . .35 ml
Distilled water .1000 ml
Dissolve the spirit blue agar in 1000 ml of water by boiling. Sterilize in autoclave for 15 minutes at 15 psi (121°C). Cool to 55°C and slowly add the 35 ml of lipase reagent, agitating to obtain even distribution. Dispense into sterile petri plates.

Tryptone Agar

Tryptone . . . . .10 g
Agar . . . . .15 g
Distilled water .1000 ml

Tryptone Broth

Tryptone . . . . .10 g
Distilled water .1000 ml

Tryptone Yeast Extract Agar

Tryptone . . . . .10 g

Yeast extract . . . . .5 g

Dipotassium phosphate . . . . .3 g

Sucrose . . . . .50 g

Agar . . . . .15 g

Water . . . . .1000 ml

pH 7.4

Identification Charts

CHART I   Interpretation of Test Results of API 20E System

Tube	Interpretation of Reactions			
	Positive		Negative	Comments
ONPG	Yellow		Colorless	(1) Any shade of yellow is a positive reaction. (2) VP tube, before the addition of reagents, can be used as a negative control.
ADH	Incubation 18–24 h	Red or Orange	Yellow	Orange reactions occurring at 36–48 hours should be interpreted as negative.
	36–48 h	Red	Yellow or Orange	
LDC	18–24 h	Red or Orange	Yellow	Any shade of orange within 18–24 hours is a positive reaction. At 36–48 hours, orange decarboxylase reactions should be interpreted as negative.
	36–48 h	Red	Yellow or Orange	
ODC	18–24 h	Red or Orange	Yellow	Orange reactions occurring at 36–48 hours should be interpreted as negative.
	36–48 h	Red	Yellow or Orange	
CIT		Turquoise or Dark Blue	Light Green or Yellow	(1) Both the tube and cupule should be filled. (2) Reaction is read in the aerobic (cupule) area.
H2S		Black Deposit	No Black Deposit	(1) H_2S production may range from a heavy black deposit to a very thin black line around the tube bottom. Carefully examine the bottom of the tube before considering the reaction negative. (2) A "browning" of the medium is a negative reaction unless a black deposit is present. "Browning" occurs with TDA-positive organisms.
URE	18–24 h	Red or Orange	Yellow	A method of lower sensitivity has been chosen. *Klebsiella, Proteus,* and *Yersinia* routinely give positive reactions.
	36–48 h	Red	Yellow or Orange	
TDA	Add 1 drop 10% ferric chloride			(1) Immediate reaction. (2) Indole-positive organisms may produce a golden orange color due to indole production. This is a negative reaction.
		Brown-Red	Yellow	
IND	Add 1 drop Kovacs' reagent			(1) The reaction should be read within 2 minutes after the addition of the Kovacs' reagent and the results recorded. (2) After several minutes, the HCl present in Kovacs' reagent may react with the plastic of the cupule resulting in a change from a negative (yellow) color to a brownish-red. This is a negative reaction.
		Red Ring	Yellow	
VP	Add 1 drop of 40% potassium hydroxide, then 1 drop of 6% alpha-naphthol.			(1) Wait 10 minutes before considering the reaction negative. (2) A pale pink color (after 10 min) should be interpreted as negative. A pale pink color appears immediately after the addition of reagents but turns dark pink or red after 10 min should be interpreted as positive.
		Red	Colorless	
				Motility may be observed by hanging drop or wet mount preparation.
GEL		Diffusion of the pigment	No diffusion	(1) The solid gelatin particles may spread throughout the tube after inoculation. Unless diffusion occurs, the reaction is negative. (2) Any degree of diffusion is a positive reaction.
GLU		Yellow or Gray	Blue or Blue-Green	**Fermentation** (Enterobacteriaceae, *Aeromonas, Vibrio*) (1) Fermentation of the carbohydrates begins in the most anaerobic portion (bottom) of the tube. Therefore, these reactions should be read from the bottom of the tube to the top. (2) A yellow color at the bottom of the tube only indicates a weak or delayed positive reaction.
MAN INO SOR RHA SAC MEL AMY ARA		Yellow	Blue or Blue-Green	**Oxidation** (Other Gram-negatives) (1) Oxidative utilization of the carbohydrates begins in the most aerobic portion (top) of the tube. Therefore, these reactions should be read from the top to the bottom of the tube. (2) A yellow color in the upper portion of the tube and a blue in the bottom of the tube indicates oxidative utilization of the sugar. This reaction should be considered positive **only** for non-Enterobacteriaceae gram-negative rods. This is a negative reaction for fermentative organisms such as Enterobacteriaceae.
GLU Nitrate Reduction	After reading GLU reaction, add 2 drops 0.8% sulfanilic acid and 2 drops 0.5% N, N-dimethylalpha-naphthylamine			(1) Before addition of reagents, observe GLU tube (positive or negative) for bubbles. Bubbles are indicative of reduction of nitrate to the nitrogenous (N_2) state. (2) A positive reaction may take 2–3 minutes for the red color to appear. (3) Confirm a negative test by adding zinc dust or 20-mesh granular zinc. A pink-orange color after 10 minutes confirms a negative reaction. A yellow color indicates reduction of nitrates to nitrogenous (N_2) state.
	NO_2	Red	Yellow	
	N_2 gas	Bubbles; Yellow after reagents and zinc	Orange after reagents and zinc	
MAN INO SOR Catalase	After reading carbohydrate reaction, add 1 drop 1.5% H_2O_2			(1) Bubbles may take 1–2 minutes to appear. (2) Best results will be obtained if the test is run in tubes which have no gas from fermentation.
		Bubbles	No bubbles	

(The text "Comments for all Carbohydrates" appears vertically spanning the GLU and MAN/INO/SOR/RHA/SAC/MEL/AMY/ARA carbohydrate rows.)

CHART II Symbol Interpretation of API 20E System

Tube	Chemical/Physical Principles	Components		Ref.
		Reactive Ingredients	**Quantity**	
ONPG	Hydrolysis of ONPG by beta-galactosidase releases yellow orthonitrophenol from the colorless ONPG; ITPG (isopropylthiogalactopyranoside) is used as inducer.	ONPG ITPG	0.2 mg 8.0 µg	12 13 14
ADH	Arginine dihydrolase transforms arginine into ornithine, ammonia, and carbon dioxide. This causes a pH rise in the acid-buffered system and a change in the indicator from yellow to red.	Arginine	2.0 mg	15
LDC	Lysine decarboxylase transforms lysine into a basic primary amine, cadaverine. This amine causes a pH rise in the acid–buffered system and a change in the indicator from yellow to red.	Lysine	2.0 mg	15
ODC	Ornithine decarboxylase transforms ornithine into a basic primary amine, putrescine. This amine causes a pH rise in the acid-buffered system and a change in the indicator from yellow to red.	Ornithine	2.0 mg	15
CIT	Citrate is the sole carbon source. Citrate utilization results in a pH rise and a change in the indicator from green to blue.	Sodium Citrate	0.8 mg	21
H₂S	Hydrogen sulfide is produced from thiosulfate. The hydrogen sulfide reacts with iron salts to produce a black precipitate.	Sodium Thiosulfate	80.0 µg	6
URE	Urease releases ammonia from urea; ammonia causes the pH to rise and changes the indicator from yellow to red.	Urea	0.8 mg	7
TDA	Tryptophane deaminase forms indolepyruvic acid from tryptophane. Indolepyruvic acid produces a brownish-red color in the presence of ferric chloride.	Tryptophane	0.4 mg	22
IND	Metabolism of tryptophane results in the formation of indole. Kovacs' reagent forms a colored complex (pink to red) with indole.	Tryptophane	0.2 mg	10
VP	Acetoin, an intermediary glucose metabolite, is produced from sodium pyruvate and indicated by the formation of a colored complex. Conventional VP tests may take up to 4 days, but by using sodium pyruvate, API has shortened the required test time. Creatine intensifies the color when tests are positive.	Sodium Pyruvate Creatine	2.0 mg 0.9 mg	3
GEL	Liquefaction of gelatin by proteolytic enzymes releases a black pigment which diffuses throughout the tube.	Kohn Charcoal Gelatin	0.6 mg	9
GLU MAN INO SOR RHA SAC MEL AMY ARA	Utilization of the carbohydrate results in acid formation and a consequent pH drop. The indicator changes from blue to yellow.	Glucose Mannitol Inositol Sorbitol Rhamnose Sucrose Melibiose Amygdalin (l +) Arabinose	2.0 mg 2.0 mg 2.0 mg 2.0 mg 2.0 mg 2.0 mg 2.0 mg 2.0 mg 2.0 mg	5 6 12
GLU Nitrate Reduction	Nitrites form a red complex with sulfanilic acid and N, N–dimethylalpha-naphthylamine. In case of negative reaction, addition of zinc confirms the presence of unreduced nitrates by reducing them to nitrites (pink-orange color). If there is no color change after the addition of zinc, this is indicative of the complete reduction of nitrates through nitrites to nitrogen gas or to an anaerogenic amine.	Potassium Nitrate	80.0 µg	6
MAN INO SOR Catalase	Catalase releases oxygen gas from hydrogen peroxide.			24

Courtesy of Anayltab Products, Plainview, N.Y.

CHART III Characterization of Gram-Negative Rods—The API 20E System

	ORGANISM	ONPG	ADH	LDC	ODC	CIT	H₂S	URE	TDA	IND	VP	GEL	GLU	MAN	INO	SOR	RHA	SAC	MEL	AMY	ARA	OXI
Escherichieae	E. coli	98.2	1.0	90.2	67.3	0	1.0	0	0	85.0	0	0	100	98.4	0.1	95.5	84.5	41.1	88.4	0.1	95.0	0
	Shigella dysenteriae	27.8	0	0	0	0	0	0	0	33.0	0	0	100	0.1	0	0	22.2	0	61.1	0	16.7	0
	Sh. flexneri	5.3	0	0	0	0	0	0	0	15.0	0	0	100	94.7	0	78.9	0	0	21.1	0	36.8	0
	Sh. boydii	5.0	0	0	0	0	0		0	20.0	0	0	100	60.0	0	53.3	1.0	0	33.3	0	66.7	0
	Sh. sonnei	96.7	0	0	80.0	0	0	0	0	0	0	0	100	99.0	0	39.9	80.0	0	50.0	0	96.7	0
	Edwardsiella tarda	0	0	99.0	99.0	0	55.0	0	0	100	0	0	100	0	0	0	0	0	50.0	0	1.1	0
Salmonelleae	Salmonella enteritidis	1.9	1.0	89.2	95.4	15.4	76.9	0	0	3.1	0	0	100	98.7	4.6	95.2	95.4	4.6	96.9	0	94.5	0
	Sal. typhi	0	0	90.0	0	0	0.1	0	0		0	0	100	99.0	0	99.0	1.8	0	100	0	27.0	0
	Sal. paratyphi A	0	0	0	100	0	0.2	0	0		0	0	100	99.0	0	99.0	99.0	0	40.0	0	80.0	0
	Arizona-S. arizonae	94.7	1.0	95.0	98.5	15.0	85.0	0	0	0	0	0	100	99.0	0	87.0	96.1	0	89.5	0	95.0	0
	Citrobacter freundii	97.0	10.0	0	60.0	10.0	81.0	0	0	6.0	0	0	100	98.0	1.0	96.0	87.0	59.0	77.0	30.0	98.0	0
	C. diversus-Levinea	97.0	10.0	0	90.0	10.0	0	0	0	91.0	0	0	100	97.0	14.5	88.0	99.0	51.0	47.0	34.0	99.0	0
	C. amalonaticus	97.0	10.0	0	95.0	10.0	0	0	0	99.0	0	0	100	97.0	0.1	93.0	99.0	29.4	53.0	80.0	93.8	0
Klebsielleae	Klebsiella pneumoniae	99.0	0	80.0	0	13.9	0	10.0	0	0	72.0	0	100	98.0	30.0	95.0	91.0	99.0	99.0	98.0	99.0	0
	K. oxytoca	98.0	0	83.0	0	13.0	0	10.0	0	100	60.0	0	100	99.0	29.0	92.0	98.0	99.0	99.0	98.0	99.0	0
	K. ozaenae	85.0	51.9	38.0	0	1.0	0	0	0	0	0	0	100	69.0	1.0	76.0	69.0	15.0	92.0	99.0	84.0	0
	K. rhinoscleromatis	0	0	0	0		0	0	0	0	0	0	100	99.0	1.0	86.0	53.0	33.0	66.0	99.0	95.0	0
	Enterobacter aerogenes	99.0	0	98.0	98.0	8.9	0	0	0	0	56.0	0	100	99.0	28.0	90.0	99.0	85.0	97.0	96.0	98.0	0
	Ent. cloacae	97.0	0	0	65.0	9.0	0	0	0	0	80.0	0	100	99.0	1.0	92.0	90.0	98.0	92.0	65.0	95.0	0
	Ent. agglomerans	90.0	0	0	0	5.4	0	0	0	50.0	20.0	0	100	99.0	1.0	80.0	60.0	33.0	70.0	70.0	95.0	0
	Ent. gergoviae	99.0	0	61.0	99.0	8.2	0	75.0	0	0	75.0	0	100	99.0	1.0	8.3	99.0	99.0	99.0	99.0	99.0	0
	Ent. sakazakii	97.0	51.6	0	59.0	8.6	0	0	0	4.0	85.0	0	100	99.0	4.0	8.5	90.0	95.0	90.0	76.0	95.0	0
	Serratia liquefaciens	85.0	0	85.0	95.0	8.9	0	1.0	0	0	50.0	60.0	100	99.0	1.0	99.0	30.0	85.0	80.7	80.0	92.9	0
	Ser. marcescens	83.0	0	88.0	94.0	8.0	0	1.0	0	0	58.0	72.0	100	96.0	1.0	97.0	2.0	98.0	37.0	72.0	18.0	0
	Ser. rubidaea	96.0	0	60.5	0.1	8.2	0	0	0	0	70.0	75.6	100	99.0	10.0	75.0	13.4	99.0	82.6	96.0	85.8	0
	Ser. odorifera 1	99.0	0	95.0	99.0	9.5	0	0	0	90.0	63.0	62.0	100	99.0	10.0	99.0	85.0	0	99.0	90.0	99.0	0
	Ser. odorifera 2	99.0	0	92.0	0	9.1	0	0	0	90.0	80.0	78.0	100	99.0	10.0	99.0	95.0	0	99.0	85.4	99.0	0
	Hafnia alvei	60.0	0	99.0	99.0	1.0	0	0	0	0	25.0	0	99.0	99.0	0	35.0	75.0	0	50.0	30.0	95.0	0

Courtesy of Anayltab Products, Plainview, N.Y.

CHART III Characterization of Gram-Negative Rods—The API 20E System cont.

	ORGANISM	ONPG	ADH	LDC	ODC	CIT	H₂S	URE	TDA	IND	VP	GEL	GLU	MAN	INO	SOR	RHA	SAC	MEL	AMY	ARA	OXI
Proteeae	Proteus vulgaris	0.5	0	0	0	4.1	75.3	91.0	95.0	75.3	0	75.3	100	0	0.1	0	0	83.0	1.0	20.0	4.0	0
	Prot. mirabilis	1.0	0	0	90.0	5.8	66.0	97.0	90.0	1.0	0	93.0	100	0	0	0	1.0	9.6	10.0	1.0	27.0	0
	Providencia alcalifaciens	0	0	0	0	9.8	0	0	95.0	94.0	0	0	100	0	0	0	0	0	0	0	25.0	0
	Prov. stuartii	1.0	0	0	0	8.5	0	0	95.0	86.0	0	0	100	0	8.0	0	0.8	3.7	34.0	0	30.0	0
	Prov. stuartii URE +	1.0	0	0	0	6.9	0	99.0	99.0	95.0	0	0	100	15.0	5.0	0	0.5	65.0	20.0	0	20.0	0
	Prov. rettgeri	1.0	0		0	7.1	0	80.0	95.0	90.0	0	0	100	85.0	1.0	30.0	40.0	5.0	0	40.0	10.0	0
	Morganella morganii	1.0	0	0	87.0	0.2	0	78.0	92.0	92.0	0	0	98.0	0	0	0	0	0	0	0	1.0	0
Yersiniae	Yersinia enterocolitica	81.0	0	0	36.0	0	0	59.0	0	54.0	0.4	0	100	99.0	1.0	95.0	9.0	78.0	40.4	31.0	76.6	0
	Y. pseudotuberculosis	80.0	0	0	0	0	0	88.0	0	0	0	0	100	94.0	0	76.0	58.0	0	5.0	0	52.0	0
	Y. pestis	93.0	0	0	0	0	0	0	0	0	1.0	0	93.0	87.0	0	56.0	0	0.6	0.6	25.0	87.0	0
Other Gram-negatives	API Group 1	99.0	0	58.8	99.0	9.2	0	0	0	99.0	0	0	100	99.0	0	75.4	82.4	82.4	94.1	97.0	94.1	0
	API Group 2	99.0	2.0	7.3	0	0	0	0	0	0	0	0	100	99.0	0	2.3	30.8	5.6	90.0	38.5	92.3	0
	Pseudomonas maltophilia	62.0	0	5.0	0	7.6	0	0	0	0	0	50.0	0.5	0	0	0	0	0	0	0	22.0	4.8
	Ps. cepacia	61.0	0	5.0	5.0	7.5	0	0	0	0	1.0	46.0	33.0	1.0	0	1.0	0	7.0	0	1.0	1.0	90.7
	Ps. paucimobilis	40.0	0	0	0	1.0	0	0	0	0	0	0	0.5	0	0	0	0	0.5	0	0	0.5	50.0
	A. calco var. anitratus	0	0	0	0	2.8	0	0	0	0	1.0	0.1	85.0	0	0	0	0	0.1	77.0	0	60.0	0
	A. calco var. lwoffii	0	0	0	0	0	0	0	0	0	0.1	0	0	0	0	0	0	0	0	0	0	0
	CDC Group VE-1	90.0	1.0	0	0	7.7	0	0	0	0	1.0	1.3	33.0	0	1.0	0	1.0	0.1	1.0	1.0	16.0	0
	CDC Group VE-2	0	0	0	0	7.9	0	0	0	0	1.0	0.1	4.5	0	1.0	0	0	0	1.0	0	5.0	0

393

CHART IV Characterization of Enterobacteriaceae—The Enterotube II System

Groups			Glucose	Gas Production	Lysine	Ornithine	H₂S	Indole	Adonitol	Lactose	Arabinose	Sorbitol	Voges-Proskauer	Dulcitol	Phenylalanine Deaminase	Urea	Citrate
ESCHERICHIEAE	Escherichia		+ 100.0	+J 92.0	d 80.6	d 57.8	−K 4.0	+ 96.3	− 5.2	+J 91.6	+ 91.3	± 80.3	− 0.0	d 49.3	− 0.1	− 0.1	− 0.2
	Shigella		+ 100.0	−A 2.1	− 0.0	∓B 20.0	− 0.0	∓ 37.8	− 0.0	−B 0.3	± 67.8	∓ 29.1	− 0.0	d 5.4	− 0.0	− 0.0	− 0.0
EDWARDSIELLEAE	Edwardsiella		+ 100.0	+ 99.4	+ 100.0	+ 99.0	+ 99.6	+ 99.0	− 0.0	− 0.0	∓ 10.7	− 0.2	− 0.0	− 0.0	− 0.0	− 0.0	− 0.0
SALMONELLEAE		Salmonella	+ 100.0	+C 91.9	+H 94.6	+I 92.7	+E 91.6	− 1.1	− 0.0	− 0.8	± 89.2	+ 94.1	− 0.0	dD 86.5	− 0.0	− 0.0	dF 80.1
		Arizona	+ 100.0	+ 99.7	+ 99.4	+ 100.0	+ 98.7	− 2.0	− 0.0	d 69.8	+ 99.1	+ 97.1	− 0.0	− 0.0	− 0.0	− 0.0	+ 96.8
	CITROBACTER	freundii	+ 100.0	+ 91.4	− 0.0	d 17.2	± 81.6	− 6.7	− 0.0	d 39.3	+ 100.0	+ 98.2	− 0.0	d 59.8	− 0.0	dw 89.4	+ 90.4
		amalonaticus	+ 100.0	+ 97.0	− 0.0	+ 97.0	− 0.0	+ 99.0	− 0.0	± 70.0	+ 99.0	+ 97.0	− 0.0	∓ 11.0	− 0.0	± 81.0	+ 94.0
		diversus	+ 100.0	+ 97.3	− 0.0	+ 99.8	− 0.0	+ 100.0	+ 100.0	d 40.3	+ 98.0	+ 98.2	− 0.0	± 52.2	− 0.0	dw 85.8	+ 99.7
PROTEEAE	PROTEUS	vulgaris	+ 100.0	±G 86.0	− 0.0	− 0.0	+ 95.0	+ 91.4	− 0.0	− 0.0	− 0.0	− 0.0	− 0.0	− 0.0	+ 100.0	+ 95.0	d 10.5
		mirabilis	+ 100.0	+G 96.0	− 0.0	+ 99.0	+ 94.5	− 3.2	− 0.0	− 2.0	− 0.0	− 0.0	∓ 16.0	− 0.0	+ 99.6	± 89.3	± 58.7
	MORGANELLA	morganii	+ 100.0	±G 86.0	− 0.0	+ 97.0	− 0.0	+ 99.5	− 0.0	− 0.0	− 0.0	− 0.0	− 0.0	− 0.0	+ 95.0	+ 97.1	−L 0.0
	PROVIDENCIA	alcalifaciens	+ 100.0	dG 85.2	− 0.0	− 1.2	− 0.0	+ 99.4	+ 94.3	− 0.3	− 0.7	− 0.6	− 0.0	− 0.0	+ 97.4	− 0.0	+ 97.9
		stuartii	+ 100.0	− 0.0	− 0.0	− 0.0	− 0.0	+ 98.6	∓ 12.4	− 3.6	− 4.0	− 3.4	− 0.0	− 0.0	+ 94.5	∓ 20.0	+ 93.7
		rettgeri	+ 100.0	∓G 12.2	− 0.0	− 0.0	− 0.0	+ 95.9	+ 99.0	d 10.0	− 0.0	− 1.0	− 0.0	− 0.0	+ 98.0	+ 100.0	+ 96.0
KLEBSIELLEAE	ENTEROBACTER	cloacae	+ 100.0	+ 99.3	− 0.0	+ 93.7	− 0.0	− 0.0	∓ 28.0	± 94.0	+ 99.4	+ 100.0	+ 100.0	d 15.2	− 0.0	± 74.6	+ 98.9
		sakazakii	+ 100.0	+ 97.0	− 0.0	+ 97.0	− 0.0	∓ 16.0	− 0.2	+ 100.0	+ 100.0	− 0.0	+ 97.0	− 6.0	− 0.0	− 0.0	+ 94.0
		gergoviae	+ 100.0	+ 93.0	+ 64.0	+ 100.0	− 0.0	− 0.0	− 0.0	∓ 42.0	+ 100.0	− 0.0	+ 100.0	− 0.0	− 0.0	+ 100.0	+ 96.0
		aerogenes	+ 100.0	+ 95.9	+ 97.5	+ 95.9	− 0.0	− 0.8	+ 97.5	+ 92.5	+ 100.0	+ 98.3	+ 100.0	− 4.1	− 0.0	− 0.0	+ 92.6
		agglomerans	+ 100.0	∓ 24.1	− 0.0	− 0.0	− 0.0	∓ 19.7	− 7.5	d 52.9	+ 97.5	d 26.3	± 64.8	d 12.9	∓ 27.6	d 34.1	d 84.2
	HAFNIA	alvei	+ 100.0	+ 98.9	+ 99.6	+ 98.6	− 0.0	− 0.0	− 0.0	d 2.8	+ 99.3	− 0.0	± 65.0	− 2.4	− 0.0	− 3.0	d 5.6
	SERRATIA	marcescens	+ 100.0	±G 52.6	+ 99.6	+ 99.6	− 0.0	−w 0.1	∓ 56.0	− 1.3	− 0.0	+ 99.1	+ 98.7	− 0.0	− 0.0	dw 39.7	+ 97.6
		liquefaciens	+ 100.0	d 72.5	± 64.2	+ 100.0	− 0.0	−w 1.8	d 8.3	d 15.6	+ 97.3	+ 97.3	∓ 49.5	− 0.0	− 0.9	dw 3.7	+ 93.6
		rubidaea	+ 100.0	dG 35.0	± 61.0	− 0.0	− 0.0	−w 2.0	± 88.0	+ 100.0	+ 100.0	− 8.0	+ 92.0	− 0.0	− 0.0	dw 4.0	± 88.0
	KLEBSIELLA	pneumoniae	+ 100.0	+ 96.0	+ 97.2	− 0.0	− 0.0	− 0.0	± 89.0	+ 98.7	+ 99.9	+ 99.4	+ 93.7	∓ 33.0	− 0.0	+ 95.4	+ 96.8
		oxytoca	+ 100.0	+ 96.0	+ 97.2	− 0.0	− 0.0	+ 100.0	± 89.0	∓ 98.7	+ 100.0	+ 98.0	+ 93.7	∓ 33.0	− 0.0	∓ 95.4	∓ 96.8
		ozaenae	+ 100.0	d 55.0	∓ 35.8	− 1.0	− 0.0	− 0.0	+ 91.8	d 26.2	+ 100.0	± 78.0	− 0.0	− 0.0	− 0.0	d 14.8	d 28.1
		rhinoscleromatis	+ 100.0	− 0.0	− 0.0	− 0.0	− 0.0	− 0.0	+ 98.0	d 6.0	+ 100.0	+ 98.0	− 0.0	− 0.0	− 0.0	− 0.0	− 0.0
YERSINIAE	YERSINIA	enterocolitica	+ 100.0	− 0.0	− 0.0	+ 90.7	− 0.0	∓ 26.7	− 0.0	− 0.0	+ 98.7	+ 98.7	− 0.1	− 0.0	− 0.0	+ 90.7	− 0.0
		pseudotuberculosis	+ 100.0	− 0.0	− 0.0	− 0.0	− 0.0	− 0.0	− 0.0	− 0.0	± 55.0	− 0.0	− 0.0	− 0.0	− 0.0	+ 100.0	− 0.0

Courtesy of Roche Diagnostics, Nutley, N.J.

E. *S. enteritidis* bioserotype Paratyphi A and some rare biotypes may be H₂S negative.

F. *S. typhi*, *S. enteritidis* bioserotype Paratyphi A and some rare biotypes are citrate-negative and *S. cholerae-suis* is usually delayed positive.

G. The amount of gas produced by *Serratia*, *Proteus* and *Providencia alcalifaciens* is slight; therefore, gas production may not be evident in the ENTEROTUBE II.

H. *S. enteritidis* bioserotype Paratyphi A is negative for lysine decarboxylase.

I. *S. typhi* and *S. gallinarum* are ornithine decarboxylase-negative.

J. The Alkalescens-Dispar (A-D) group is included as a biotype of *E. coli*. Members of the A-D group are generally anaerogenic, non-motile and do not ferment lactose.

K. An occasional strain may produce hydrogen sulfide.

L. An occasional strain may appear to utilize citrate.

Courtesy of Anayltab Products, Plainview, N.Y.

CHART V Reaction Interpretations for API Staph-Ident

MICROCUPULE		INTERPRETATION OF REACTIONS		
NO.	SUBSTRATE	POSITIVE	NEGATIVE	COMMENTS AND REFERENCES
1	PHS	Yellow	Clear or straw-colored	A positive result should be recorded only if significant color development has occurred.(3)
2	URE	Purple to Red-Orange	Yellow or Yellow-Orange	Phenol red has been added to the urea formulation to allow detection of alkaline end products resulting from urea utilization.(1)
3	GLS	Yellow	Clear or straw-colored	A positive result should be recorded only if significant color development has occurred.
4 5 6 7	MNE MAN TRE SAL	Yellow or Yellow-Orange	Red or Orange	Cresol red has been added to each carbohydrate to allow detection of acid production if the respective carbohydrates are utilized. (1,7)
8	GLC	Yellow	Clear or straw-colored	A positive result should be recorded only if significant color development has occurred.
9	ARG	Purple to Red-Orange	Yellow or Yellow-Orange	Phenol red has been added to the arginine formulation to allow detection of alkaline end products resulting from arginine utilization.(1)
10	NGP	Add 1–2 drops of STAPH-IDENT REAGENT Plum-Purple (Mauve)	Yellow or colorless	Color development will begin within 30 seconds of reagent addition. (1,5)

Courtesy of Anayltab Products, Plainview, N.Y.

Abbreviation	Test
PHS	Phosphatase
URE	Urea utilization
GLS	β-Glucosidase
MNE	Mannose utilization
MAN	Mannitol utilization
TRE	Trehalose utilization
SAL	Salicin utilization
GLC	β-Glucuronidase
ARG	Arginine utilization
NGP	β-Galactosidase

CHART VI   Biochemistry of API Staph-Ident Tests

MICROCUPULE		CHEMICAL/PHYSICAL PRINCIPLES	REACTIVE INGREDIENTS	QUANTITY
NO.	SUBSTRATE			
1	PHS	Hydrolysis of p-nitrophenyl-phosphate, disodium salt, by alkaline phosphatase releases yellow paranitrophenol from the colorless substrate.	p-nitrophenyl-phosphate, disodium salt	0.2%
2	URE	Urease releases ammonia from urea; ammonia causes the pH to rise and changes the indicator from yellow to red.	Urea	1.6%
3	GLS	Hydrolysis of p-nitrophenyl-β-D-glucopyranoside by β-glucosidase releases yellow para-nitrophenol from the colorless substrate.	p-nitrophenyl-β-D-glucopyranoside	0.2%
4	MNE	Utilization of carbohydrate results in acid formation and a consequent pH drop. The indicator changes from red to yellow.	Mannose	1.0%
5	MAN		Mannitol	1.0%
6	TRE		Trehalose	1.0%
7	SAL		Salicin	1.0%
8	GLC	Hydrolysis of p-nitrophenyl-β-D-glucuronide by β-glucuronidase releases yellow para-nitrophenol from the colorless substrate.	p-nitrophenyl-β-D-glucuronide	0.2%
9	ARG	Utilization of arginine produces alkaline end products which change the indicator from yellow to red.	Arginine	1.6%
10	NGP	Hydrolysis of 2–naphthol-β-D-galactopyranoside by β-galactosidase releases free β-naphthol which complexes with STAPH-IDENT REAGENT to produce a plum-purple (mauve) color.	2–naphthol-β-D-galactopyranoside	0.3%

Courtesy of Anayltab Products, Plainview, N.Y.

CHART VII   API Staph-Ident Profile Register*

Profile	Identification		Profile	Identification	
0 040	STAPH CAPITIS		4 700	STAPH AUREUS	COAG +
0 060	STAPH HAEMOLYTICUS			STAPH SCIURI	COAG −
0 100	STAPH CAPITIS		4 710	STAPH SCIURI	
0 140	STAPH CAPITIS		5 040	STAPH EPIDERMIDIS	
0 200	STAPH COHNII		5 200	STAPH SCIURI	
0 240	STAPH CAPITIS		5 210	STAPH SCIURI	
0 300	STAPH CAPITIS		5 300	STAPH AUREUS	COAG +
0 340	STAPH CAPITIS			STAPH SCIURI	COAG −
0 440	STAPH HAEMOLYTICUS		5 310	STAPH SCIURI	
0 460	STAPH HAEMOLYTICUS		5 600	STAPH SCIURI	
0 600	STAPH COHNII		5 610	STAPH SCIURI	
0 620	STAPH HAEMOLYTICUS		5 700	STAPH AUREUS	COAG +
0 640	STAPH HAEMOLYTICUS			STAPH SCIURI	COAG −
0 660	STAPH HAEMOLYTICUS		5 710	STAPH SCIURI	
1 000	STAPH EPIDERMIDIS		5 740	STAPH AUREUS	
1 040	STAPH EPIDERMIDIS		6 001	STAPH XYLOSUS	XYL + ARA +
1 300	STAPH AUREUS			STAPH SAPROPHYTICUS	XYL − ARA −
1 540	STAPH HYICUS (An)		6 011	STAPH XYLOSUS	
1 560	STAPH HYICUS (An)		6 021	STAPH XYLOSUS	
2 000	STAPH SAPROPHYTICUS	NOVO R	6 101	STAPH XYLOSUS	
	STAPH HOMINIS	NOVO S	6 121	STAPH XYLOSUS	
2 001	STAPH SAPROPHYTICUS		6 221	STAPH XYLOSUS	
2 040	STAPH SAPROPHYTICUS	NOVO R	6 300	STAPH AUREUS	
	STAPH HOMINIS	NOVO S	6 301	STAPH XYLOSUS	
2 041	STAPH SIMULANS		6 311	STAPH XYLOSUS	
2 061	STAPH SIMULANS		6 321	STAPH XYLOSUS	
2 141	STAPH SIMULANS		6 340	STAPH AUREUS	COAG +
2 161	STAPH SIMULANS			STAPH WARNERI	COAG −
2 201	STAPH SAPROPHYTICUS		6 400	STAPH WARNERI	
2 241	STAPH SIMULANS		6 401	STAPH XYLOSUS	XYL + ARA +
2 261	STAPH SIMULANS			STAPH SAPROPHYTICUS	XYL − ARA −
2 341	STAPH SIMULANS		6 421	STAPH XYLOSUS	
2 361	STAPH SIMULANS		6 460	STAPH WARNERI	
2 400	STAPH HOMINIS	NOVO S	6 501	STAPH XYLOSUS	
	STAPH SAPROPHYTICUS	NOVO R	6 521	STAPH XYLOSUS	
2 401	STAPH SAPROPHYTICUS		6 600	STAPH WARNERI	
2 421	STAPH SIMULANS		6 601	STAPH SAPROPHYTICUS	XYL − ARA −
2 441	STAPH SIMULANS			STAPH XYLOSUS	XYL + ARA +
2 461	STAPH SIMULANS		6 611	STAPH XYLOSUS	
2 541	STAPH SIMULANS		6 621	STAPH XYLOSUS	
2 561	STAPH SIMULANS		6 700	STAPH AUREUS	
2 601	STAPH SAPROPHYTICUS		6 701	STAPH XYLOSUS	
2 611	STAPH SAPROPHYTICUS		6 721	STAPH XYLOSUS	
2 661	STAPH SIMULANS		6 731	STAPH XYLOSUS	
2 721	STAPH COHNII (SSP1)		7 000	STAPH EPIDERMIDIS	
2 741	STAPH SIMULANS		7 021	STAPH XYLOSUS	
2 761	STAPH SIMULANS		7 040	STAPH EPIDERMIDIS	
3 000	STAPH EPIDERMIDIS		7 141	STAPH INTERMEDIUS (An)	
3 040	STAPH EPIDERMIDIS		7 300	STAPH AUREUS	
3 140	STAPH EPIDERMIDIS		7 321	STAPH XYLOSUS	
3 540	STAPH HYICUS (An)		7 340	STAPH AUREUS	
3 541	STAPH INTERMEDIUS (An)		7 401	STAPH XYLOSUS	
3 560	STAPH HYICUS (An)		7 421	STAPH XYLOSUS	
3 601	STAPH SIMULANS	NOVO S	7 501	STAPH INTERMEDIUS (An)	COAG +
	STAPH SAPROPHYTICUS	NOVO R		STAPH XYLOSUS	COAG −
4 060	STAPH HAEMOLYTICUS		7 521	STAPH XYLOSUS	
4 210	STAPH SCIURI		7 541	STAPH INTERMEDIUS (An)	
4 310	STAPH SCIURI		7 560	STAPH HYICUS (An)	
4 420	STAPH HAEMOLYTICUS		7 601	STAPH XYLOSUS	
4 440	STAPH HAEMOLYTICUS		7 621	STAPH XYLOSUS	
4 460	STAPH HAEMOLYTICUS		7 631	STAPH XYLOSUS	
4 610	STAPH SCIURI		7 700	STAPH AUREUS	
4 620	STAPH HAEMOLYTICUS		7 701	STAPH XYLOSUS	
4 660	STAPH HAEMOLYTICUS		7 721	STAPH XYLOSUS	
			7 740	STAPH AUREUS	

*Date of Publication: March, 1984
Courtesy of Anayltab Products, Plainview, N.Y.

The Streptococci: Classification, Habitat, Pathology, and Biochemical Characteristics

To fully understand the characteristics of the various species of medically important streptococci, this appendix has been included as an adjunct to Exercise 53. The table of streptococcal characteristics on this page is the same one that is shown on page 329 of Exercise 53. It is also the basis for much of the discussion that follows.

The first system that was used for grouping the streptococci was based on the type of hemolysis and was proposed by J. H. Brown in 1919. In 1933, R. C. Lancefield proposed that these bacteria be separated into groups A, B, C, and so on, on the basis of precip-

itation-type serological testing. Both hemolysis and serological typing still play predominant roles today in our classification system. Note below that the Lancefield groups are categorized with respect to the type of hemolysis that is produced on blood agar.

BETA HEMOLYTIC GROUPS

Using a streak-stab technique, a blood agar plate is incubated aerobically at 37° C for 24 hours. Isolates that have colonies surrounded by clear zones completely free of red blood cells are characterized as

TABLE E.1 Physiological Tests for Streptococcal Differentiation

GROUP	Type of Hemolysis	Bacitracin Susceptibility	CAMP Reaction or Hippurate Hydrolysis	SXT Sensitivity	Bile Esculin Hydrolysis	Tolerance to 6.5% NaCl	Optochin Susceptibility	Bile Solubility
Group A S. pyogenes	beta	+	−	R	−	−	−	−
Group B S. agalactiae	beta	−*	+	R	−	±	−	−
Group C S. equi S. equisimilis S. zooepidemicus	beta	−*	−	S	−	−	−	−
Group D** (enterococci) S. faecalis S. faecium etc.	alpha beta none	−	−	R	+	+	−	−
Group D** (nonenterococci) S. bovis etc.	alpha none	−	−	R/S	+	−	−	−
Viridans S. mitis S. salivarius S. mutans etc.	alpha none	−*	−*	S	−	−	−	−
Pneumococci S. pneumoniae	alpha	±	−		−	−	+	+

*Exceptions occur occasionally   **See comments on pp. 363 and 364 concerning correct genus.

Note: R = resistant; S = sensitive; blank = not significant

being *beta hemolytic*. Three serological groups of streptococci fall in this category: groups A, B, and C; a few species in group D are also beta hemolytic.

Group A Streptococci

This group is represented by only one species: *Streptococcus pyogenes*. Approximately 25% of all upper respiratory infections (URIs) are caused by this species; another 10% of URIs are caused by other streptococci; most of the remainder (65%) are caused by viruses. Since no unique clinical symptoms can be used to differentiate viral from streptococcal URIs, and since successful treatment relies on proper identification, it becomes mandatory that throat cultures be taken in an attempt to prove the presence or absence of streptococci. It should be added that if streptococcal URIs are improperly treated, serious sequelae such as pneumonia, acute endocarditis, rheumatic fever, and glomerularnephritis can result.

S. pyogenes is the only beta hemolytic streptococcus that is primarily of *human origin*. Although the pharynx is the most likely place to find this species, it may be isolated from the skin and rectum. Asymptomatic pharyngeal and anal carriers are not uncommon. Outbreaks of postoperative streptococcal infections have been traced to both pharyngeal and anal carriers among hospital personnel.

These coccoidal bacteria (0.6–1.0 μm diameter) occur as pairs and as short to moderate-length chains in clinical specimens; in broth cultures, the chains are often longer.

When grown on blood agar, the colonies are small (0.5 mm dia.), transparent to opaque, and domed; they have a smooth or semimatte surface and an entire edge; complete hemolysis (beta-type) occurs around each colony, usually two to four times the diameter of the colony.

S. pyogenes produces two hemolysins: streptolysin S and streptolysin O. The beta-type hemolysis on blood agar is due to the complete destruction of red blood cells by the streptolysin S.

There is no group of physiological tests that can be used with *absolute* certainty to differentiate *S. pyogenes* from other streptococci; however, if an isolate is beta hemolytic and sensitive to bacitracin, one can be 95% certain that the isolate is *S. pyogenes*. The characteristics of this organism are the first ones tabulated in table E.1.

Group B Streptococci

The only recognized species of this group is *S. agalactiae*. Although this organism is frequently found in milk and associated with *mastitis in cattle,* the list of human infections caused by it is as long as the one for *S. pyogenes*: abscesses, acute endocarditis, impetigo, meningitis, neonatal sepsis, and pneumonia

are just a few. Like *S. pyogenes*, this pathogen may also be found in the pharynx, skin, and rectum; however, it is more likely to be found in the genital and intestinal tracts of healthy adults and infants. It is not unusual to find the organism in vaginal cultures of third-trimester pregnant women.

Cells are spherical to ovoid (0.6–1.2 μm dia) and occur in chains of seldom less than four cells; long chains are frequently present. Characteristically, the chains appear to be composed of paired cocci.

Colonies of *S. agalactiae* on blood agar often produce double zone hemolysis. After 24 hours incubation, colonies exhibit zones of beta hemolysis. After cooling, a second ring of hemolysis forms, which is separated from the first by a ring of red blood cells.

Reference to table E.1 emphasizes the significant characteristics of *S. agalactiae*. Note that this organism gives a positive CAMP reaction, hydrolyzes hippurate, and is not (usually) sensitive to bacitracin. It is also resistant to SXT. Presumptive identification of this species relies heavily on a positive CAMP test or hippurate hydrolysis, even if beta hemolysis is not clearly demonstrated.

Group C Streptococci

Three species fall in this group: *S. equisimilis, S. equi,* and *S. zooepidemicus*. Although all of these species may cause human infections, the diseases are not usually as grave as those caused by groups A and B. Some group C species have been isolated from impetiginous lesions, abscesses, sputum, and the pharynx. There is no evidence that they are associated with acute glomerularnephritis, rheumatic fever, or even pharyngitis.

Presumptive differentiation of this group from *S. pyogenes* and *S. agalactiae* is based primarily on (1) resistance to bacitracin, (2) inability to hydrolyze hippurate or bile esculin, and (3) a negative CAMP test. There are other groups that have some of these same characteristics, but they will not be studied here. Tables 12.16 and 12.17 on page 1049 of *Bergey's Manual*, vol. 2, provide information about these other groups.

ALPHA HEMOLYTIC GROUPS

Streptococcal isolates that have colonies with zones of incomplete lysis around them are said to be **alpha hemolytic.** These zones are often greenish; sometimes they are confused with beta hemolysis. *The only way to be certain that such zones are not beta hemolytic is to examine the zones under* 60 × *microscopic magnification.* Figure 53.4, page 326, illustrates the differences between alpha and beta hemolysis. If some red blood cells are seen in the zone, the isolate is classified as being alpha hemolytic.

The grouping of streptococci on the basis of alpha hemolysis is not as clear-cut as it is for beta hemolytic groups. Note in table E1 that the bottom four groups that have alpha hemolytic types may also have beta hemolytic or nonhemolytic strains. Thus, we see that hemolysis in these four groups can be a misleading characteristic in identification.

Alpha hemolytic isolates from the pharynx are usually *S. pneumoniae*, viridans streptococci, or group D. Our primary concern here in this experiment is to identify isolates of *S. pneumoniae*. To accomplish this goal, it will be necessary to differentiate any alpha hemolytic isolate from group D and viridans streptococci.

Streptococcus pneumoniae
(Pneumococcus)

This organism is the most frequent cause of bacterial pneumonia, a disease that has a high mortality rate among the aged and debilitated. It is also frequently implicated in conjunctivitis, otitis media, pericarditis, subacute endocarditis, meningitis, septicemia, empyema, and peritonitis. Thirty to 70% of normal individuals carry this organism in the pharynx.

Spherical or ovoid, these cells (0.5–1.25 μm dia) occur typically as pairs, sometimes singly, often in short chains. Distal ends of the cells are pointed or lancet-shaped and are heavily encapsulated with polysaccharide on primary isolation.

Colonies on blood agar are small, mucoidal, opalescent, and flattened with entire edges surrounded by a zone of greenish discoloration (alpha hemolysis). In contrast, the viridans streptococcal colonies are smaller, gray to whitish gray, and opaque with entire edges.

Presumptive identification of *S. pneumoniae* can be made with the optochin and bile solubility tests. On the optochin test, the pneumococci exhibit sensitivity to ethylhydrocupreine (optochin). With the bile solubility test, pneumococci are dissolved in bile (2% sodium desoxycholate). Table E.1 reveals that except for bacitracin susceptibility ($\pm$), *S. pneumoniae* is negative on all other tests used for differentiation of streptococci.

Viridans Group

Streptococci that fall in this group are primarily alpha hemolytic; some are nonhemolytic. Approximately 10 species are included in this group. All of them are highly adapted parasites of the upper respiratory tract. Although usually regarded as having low pathogenicity, they are opportunistic and sometimes cause serious infections. Two species (*S. mutans* and *S. sanguis*) are thought to be the primary cause of dental caries, since they have the ability to form dental plaque. Viridans streptococci are implicated more often than any other bacteria in subacute bacterial endocarditis.

When it comes to differentiation of bacteria of this group from the pneumococci and enterococci, we will use the optochin, bile solubility, and salt-tolerance tests. See table E.1.

Group D Streptococci (Enterococci)

Members of this group are, currently, considered by most taxonomists to belong to the genus *Enterococcus*. During the preparation of volume 2 of *Bergey's Manual*, Schleifer and Kilper-Balz presented conclusive evidence that *S. faecalis*, *S. faecium*, and *S. bovis* were so distantly related to the other groups of streptococci that they should be transferred to another genus. Since the term *Enterococcus* had been previously suggested by others, Schleifer and Kilper-Balz recommended that this be the name of a new genus to include all of the Group D streptococci, nonenterococci included. The fact that these papers came too late for *Bergey's Manual* to include this new genus caused the genus *Streptococcus* to be retained. To avoid confusion in our use of *Bergey's Manual*, we have retained the same terminology used in *Bergey's Manual*.

The enterococci of serological group D may be alpha hemolytic, beta hemolytic, or nonhemolytic. The principal species of this enterococcal group are *S. faecalis*, *S. faecium*, *S. durans*, and *S. avium*.

Subacute endocarditis, pyelonephritis, urinary tract infections, meningitis, and biliary infections are caused by these organisms. All five of these species have been isolated from the intestinal tract. Approximately 20% of subacute bacterial endocarditis and 10% of urinary tract infections are caused by members of this group. Differentiation of this group from other streptococci in systemic infections is mandatory because *S. faecalis*, *S. faecium*, and *S. durans* are resistant to penicillin and require combined antibiotic therapy.

Since *S. faecalis* can be isolated from many food products (not connected with fecal contamination), it can be a transient in the pharynx and show up as an isolate in throat cultures. Morphologically, the cells are ovoid (0.5–1.0 μm dia) occurring as pairs in short chains. Hemolytic reactions of *S. faecalis* on blood agar will vary with the type of blood used in the medium. Some strains produce beta hemolysis on agar with horse, human, and rabbit blood; on sheep blood agar the colonies will always exhibit alpha hemolysis. Other streptococci are consistently either beta, alpha, or nonhemolytic.

Cells of *S. faecium* are morphologically similar to *S. faecalis* except that motile strains are often encountered. A strong alpha-type hemolysis is usually seen around colonies of *S. faecium* on blood agar.

Although presumptive differentiation of group D enterococcal streptococci from groups A, B, and C is not too difficult with physiological tests, it is more laborious to differentiate the individual species within group D. As indicated in table E.1, the enterococci (1) hydrolyze bile esculin, (2) are CAMP negative, and (3) grow well in 6.5% NaCl broth.

Differentiation of the five species within this group involves nine or ten physiological tests.

Group D Streptococci (Nonenterococci)

The only medically significant nonenterococcal species of group D is *S. bovis*. This organism is found in the intestinal tract of humans as well as in cows, sheep, and other ruminants. It can cause meningitis, subacute endocarditis, and urinary tract infections. On blood agar, the organism is usually alpha hemolytic; occasionally, it is nonhemolytic. The best way to differentiate it from the group D enterococci is to test its tolerance to 6.5% NaCl. Note in table E.1 that *S. bovis* will not grow in this medium, but all enterococci will.

Reading References

General Information

Alcamo, I. Edward. *Fundamentals of Microbiology,* 5th ed. Reading, Mass.: Addison-Wesley, 1997.

Atlas, R. M., and Bartha, R. *Microbial Ecology: Fundamentals and Applications,* 3rd ed. Menlo Park, Calif.: Benjamin/Cummings Publishing, 1993.

Baron, Samuel, ed. *Medical Microbiology,* 4th ed. Reading, Mass.: Addison-Wesley, 1996.

Brock, Thomas D. *Robert Koch: A Life in Medicine and Bacteriology.* Herndon, Va.: ASM Press, 1999.

Brogden, Kim A., et al. *Virulence Mechanisms of Bacterial Pathogens.* Herndon, Va.: ASM Press, 2000.

Brun, Yves V., and Shimkets, L. J. *Prokaryotic Development.* Herndon, Va.: ASM Press, 2000.

Burlage, Robert S., Atlas, R., Stahl, D., Geesey, G., and Saylor, G. *Techniques in Microbial Ecology.* Cary, N.C.: Oxford University Press, 1998.

Collier, Leslie H., et al. *Topley and Wilson's Microbiology and Microbial Infections.* Six Volumes. Herndon, Va.: ASM Press, 1998.

Colwell, Rita R. *Nonculturable Microorganisms in the Environment.* Herndon, Va.: ASM Press, 2000.

Doyle, Michael P., et al. *Food Microbiology: Fundamentals and Frontiers.* Herndon, Va.: ASM Press, 1997.

Dubos, Rene. *Pasteur and Modern Science.* Paperback. Herndon, Va.: ASM Press, 1998.

Flint, S. J., et al. *Principles of Virology: Molecular Biology, Pathogenesis, and Control.* Herndon, Va.: ASM Press, 1999.

Gerhardt, Philipp, et al. *Methods for General and Molecular Bacteriology.* Herndon, Va.: ASM Press, 1997.

Hurst, Christon J., et al. *Manual of Environmental Microbiology.* Herndon, Va.: ASM Press, 1997.

Jakoby, W. B. *Methods in Enzymology.* New York: Academic Press, 1987.

Karam, Jim D., et al. *Molecular Biology of Bacteriophage T-4.* Herndon, Va.: ASM Press, 1994.

Lacey, Alan J. *Light Microscopy in Biology.* Cary, N.C.: Oxford University Press, 1999.

Lederberg, Joshua, et al. *Encyclopedia of Microbiology.* New York: Academic Press, 1992.

Lovley, Derek R. *Environmental Microbe-Metal Interactions.* Herndon, Va.: ASM Press, 2000.

Madigan, Michael T., and Marrs, Barry L. *Extremophiles.* New York: Scientific American Vol. 276, Number 4: pp. 82–87, 1997.

Madigan, Michael T., Martinko, John M., and Parker, Jack. *Brock Biology of Microorganisms,* 8th ed. Englewood Cliffs, N.J.: Prentice-Hall, 1997.

Mobley, Harry L. T., and Warren, John W. *Urinary Tract Infections.* Herndon, Va.: ASM Press, 1995.

Needham, Cynthia, et al. *Intimate Strangers: Unseen Life on Earth.* Herndon, Va.: ASM Press, 2000.

Pelczar, M. J., and Chan, E. C. *Microbiology,* 5th ed. New York: McGraw-Hill, 1993.

Prescott, Lansing M., Harley, John P., and Klein, Donald A. *Microbiology,* 4th ed. New York: McGraw-Hill, 1999.

Rose, Noel R. *Manual of Clinical Laboratory Immunology,* 5th ed. Herndon, Va.: ASM Press, 1997.

Rosenburg, Eugene. *Microbial Ecology and Infectious Disease.* ASM Press, 1999.

Salyers, Abigail A., and Whitt, D. D. *Bacterial Pathogenesis.* Herndon, Va.: ASM Press, 1994.

Smith, A. D., et al. *Oxford Dictionary of Biochemistry and Molecular Biology.* Cary, N.C.: Oxford University Press, 1997.

Snyder, Larry, and Champness, Wendy. *Molecular Genetics of Bacteria.* Herndon, Va.: ASM Press, 1997.

Talaro, K., and Talaro, A. *Foundations in Microbiology,* 3rd ed. Dubuque, IA; New York: McGraw-Hill, 1999.

Tortora, Gerard J., Funke, B. R., and Case, C. L. *Microbiology: An Introduction,* 6th ed. Menlo Park, Calif.: Benjamin/Cummings Publishing, 1999.

Volk, W.A., and Wheeler, M. F. *Basic Microbiology,* 6th ed. Reading, Mass.: Addison-Wesley, 1996.

Walker, Graham C., and Kaiser, Dale. *Frontiers in Microbiology: A Collection of Minireviews from the Journal of Bacteriology.* Herndon, Va.: ASM Press, 1993.

White, David. *The Physiology and Biochemistry of Prokaryotes,* 2nd ed. Cary, N.C.: Oxford University Press, 1999.

Laboratory Procedures

American Type Culture Collections. *Catalog of Cultures,* 8th ed. Rockville, Md. n.d.

Atlas, R. M., and Snyder, J. W. *Handbook of Media for Clinical Microbiology.* Boca Raton, Fla.: CRC Press, 1996.

Chart, Henrik. *Methods in Practical Laboratory Bacteriology.* Boca Raton, Fla.: CRC Press, 1994.

Difco Laboratory Staff. *Difco Manual of Dehydrated Culture Media and Reagents,* 11th ed. Detroit, Mich.: Difco Laboratories, 1998.

Flemming, D. O., Richardson, J. H., Tulis, J. J., and Vesley, D. *Laboratory Safety: Principles and Practices,* 2nd ed. Herndon, Va.: ASM Press, 1995.

Garcia, Lynne S., and Brukner, David A. *Diagnostic Medical Parasitology,* 3rd ed. Herndon, Va.: ASM Press, 1996.

Isenberg, Henry D., et al. *Clinical Microbiology Procedures Handbook,* Vols. 1 and 2. Herndon, Va.: ASM Press, 1992.

Miller, Michael J. *A Guide to Specimen Management in Clinical Microbiology*, 2nd ed. Herndon, Va.: ASM Press, 1998.

Murray, Patrick R., et al. *Manual of Clinical Microbiology*, 6th ed. Herndon, Va.: ASM Press, 1997.

Murray, Patrick R., et al. *Manual of Clinical Microbiology*, 7th ed. Herndon, Va.: ASM Press, 1999.

Norris, John R., and Ribbons, D. W. *Methods in Microbiology*, Vols. 23 and 24. New York: Academic Press, 1991.

Shapton, D. A., and Shapton, N.F. *Principles and Practices of Safe Processing of Food.* New York: Academic Press, 1994.

Smith, Robert F. *Microscopy and Photomicrography.* Boca Raton, Fla.: CRC Press, 1994.

Identification of Microorganisms

Balows, Albert, et al. *Manual of Clinical Microbiology*, 5th ed. Bethesda, Md.: American Society for Microbiology, 1991.

Chandler, Francis W., and Watts, John C. *Pathologic Diagnosis of Fungal Infections.* Chicago, Ill.: American Society of Clinical Pathologists, 1987.

Fischetti, Vincent A., et al. *Gram-Positive Pathogens.* Herndon, Va.: ASM Press, 2000.

Goodfellow, M., and O'Donnell, A. G. *Handbook of New Bacteria and Systematics.* New York: Academic Press, 1993.

Holt, John G., Kreig, N. R., et al. *Bergey's Manual of Systematic Bacteriology*, vol. 1, 4th ed. Baltimore, Md.: Williams & Wilkins, 1984.

Jahn, Theodore L., et al. *Protozoa*, 2nd ed. Dubuque, Ia.: WCB/McGraw-Hill, 1978.

Lapage, S. P., et al. *International Code of Nomenclature of Bacteria.* Herndon, Va.: ASM Press, 1990.

Larone, Davise. *Medically Important Fungi: A Guide to Identification.* Herndon, Va.: ASM Press, 1995.

Piggot, Patrick J., et al. *Regulation of Bacterial Differentiation.* Herndon, Va.: ASM Press, 1993.

Skerman, V. B. D., and Sneath, P. H. A. *Approved Lists of Bacterial Names.* Herndon, Va.: ASM Press, 1998.

Sneath, Peter H. A., et al. *Bergey's Manual of Systematic Bacteriology*, vol. 2. Baltimore, Md.: Williams & Wilkins, 1986.

Staley, James T., et al. *Bergey's Manual of Systematic Bacteriology*, vol. 3. Baltimore, Md.: Williams & Wilkins, 1989.

Sanitary and Medical Microbiology

Balows, Albert et al. *Manual of Clinical Microbiology*, 5th ed. Herndon, Va.: ASM Press, 1991.

Flemming, D. O., Richardson, J. H., Tulis, J. J., and Vesley, D. *Laboratory Safety: Principles and Practices*, 2nd ed. Herndon, Va.: ASM Press, 1995.

Greenberg, Arnold E., et. al. *Standard Methods for the Examination of Water and Wastewater*, 19th ed. Washington, D.C.: American Public Health Association, 1995.

Jay, James M. *Modern Food Microbiology*, 5th ed. New York: Chapman-Hall, 1996.

Kneip, Theodore, and Crable, John V. *Methods for Biological Monitoring: A Manual for Assessing Human Exposure to Hazardous Substances.* Washington, D.C.: American Public Health Association, 1988.

Marshall, Robert T. *Standard Methods for the Examination of Dairy Products*, 16th ed. Washington, D.C.: American Public Health Association, 1992.

Miller, Michael J. *A Guide to Specimen Management in Clinical Microbiology*, 2nd ed. Herndon, Va.: ASM Press, 1998.

Ray, Bibek. *Fundamental Food Microbiology.* Boca Raton, Fla.: CRC Press, 1996.

Vanderzant, Carl, and Splittstoesser, Don. *Compendium of Methods for the Microbiological Examination of Foods*, 3rd ed. Washington, D.C.: American Public Health Association, 1992.

Index

ABO blood types, 357–61
acetone, 6
acid-fast bacteria, 103
acid-fast staining, 103–04, 222
acidic dyes, 87
acido-thermophiles, 30
acid reaction, and litmus milk test, 243
adonitol, and Enterotube II system, 267
adsorption, and lytic cycle, 153
Actinastrum, 36
Actinobacter, 266
aerobic bacteria, 227
aerotolerant organisms, 125
agar, 116, 118. *See also* agar culture method; agar plates; blood agar;
    cornmeal agar; Hektoen Enteric (HE) agar; Kligler's iron agar;
    MacConkey agar; mannitol salt agar (MSA); Mueller-Hinton II
    agar; Sabouraud's agar; spirit blue agar; tryptone glucose yeast
    extract agar; Xylose Lysine Desoxycholate (XLD) agar
agar culture method, 143
agar plates, 62–64, 70
agglutination, 341, 347, 357
Agmenellum, 39
Alcaligenes, 252
    A. faecalis, 181
alcohol, as antiseptic, 197–98
alcoholic beverages, and fermentation, 307–08
algae, 33–38
alkaline reaction, and litmus milk test, 243
alkaline reversion, 338
allophycocyanin, 38
alpha-prime hemolytic organisms, 326
alpha-type inoculations, 327
Alternaria, 50, 52, 54
Amastigomycota, 49, 51
amino acids, 181
ammonia, 237
amoebas, 32
amoeboid movement, 32
amplitude objects, 15
amylases, 237
Anabaena, 39
anabolism, 229
Anacystis, 39
anaerobes, 125–28, 227
anaerobic respiration, 125
Analytab Products, 255, 257, 279
Analytical Profile Index, and API 20E System, 257, 259, 260
Ankistrodesmus, 36
annular ring, of phase-contrast microscope, 16, 18, 19, 20
antibiotics
    genetics of bacterial resistance, 382, 422
    sensitivity testing, 201–04
antigens, 341, 347
antimicrobic sensitivity testing, 201–04
antiseptics, 197–98, 207–08. *See also* control; disinfectants; sterilization
antiserum, 341
Aphanizomenon, 39
Apicomplexa, 32
API Staph-Ident System, 279–81, 319
API 20E System, 257–60
applied microbiology, 285

alcohol fermentation, 307–08
    bacterial counts of foods, 287–90
    membrane filter method, 297–99
    qualitative tests for bacteriological examination of water, 291–96
    reductase test, 301–03
    spoilage of canned food, 305–07
arabinose, and Enterotube II system, 267
arborescent growth, 226
Archaea, 29, 30, 173, 185
arginine dihydrolase, 274
Arthrobacter and *A. globiformis,* 251
Arthrospira, 39
arthrospores, 50
Ascomycetes, 51
Ascomycotina, 51
ascospores, 51
asepsis and aseptic technique, 57, 59–65
aseptate hyphae, 49
asexual spores, 49–50, 51
Aspergillus, 52, 54
    A. niger, 54
Asterionella, 37
autoclave, 99, 117, 121–22
autotrophs, 115

Bacillus, 99, 250
    B. coagulans, 305
    B. megaterium
        gram stain, 97, 98
        lethal effects of temperature, 177–78
        negative staining, 71
        spore staining, 99
        ultraviolet light, 189, 190
    B. stearothermophilus, 173–74, 305
    B. subtilis, 100, 237, 238
bacitracin, 328
bacteria. *See also* identification; microorganisms; *Staphylococcus;*
    Streptococcus
    diversity, 285
    nutritional requirements, 115–16
    survey, 29, 30, 38
    ubiquity, 45–46
    viruses compared to, 151
bacterial counts, 287–90
bacteriochlorophyll, 38
bacteriological examination, and qualitative tests of water, 291–96
bacteriophages, 151, 152, 155–57, 167. *See also* phages
bacteriostatic agents, 207–08
base plate, of virus, 151, 155
basic dyes, 87
basic fuchsin, 87
Basidiomycetes and basidiomycotina, 51
basidiospores, 51
basophils, 351, 352
Batrachospermum, 35
beaded growth, 226
Becton-Dickinson, 255, 275
Bergey's Manual of Systematic Bacteriology, 87, 217, 219, 225, 249–53,
    315
beta hemolytic stabs, 325

- On Emb agar Coliforms - produce chalk center.
- Endo agar → Red.

Completed test:
- → Nutrient agar slant → Durham tube.
- Agar reveals that we have gram-negative, non-spore forming rod.

Confirmed test
- Levine EMB agar contains methylene blue.
- Colonies of E. coli and E. aerogenes can be differentiated on the basis of size and presence of greenish metallic sheen.
- Endo agar contains a fuchsin sulfite indicator.

- Check water for these pathogens → Vibrio Cholerae, Salmonella typhi, and Shigella dysenteriae.

- E.coli → found in the Human intestine; but is not found in soil or water.

- E.coli - good indicator of fecal contamination.
  1) occurs primarily in the intestines of humans.
  2) organism can be easily identified by microbiological test
  3) it is not as fastidious as the intestinal pathogens, it survives longer in water samples

- E.coli and Enterobacter aerogenes are designed as coliforms;

→ Coliforms - gram negative, facultat___ forming rods that ferment lacto___ in 48 hours at 35°C.

- Streptococcus fecalis → a___

- Three tests to det___

Presumptive
9 t___